Wind Turbines - Advances and Challenges in Design, Manufacture and Operation

Edited by Karam Maalawi

Published in London, United Kingdom

Wind Turbines - Advances and Challenges in Design, Manufacture and Operation
http://dx.doi.org/10.5772/intechopen.97967
Edited by Karam Maalawi

Contributors
Lijun Cai, Karthik Jayanarasimhan, Vignesh Subramani-Mahalakshmi, Mohamed Mahran Kasem, Tefera Kitaba, Karam Youssef Maalawi, Vitaly Beresnevich, Janis Viba, Martins Irbe, Anubhav Jain, Enas Raafat Maamoun Shouman, Girmaw Teshager Bitew, Minxiao Han, Laurent Barthélemy, Alice Elizabeth Gonzalez, José Cataldo-Ottieri, John Campbell, Balaraman Kannan, Bastin Jeyaraj, Jigneshkumar P. Pramodbhai Desai, Ramesh Kumar Behara, Kavita Behara

First published in London, United Kingdom, 2022 by IntechOpen
IntechOpen is the global imprint of INTECHOPEN LIMITED, registered in England and Wales, registration number: 11086078, 5 Princes Gate Court, London, SW7 2QJ, United Kingdom

British Library Cataloguing-in-Publication Data
A catalogue record for this book is available from the British Library

Additional hard and PDF copies can be obtained from orders@intechopen.com

Wind Turbines - Advances and Challenges in Design, Manufacture and Operation
Edited by Karam Maalawi
p. cm.
Print ISBN 978-1-80355-492-1
Online ISBN 978-1-80355-493-8
eBook (PDF) ISBN 978-1-80355-494-5

For EU product safety concerns:
IN TECH d.o.o., Prolaz Marije Krucifikse Kozulić 3, 51000 Rijeka, Croatia,
info@intechopen.com or visit our website at intechopen.com.

Meet the editor

Karam Maalawi is a Professor of Aeronautics and Mechanics at the Mechanical Engineering Department, Institute of Engineering Research and Renewable Energy, National Research Centre, Cairo, Egypt. He has published extensively in the field of structural optimization and wind turbine design and performance. Dr. Maalawi has contributed to several research projects regarding aerospace engineering technology, wind turbine structures, and renewable energy applications. For his outstanding contributions to his profession, the National Research Centre has recognized him with two awards for engineering sciences. He has been presented with the Albert Nelson Marquis Lifetime Achievement Award and endorsed by Marquis Who's Who as a leader in the aerospace engineering industry.

Contents

Preface

Nowadays, intensive research and collaborative projects dealing with the installation of large wind turbines are taking place in many countries. The goal of this work is the decarbonization of electricity grid systems through the exploitation of wind power. In this context, modern and larger offshore wind turbines with power capacity reaching 15 MW and rotors of more than 230 meters in diameter are under continuous development to minimize the unit cost of energy production.

This book presents research in wind turbine design, manufacture, and operation. Its main topics include basic aspects of wind turbine design, low-specific power turbines, offshore wind industry and floating wind turbines, wind measurement and forecasting models, design and manufacturing of rotor blades, manufacture of power transmission bearings, and challenges in control strategies and computational aerodynamics.

Chapter 1 outlines the major design aspects of wind turbines. It analyzes the main structures to build a general optimization model. It defines and measures the most significant design objectives as well as design environment and constraints. It also identifies and discusses all effective system design variables and parameters. Several design alternatives are considered to see how the various design criteria are affected in each case.

Chapter 2 describes wind turbine generator protection based on fault current and voltage analysis, which can identify the instantaneous operation, delay operation, or immune operation. The fault scenarios are explained using simulation results on the wind turbine generator system modeled in a software environment.

Chapter 3 discusses the technological advantages provided by the low specific power (SP) turbine synthesized close to a target SP of 100 W/m2, which was determined by ground-based measurements. The findings imply that low-specific power wind turbines can improve the capacity utilization factor, lower the cost of energy, boost the value of wind, and better utilize the transmission system in all wind situations, albeit at varying magnitudes, according to the given results.

Chapters 4–9 focus on offshore wind farms, which are forecast to reach power capacity near 1000 GW by 2050 and will be a crucial part of the transition to net-zero emission. Chapter 4 gives insight into the forthcoming challenges and highlights potential solutions to make wind farms more self-reliant. It is concluded that the advent of a new class of converter-based power modules viz. grid forming will support the stability of offshore wind power plants and provide the needed services to ensure reliable and secure operation of future decarbonized electric networks.

Chapter 5 describes a new method for operating an offshore wind farm (OWF) with a diode-rectifier unit (DRU). A phase shifting transformer (PST) is applied on the onshore side of the medium-voltage submarine cable to avoid uncontrolled current flow through the cable. The application of PST is to ensure the smooth black-start and stable operation of the OWF and DRU-HVDC links. Both static and dynamic behaviours of the proposed method are presented, and the simulation results validate the given mathematical model.

Chapter 6 deals with the optimization of a crew transfer vessel (CTV) against an offshore wind turbine. It presents a proposal for improving the CTV by promoting the use of surface effect ships (SES) to minimize their heave under the constraint of fuel consumption. The study shows that another possible axis of development would be to design an additional wall to existing boat landings, providing sheltered water.

Chapter 7 discusses the prediction of sound pressure levels from a wind farm. The main characteristics are taking into account atmospheric stability to determine the acoustic power of the wind turbines; describing the generation of noise along the blades due to turbulent phenomena; and calculating the noise propagation at different distances from the tower taking into account atmospheric absorption, turbulence energy dissipation, and geometric divergence.

Chapter 8 addresses the different models for wind power forecasting, which is becoming increasingly significant due to the high penetration of wind power in the energy grid. Two main groups related to wind speed prediction are considered. The first group is based upon analysis of historical time series of wind, and the second uses forecasted values from a numerical weather prediction (NWP) model as an input. It is concluded that future research should focus on the following areas: implementation of artificial intelligence approaches to increase forecast accuracy; new strategies dealing with complicated terrain; more research into hybrid methods to combine physical and statistical approaches for achieving good results in both long-term and short-term prediction; and additional research into NWP models designed for use in an offshore environment.

Chapter 9 presents a droop-fed direct power control strategy in the variable speed pumped storage (VSPS) of a wind farm grid integration system. Modeling of the system is carried out based on the phasor model technique. The frequency spectrum analysis approach is used in the VSPS system for determining the dynamic performances of the grid in case of wind power fluctuation compensation and contingencies and validated in the MATLAB/Simulink platform. Results show that the frequency spectrum analysis method is effective for determining the wind power fluctuation and stability requirement in the large power system.

Chapter 10 focuses on the influences of the integration of wind power generators into power grid systems. The rotor side converter controller is used for active and reactive power control by controlling rotor current and the speed control of the Doubly Fed Induction Generator (DFIG). The control scheme and the simulation mode controller employed for the study assure the wind generator supplying the grid at varying wind speeds behaves like a synchronous generator at a 0 Hz rotor frequency.

Chapter 11 presents the fundamentals of wind turbine aerodynamics and the related terminologies to the design of blades. It discusses the three-dimensional computational aerodynamics of the blades as well as the design of vortex generators as an effective passive control device for airflow. The optimum dimensions and arrangement of these devices along the blade span are also studied aiming at the increase of the power output. It is concluded that integrating vortex generators in wind turbines is the next giant leap in aerodynamic research.

Chapter 12 discusses some techniques for the design and manufacturing of wind turbine blades, including the appropriate selection of the airfoil type, the design optimization methods, and manufacturing techniques. It highlights the superiority of using chord-wise and span-wise stiffeners to increase the stiffness of the skin of carbon fiber wind turbine blades. These stiffeners are not bonded externally to the skin, but rather they are layers of carbon fibers that are buried inside the skin of the wind turbine blades.

Chapter 13 describes the author's experience of a lifetime of casting metals and how the casting technique relates to the quality of the metal and offers answers to engineering failures. It is a concern that the failure of wind turbine bearings continues, on occasions, to defy substantial metallurgical efforts. It is proposed that there is good reason to identify the casting process as the generator of pervasive defects, which the author calls bifilms. These defects originate from the casting process during the pouring of the liquid steel. They exist in finished steel components as a substantial population of cracks. These pre-existing cracks are usually the initiators of fatigue failure as well as other failure modes.

Chapter 14 discusses wind turbine characteristics to design low-power rating generators, which is necessary for remote and rural electrification. The generator specifications have been obtained from wind turbine models such as torque, speed, and power. The rotor design reducing q-axis inductance of this generator is analyzed and the relationship between generated EMF voltage and torque with the change of time is evaluated. The effects of stator resistance on electromagnetic torque with a variation of power angle are also considered.

Chapter 15 discusses the optimization of operational safety and efficiency of wind energy conversion equipment. The proposed method involves flat blades or space prisms that perform translation motion due to the interaction with airflow. Theoretical results obtained are used in the design of new devices for energy extraction from airflow. Models of wind energy conversion devices equipped with one vibrating blade are developed (quasi-translatory blade's motion model; model with vibrating blade equipped with crank mechanism). Operation of the system due to the action of airflow is simulated with computer programs. Possibilities to obtain energy with generators of different characteristics, using mechatronic control, have been studied. The effect of wind flow with a constant speed and with a harmonic or polyharmonic component is considered.

Finally, it is a great pleasure for me to take this opportunity to express my gratitude and thanks to all the contributing authors. I also wish to express my gratitude for the help and support of the staff at IntechOpen, particularly Author Service Manager Ms. Elena Vračarić.

Karam Maalawi
Professor of Aeronautics and Mechanics,
Department of Mechanical Engineering,
Institute of Engineering Research and Renewable Energy,
National Research Centre,
Cairo, Egypt

Chapter 1

Introductory Chapter: General Design Aspects of Horizontal-Axis Wind Turbines

Karam Maalawi

1. Introduction

Numerous research contributions in developing wind industry technologies worldwide have been initiated since the oil crisis in 1973, and various configurations of wind turbines and large-scale wind farms have been installed in many places. These clean energy sources can make a substantial and economically competitive contribution to the future energy needs.

Irrespective of the specific application, a wind turbine system design should be based on the cost-effective production of energy. The main objective should be based on the minimum cost of energy depending on the rotor diameter, rated power as well as the wind characteristics for a given site. The economic feasibility of large-scale wind turbines operated as a part of electrical power systems has been considered by H. M. Bae [1]. In this paper, the design variables were taken to be the rotor diameter, rated power, and number of the installed machines. Maximization of the total net value of the generated power, which is equal to the annual expected fuel cost savings minus the total cost of the system, was taken as the main system objective. Power was considered as constraint rather than design objective. Hansen [2] addressed optimum blade shapes for maximizing the power coefficient of the rotor. He presented a method to obtain the optimum blade chord and twist distributions for better aerodynamic performance. Another important consideration in the design of wind energy generator systems is to reduce vibration without increasing structural weight. This is because the economics require that large wind turbines operate reliably for long periods of time while subject to significant vibratory loads [3, 4].

In this chapter, the wind turbine will be analyzed as a system in order to build a general model for its structural design optimization. The most significant design objectives as well as design environment and constraints are defined and measured. All effective system design variables and parameters are identified and discussed. Several design alternatives will be considered to see how the various design criteria are affected in each case.

2. System definition and main function

A wind turbine can be defined as a device that converts the wind's kinetic energy into useful mechanical power. This produced power can be exploited in many applications such as:

a. Corn grinding and wood sawing.

b. Battery charging.

c. Water pumping (e.g. irrigation and agricultural purposes).

d. Domestic use (e.g. heating and illumination).

e. Electricity producing (wind/diesel systems, wind farm, and utility operation).

The present chapter focuses on horizontal-axis wind turbines (HAWTs) utilized for electricity generation.

3. System components and subcomponents

To carry out its intended function, a wind turbine system must have the following main subsystems:

a. *Momentum exchange device (main wind rotor)*: This consists of rotating aerodynamical surfaces, called the blades that provide the main driving aerodynamic forces. The blades are mounted on a rotting hub/shaft assembly.

b. *Power transmission mechanism (power train unit)*:

This is composed of the following subcomponents:

- Shaft/bearing assembly
- Speed-up mechanism (gearbox)
- Braking system
- Energy utilizer (generator)

c. *Control system* for changing the blade setting angle to limit the turbine output at high wind speed and for yawing the rotor so that it may face the wind properly.

d. *Casing unit (Nacelle)* for housing the power train and control units. It is the interfacing device that connects these units together with the rotor and the supporting tower structure.

e. *Tower,* which supports the above units and elevates the rotor above the earth's boundary layer.

f. *Foundation*, which provides firm fixation of the system to the ground.

A typical horizontal-axis wind turbine system is shown in **Figure 1**.

DOI: http://dx.doi.org/10.5772/intechopen.106330

Figure 1.
Horizontal-axis wind turbine system components.

4. Design variables

The definition of wind turbine variables and parameters is of great importance in formulating a design optimization model. Actually, a wind turbine is a complicated network composed of thousands of interconnected elements. However, a breakdown of the system may help identify the most important design variables for each subsystem.

The following list shows the main design variables of wind turbine subsystems, with emphasis on variables related to the blade and tower structures.

4.1 Rotor variables

a. Main variables

- Diameter (rotor size)
- Location with respect to tower (e.g., upwind or downwind rotor)
- Number of blades (one, two, or three)

b. Blade variables

- General layout (length – chord and twist distributions – coning angle).
- Cross section (airfoil type – construction).
- Blade-Hub connection (semi-articulated – fully articulated – hingeless).

- Material of construction (aluminum – steel – fiberglass composites, etc.).

c. Rotor support (hub/shaft) variables

- Type of hub (fixed – teetered)
- Type of bearings
- Shaft construction and dimensions

Type of mounting of rotor shaft onto gearbox.

4.2 Power transmission variables

a. Gearbox

- Type (e.g., parallel shaft – planetary – ...).
- Size and type of casing
- Gear ratio

b. Generator

- Type (induction – synchronous – permanent magnet – .....).
- Mechanical and performance characteristics (torque-speed relation, rated power, ...)

4.3 Control system variables

a. Rotor yaw control

b. Blade pitch control

4.4 Nacelle variables

Configuration – dimensions – construction – mounting – material

4.5 Tower variables

a. Type (truss – tubular – guyed pole)

b. Layout parameters (height – width)

c. Type of cross section

d. Material of construction (steel – concrete – ...)

5. System objectives

A successful wind turbine design should ensure efficient, safe, and economic operation of the machine. It should provide easy access for maintenance and easy transportation and erection of the system components and subcomponents. Good designs should incorporate esthetic features of the overall machine shape. In fact, there are no simple criteria for measuring the above set of objectives. However, it should be recognized that the success of structural design ought to be judged by the extent to which the wind turbine main function is achieved.

5.1 Cost of energy production

The effectiveness of the design should be based on the end-product economics; i.e., the cost of energy produced. This may be expressed on an annual basis as:

$$\text{Minimize; Unit energy cost} = \frac{\text{Total annual cost}}{\text{Annual energy produced}} \quad (\$/\text{Kw.h}) \tag{1}$$

In Ref. [5], it was demonstrated that designs of large wind turbines are projected to be cost competitive for utility applications when produced in quantity. The cost of electricity produced can be decreased when operated at sites with a mean annual wind speed of about 6.5 m/s at 10 m height.

5.1.1 Annual cost

The main cost items of a wind turbine are incurred in the following major stages:

a. Initial capital cost

b. Operation, maintenance, and repair costs

c. Other cost items

A breakdown of cost components of each stage is shown in **Figure 2**.

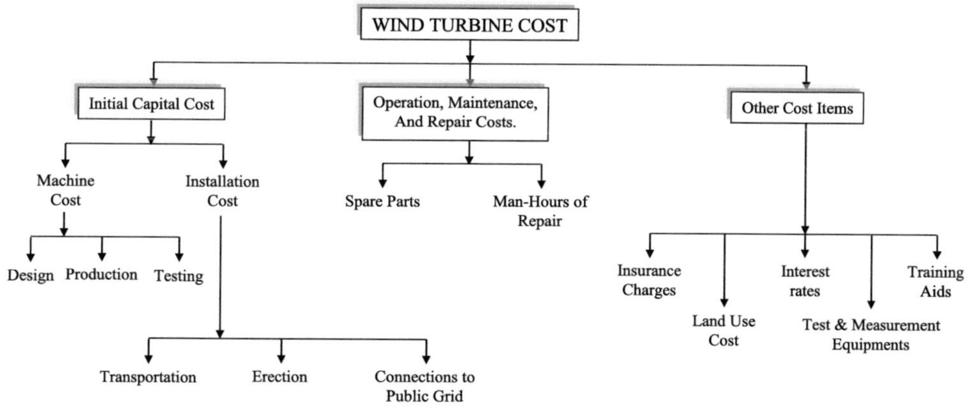

Figure 2.
Main cost items of a wind turbine.

Capital cost analysis depends on the development of statistical cost estimates, which relate the various design parameters and variables of the turbine to its total capital cost or its subsystem costs. The most significant design variables that have a bearing on wind turbine system costs are:

a. Rotor size

b. Rated power

c. Rated wind speed

d. Expected service life of the machine

e. Quantity of production

f. Type and material of construction of the various components

g. Degree of utilization of the machine

h. Type of generator and power transmission systems

For large-scale machines, **Figure 3**, taken from Ref. [1], shows typical machine cost as a function of rated power for different rated wind speeds and rotor sizes. The curves were determined by interpolating statistical cost estimates and shown on a logarithmic scale.

The initial capital is transferred to annual rates by multiplying with annualization factor (charge rate), which depends on the interest rates and machine life. Operation and maintenance costs are usually given as a fraction of the total capital. They are greatly influenced by how easy it is to exchange components for maintenance and repair.

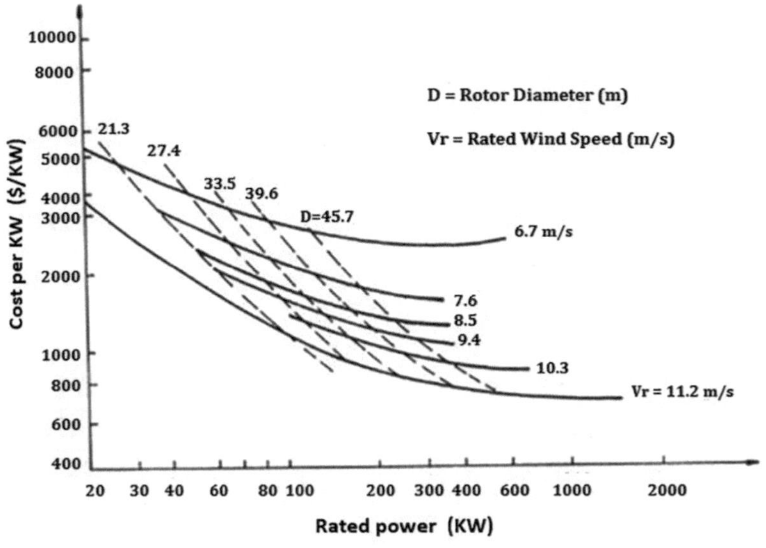

Figure 3.
Wind turbine machine cost as a function of rated power [1].

5.1.2 Annual energy productivity

The annual energy yield of a wind turbine is readily defined as the total number of kilowatt hours (Kwh) actually produced by the machine installation in a year (8760 hours). It depends very much on the site wind characteristics and machine performance characteristics. W. R. Powell [6] derived the following expression for the annual energy, E:

$$E = 8.760\ P_r \left[\frac{Exp(-\hat{v}_{in}^2) - Exp(-\hat{v}_r^2)}{\hat{v}_r^2 - \hat{v}_{in}^2} - Exp(-\hat{v}_{ou}^2) \right] * \text{availability factor (Kwh)} \quad (2)$$

where P_r is the rated power, and the term between brackets is called the capacity factor, which is given by the ratio of the average output power to the maximum rated power. All wind speed terms are described in a non-dimensional form, $\hat{V} = \frac{V}{\overline{V}}\sqrt{\frac{\pi}{4}}$, and are defined as:

V_{in}: Cut-in speed at which the machine starts to develop power

V_{out}: Cut-out speed at which the machine shuts down in high winds

$\overline{V}$: Mean wind speed in a year

The availability factor accounts for the availability of the wind turbine for service in the period in which the wind speed is in its operating range. Powell's expression was based on a Rayleigh wind distribution and a quadratic power-speed curve.

In general, wind machines with higher rated to cut-in speed ratios can both produce more energy and have higher capacity factors, but, unfortunately, they cost more. The selection of optimum rated to mean wind speed ratio is also a compromise. High capacity factors are available at low rated speeds, but less energy will be produced. The rated speed depends on the specific load application and rotor size, while the cut-in speed depends on the mechanical and power transmission system design. Variable pitch machines can adjust the blade angles to the wind in order to capture more energy over a wide range of wind speeds. However, cost will be incurred in the needed control systems.

The maximization of the annual energy production may be attained by maximization of the rotor power coefficient, C_p. Several authors have studied optimum blade shapes for maximizing C_p, which, for a prescribed value of the design tip-speed ratio (rotational speed*radius/wind speed), depends on the following design variables:

- Type of airfoil section (C_L/C_D ratio)
- Blade configuration and chord distribution
- Blade twist variation
- Number of blades (i.e., rotor solidity)

Optimization results show that:

a. The higher the lift-to-drag ratio, the better the aerodynamic performance of the machine.

b. The greater the blade number, the better the performance.

c. There is an optimum value for the power coefficient at a certain tip-speed ratio, called the design tip-speed ratio.

5.2 Weight considerations

An improved technology would result in a lightweight design, which performs the intended function efficiently. Lightweight also furthers some other objectives such as lower cost and better performance characteristics. Therefore, minimization of structural weight can be taken as a useful criterion for measuring the success of a wind turbine design. This would include both the tower and rotating blades as they are the main structural components of the machine. The component's weight depends on the material of construction, dimensions, and configuration.

5.3 Fatigue life

The fatigue life of the major structural components must be adequate to allow the production of enough energy to balance the initial investment. Approximately half the failures caused by fatigue occurred in the rotor assembly. This is expected because the rotor is the primary structure, which transfers wind loads to other structural components.

The design variables necessary for predicting fatigue life may be classified as follows [7]:

a. Wind speed characteristics of the site

b. Material strength levels and safety factors

c. Statistical distribution of stress levels

d. Choice and definition of applied loads, which are stratified in **Figure 4**.

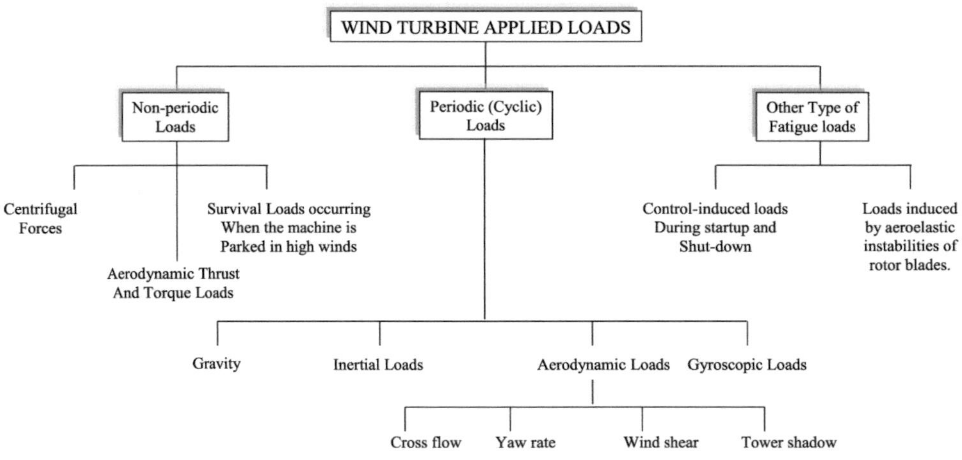

Figure 4.
Definition of wind turbine loads.

5.4 Design for minimum vibration

The reduction or control of the vibration of wind turbine structural components is an important design consideration. Vibration can greatly influence the commercial acceptance of the machine because of its adverse effects on performance, cost, stability, fatigue life, and noise. Such undesired effects become more pronounced in the case of large horizontal-axis wind turbines [8], which have the unique feature of slender rotating blades mounted on flexible tall towers.

When the machine is operating, the rotating blades of the main rotor are the prime source of vibration, which is then transmitted to the supporting tower structure primarily through a time-dependent shearing force at the hub. The forcing frequencies are integer multiples of the rotation rate. A common way to present natural frequency data and look for possible resonances is to plot the Campbell diagram as shown in **Figure 5**.

The intersection of one of the radial lines with one of the system natural frequency curves indicates a potential for resonant vibration near the rotor speed at the intersection point.

A good design philosophy for vibration reduction is to separate the natural frequencies of the structure from the harmonics of air loads or other excitation. This would avoid resonance where large amplitudes of vibration could severely damage the structure. Frequency placement is one of the techniques that have been used for reducing helicopter rotor blade vibrations [9]. The mass and stiffness distributions of the blades are to be tailored in such a way to give a predetermined placement of blade natural frequencies. Frequency placement can also help in controlling the forced

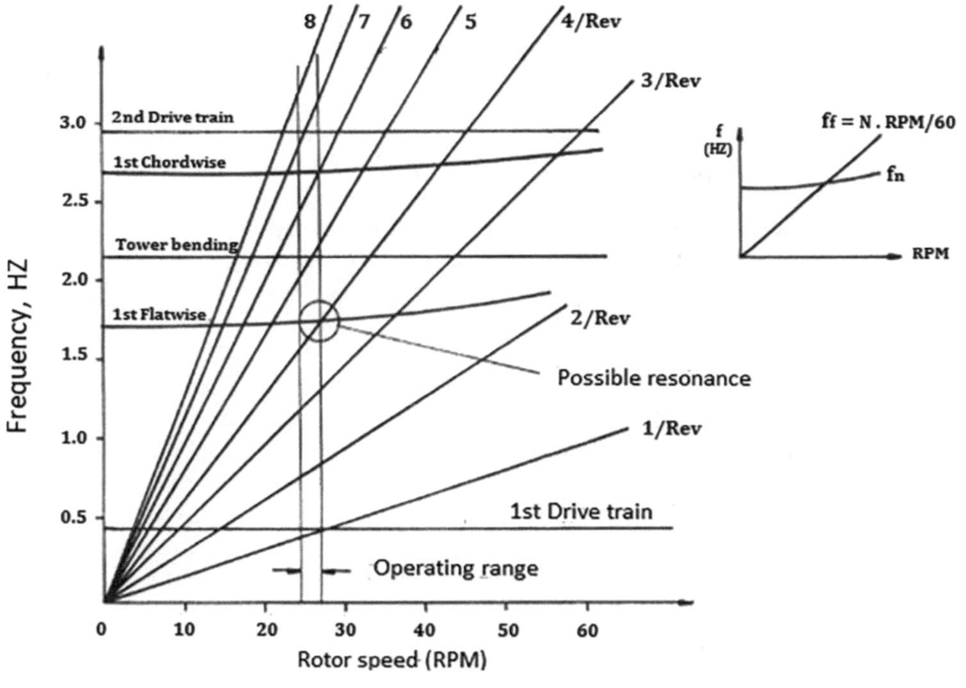

Figure 5.
Campbell diagram for a two-bladed wind turbine.

response of the blade. Another way of vibration reduction is to minimize the induced shearing forces transmitted to the supporting structure by the rotating blades.

5.5 Noise reduction

Wind turbine design may also be judged by its noise annoyance potential perceived by the nearby residents in both indoor and outdoor environments. The main sources of sound radiated from a wind machine are summarized in the following subsections.

5.5.1 Mechanical noise sources

These are mainly associated with the power-transmission system operation. This noise depends on the types and sizes of the gear box, generator, and bearings, and their mechanical and performance characteristics.

5.5.2 Aerodynamic noise sources

These are mainly associated with rotation of the blades in the surrounding air. It comprises three major components:

a. *Rotational noise* produced by the steady thrust and in plane torque loads acting on the blades. This noise is characterized by a large number of discrete frequency bands [10], which are harmonically related to the blade passage frequency. As a result of the low rotational speed of wind turbines, the associated acoustic energy resides in the low-frequency and sub-audible ranges ($\leq$ 20 HZ). It was shown that the acoustic pressure depends on the following design variables:

Position of the receiver – wind velocity – R.P.M – diameter – number of blades – airfoil type – plan form geometry of the blades – coning and pitch angles of the blades.

b. *Ground shear noise* produced by the unsteady blade loads that vary with the position around the rotor disk as a consequence of wind shear effect. The developed sound is characterized by low-frequency patterns and is largely dependent upon the shape of the wind shear profile.

c. *Impulsive noise* due to passage of the blades through the tower wake. It is identified with short, transient fluctuations in the radiated acoustic field. This is the most annoying source of noise because of its high degree of coherence and radiation efficiency. It depends, to a great extent, on the structural configuration of the tower.

d. *Impulsive noise* caused by cross-flow unsteadiness, which is identified by low-frequency sound radiation.

e. *Vortex noise* generated by vortices in the rotor wake, which are shed by the blades during rotation. This noise is characterized by largely incoherent radiation over a wide frequency range.

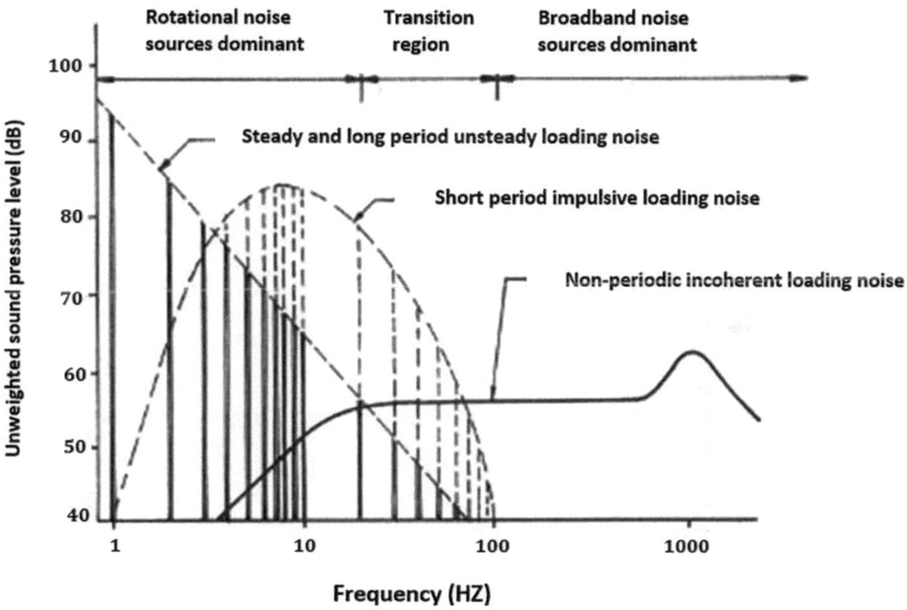

Figure 6.
Wind turbine noise spectrum characteristics [10].

It has been demonstrated that noise due to steady and unsteady aerodynamic loading arising from wind shear does not substantially contribute to the acoustic signal from wind machines. On the other hand, it was shown that community annoyance associated with turbine operations was related to coherent impulsive noise and the subsequent coupling of acoustic energy with residential structures. **Figure 6**, taken from Ref. [10], summarizes the acoustic pressure spectrum associated with large wind turbines for dominate noise sources as a function of frequency.

5.5.3 Noise caused by vibrations of structural components

Sound can be radiated from a wind turbine as a consequence of tower and blade vibrations. The efforts aiming at bringing structural vibrations to a minimum decrease this noise source automatically.

6. System environment and constraints

There are many limitations that restrict wind turbine design, manufacturing, and operation. The most significant among these are given below:

a. Type of application (e.g. electricity generation)

b. Site conditions (location, wind speed characteristics, wind shear, turbulence level, available area, transportation, soil conditions, local electricity system, etc.)

c. Project budget and financial limitations

d. Technological and manufacturing limitations

e. Manpower skills and design experience

f. Availability of certain material types

g. Safety, strength, and stiffness requirements

h. Dynamic and stability requirements of the whole structure

i. Performance requirements

The problem of wind turbine system optimization is that of finding values of the design variables, which best achieve the system objectives and, in the meantime, satisfy all design constraints.

7. Design alternatives and solutions

There are tremendous differences among horizontal axis wind turbines, depending on the size of the rotor and the specific energy application. However, the differences become contained in general design categories for turbines operating in the same environment and for the same application. Based on the selected design objectives, it is possible to identify a number of design solutions that are governed by the choice of the main design variables. **Table 1** gives some of the alternatives that concern the blades and tower designs.

Design solution		Effects on system objectives
(1) Rotor Size	(a) Small	Low energy productivity, light weight, low vibration levels, long machine life.
	(b) Large	High energy, heavy weight, high cost, high vibrations and fatigue loads.
(2) Rotor Position	(a) Upwind	Low noise and long fatigue life (No tower shadow effects), cost of control increases (requires yaw drive), has tower clearance problem.
	(b) Downwind	Noisy (tower shadow), reduce cost of control (free yaw), No tower-blade clearance problems.
(3) No. of Blades	(a) One	Less popular, noisy (unbalanced forces), less energy capture, cheaper, easy to erect.
	(b) Two	Reduce cost of transmission, long fatigue life and low vibrations (with teetered hub), cost of hub construction is high.
	(c) Three	More energy capture, more blade weight, low cost of hub construction, balanced gyroscopic forces, more esthetic.
(4) Blade Coning	(a) Without Precone	More energy, high cyclic loads (shorter fatigue life).
	(b) With Precone	Light weight, long fatigue life (reduce bending loads at blade root).

Design solution		Effects on system objectives
(5) Blade Material	(a) Wood Epoxy	Light weight, long life, high tooling cost, low stiffness.
	(b) Glass-Reinforced Polyester (GRP)	Lightweight, high tooling costs, possible to form complex shapes.
	(c) Steel	Low tooling costs-heavier weight, high stiffness, established technology.
	(d) Aluminum	Light weight, high stiffness.
(6) Tower type	(a) Lattice (Truss)	Cheap-easy transportation and erection – high strength and stiffness. Poor visual appeal-external access to nacelle – high tower shadow effects.
	(b) Tubular	Expensive – more esthetic – less tower shadow effects – internal access to nacelle.
	(c) Guyed tower	Low weight – low cost – high vibration (very soft) – less popular.
	(d) Concrete	Can be economical for large wind turbines.

Table 1.
Some alternatives for wind turbine blade and tower designs.

Author details

Karam Maalawi
Mechanical Engineering Department, National Research Centre, Cairo, Egypt

*Address all correspondence to: maalawi@netscape.net

References

[1] Bae HM, Devine MD. Optimization models for the economic design of wind power systems. Solar Energy. 1978;**20**: 469-481

[2] Hansen MO. Aerodynamics of Wind Turbines. New York, NY, USA: Routledge; 2015

[3] Chattopadhyay A, Walsh JL. Minimum weight design of rotorcraft blades with multiple frequency and stress constraints. AIAA. 1990;**28**(3): 565-567

[4] Pritchard JI, Adelman HM. Optimal placement of tuning masses for vibration reduction in helicopter rotor blades. AIAA. 1990;**28**(2):309-315

[5] Thomas RL, Robbins WH. Large wind turbine projects in the United States wind energy programs. Journal of Industrial Aerodynamics. 1980;**5**: 323-335

[6] Powell WR. An analytical expression for the average output power of a wind turbine. Journal of Solar Energy. 1981;**26**: 77-80

[7] Grujicic M, Arakere G, Subramanian E, Sellappan V, Vallejo A, Ozen M. Structural-response analysis, fatigue-life prediction, and material selection for 1 MW horizontal-axis wind-turbine blades. Journal of Materials Engineering and Performance. 2010;**19**:790-801

[8] Spera DA. Wind Turbine Technology. New York, NY, USA: American Society of Mechanical Engineers; 1994

[9] Peter DA, Rossow MP, Ko T. Design of helicopter rotor blades for optimum dynamic characteristics. Journal of Computer and Mathematics with Applications. 1986;**12A**(1):85-109

[10] Martinez R, Widnall SE, Harris WL. Predictions of low-frequency and impulsive sound radiation from horizontal-axis wind turbines. Journal of Solar Energy Engineering. 1982;**104**: 124-130

Chapter 2

An Algorithm for Default Detection of Wind Turbine Generators

Jigneshkumar P. Desai

Abstract

The protection of the wind turbine generator (WTG) required discrimination between internal and parallel WTG faults. Furthermore, it must discriminate the fault of its feeder line and parallel feeder line. This chapter describes the protection of wind turbine generators based on fault current and voltage analysis, which can identify the instantaneous operation, delay operation, or immune operation. A proposed Algorithm based digital relay is presented to provide all the different fault detection in a single unit suitable for internal and external fault protection of wind turbine generator. The main challenge to this scheme is that fault resistance may wrongly operate the scheme in some rare conditions. The phase angle of negative sequence current components determines the type of fault. The algorithm used negative sequence current and voltage to positive sequence current and voltage ratio, which is less than the set value in case of external fault. The fault seniors have been explained using simulation results on the wind turbine generator system modeled in a software environment.

Keywords: internal fault, feeder, positive sequence current, relay, wind turbine generator

1. Introduction

The parts of the wind turbine generator are shown in **Figure 1**. Blades are connected to this rotor hub. By rotating the motor shaft angle, one can turn the blade's direction, resulting in a change in mechanical power. This rotational mechanical energy rotates the rotor, and by using a gearbox, speed can be changed. By changing the speed, torque will be changed. The frequency of the generated voltage depends on the speed and number of poles. The variable frequency is converted into the constant frequency using a power converter. At this stage, two converters are used. One is an AC-DC converter, and the second is a DC-AC converter called a back-to-back converter. This is often called a gear-less wind turbine generator [2]. As electrical technology is very advanced, mechanical energy to electrical energy can be converted with different machines. Based on this machine used, the wind turbine generators are classified. The most common challenges for the wind turbine are as follows: (1) highly variable wind power injection into the grid, (2) increased penetration of wind energy, (3) Electrically weak distribution network, and (4) heavy reactive power burden by Induction generator (IG).

The classification of wind energy conversion system (WECS) is shown in **Figure 2**. Squirrel cage induction generators (SCIG) are a traditional method, but one cannot

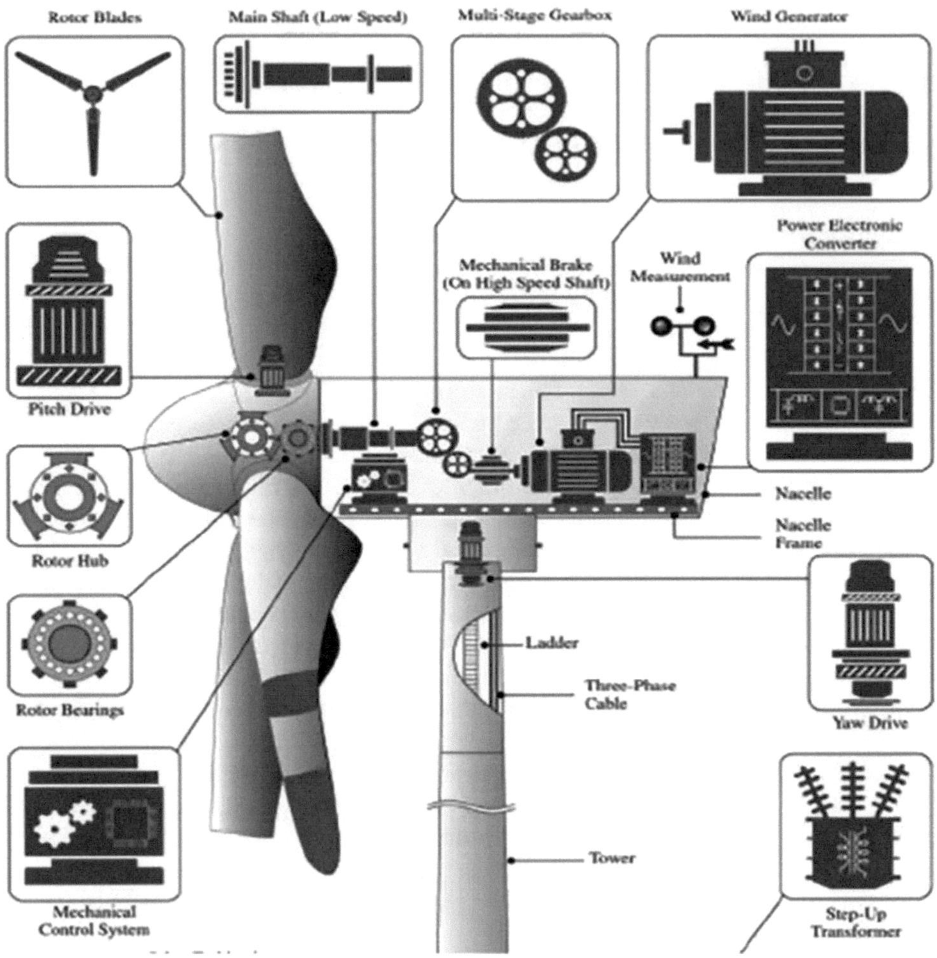

Figure 1.
Parts of wind turbine generator adapted from [1].

get maximum power at different wind speeds. The SCIG is generally known as a fixed-speed wind turbine. At variable rates, wind turbine generators are two types which are gear-less and with gear. Gear-less wind turbine generators may be running at a slower speed, but one can change the number of poles. The wound rotor synchronous generator (WRSG) and permanent magnet synchronous generator (PMSG) is the gear-less wind turbine. The wound rotor synchronous generator (WRSG) and permanent magnet synchronous generator (PMSG) maybe with gear also [3]. Doubly fed induction generator (DFIG), SCIG, wound rotor induction generator (WRIG) with variable rotor resistance also come under the gear category.

2. Grid connected operation of SCIG

The SCIG has the following main parts: (1) Gear Box, (2) cage induction generator, (3) soft starter, and (4) Capacitor for power factor compensation.

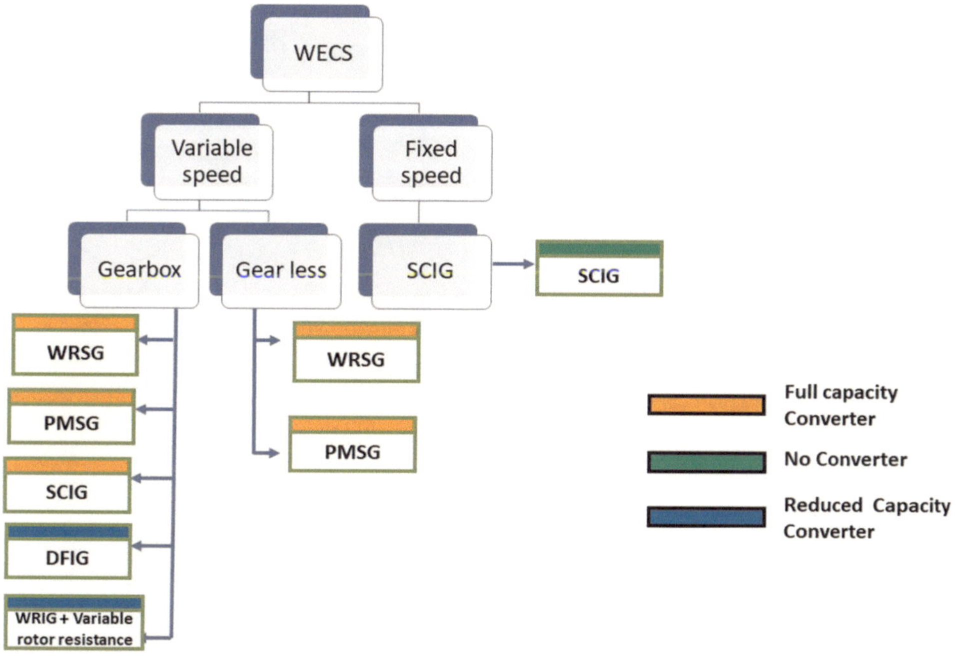

Figure 2.
Classification of wind turbine generator.

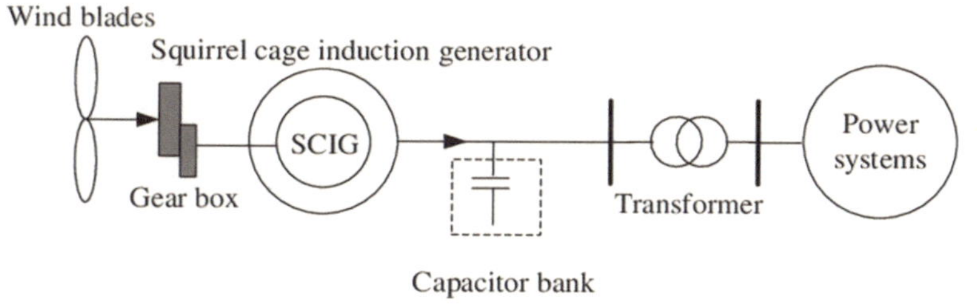

Figure 3.
Fixed speed SCIG wind turbines from [4].

The SCIG is a very simple structure of WTG used in the system. SCIG is directly connected to the grid using a starter and transformer. A soft starter is available in starting only to limit high inrush current. This is the most widely used structure worldwide due to less maintenance and simple design. The main disadvantage is that full power cannot be extracted from the grid. The SCIG needs high reactive power, which the capacitor bank locally supplies, as shown in **Figure 3**. The machine is run above synchronous speed using pitch control.

3. Configuration of wind farm

A wind farm composed of six 1.5 MW wind turbines is connected to a 25 kV distribution system that exports electricity to a 120 kV network via a 25 km long feeder from a 25 kV bus 4. Three 1.5 MW wind turbines pairs simulate the 9-MW wind farm. Wind turbines use squirrel cage induction generators (SCIG) [5].

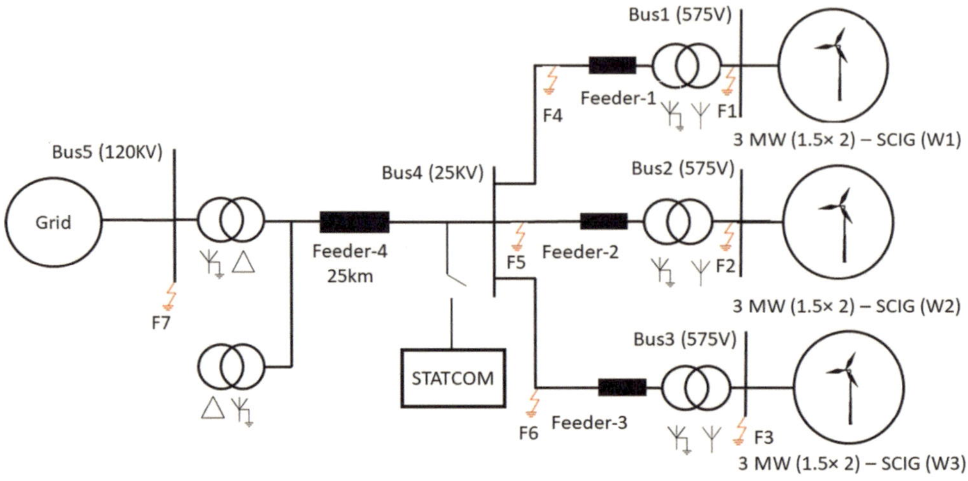

Figure 4.
9 MW—Wind farm connected to the grid.

Figure 4 shows the system considered for protection in which three wind turbine generators are connected to the grid.

The stator winding is connected directly to the 60 Hz grid, and a controllable-pitch windmill drives the rotor [6, 7]. The pitch angle is controlled to limit the generator's output power to its nominal value for winds exceeding the little velocity (9 m/s). A protection system is installed at each wind generator from W1 to W3, which measures voltage, current, and speed. Reactive power absorbed by the IGs is partly compensated by capacitor banks connected at each wind turbine low voltage bus [8–10]. The rest of the reactive power required to maintain the 25-kV voltage at bus B4 close to 1 pu is provided by a 3-Mvar STATCOM with a 3% droop setting. Modeling of Wind Turbine Generator is carried in MATLAB Software [11]. The data of wind turbine generator modeling is shown in Appendix A.

4. Protection of wind turbine generator

The digital protection system installed on W1 to W3 consist of following protections covers in single digital relay for wind turbine generator [12].

1. Instantaneous AC overcurrent

2. Positive-sequence AC overcurrent

3. Unbalance AC current

4. Positive-sequence under voltage

5. Positive-sequence under voltage

6. Negative-sequence unbalanced voltage

7. Zero-sequence unbalanced voltage

Fault position	Relay 1 operation	Relay 2 operation	Relay 3 operation
F1	Instantaneous	Non-operation	Non-operation
F2	Non-operation	Instantaneous	Non-operation
F3	Non-operation	Non-operation	Instantaneous
F4	Instantaneous	Instantaneous	Instantaneous
F5	Instantaneous	Instantaneous	Instantaneous
F6	Instantaneous	Instantaneous	Instantaneous
F7	Delayed	Delayed	Delayed

Table 1.
Desired operation of digital relay R1, R2, and R3 at W1, W2, and W3 respectively.

The desired protection for relay 1 at W1, relay 2 at W2, and relay 3 at W3 are shown in **Table 1** for different fault locations.

4.1 Protection algorithm

The relay R1, R2, and R3 are located at Bus 1, Bus 2, and Bus 3. The algorithm is explained in this section by considering Relay R1 to protect W1 against internal faults F1. For faults F4, F5 and F6 at POC at Bus 4, R1 reacts instantaneously as both these three faults impact the W1 directly. On the other hand, the relay R1 remains stable for F2, F3 and F7 and maybe operate as a backup to the primary relay addressing these faults. Here, F2 and F3 are parallel feeder faults considered external faults for F1. F7 is a external fault in the grid system. The protection algorithm for such desired operation as per **Table 1** is shown in **Figure 5**. As per the Algorithm, relay R1 measured three-phase voltage and current V_{abc} and I_{abc} with the help of PT and CT in the beginning [13]. Using the symmetrical component method, Positive sequence, negative

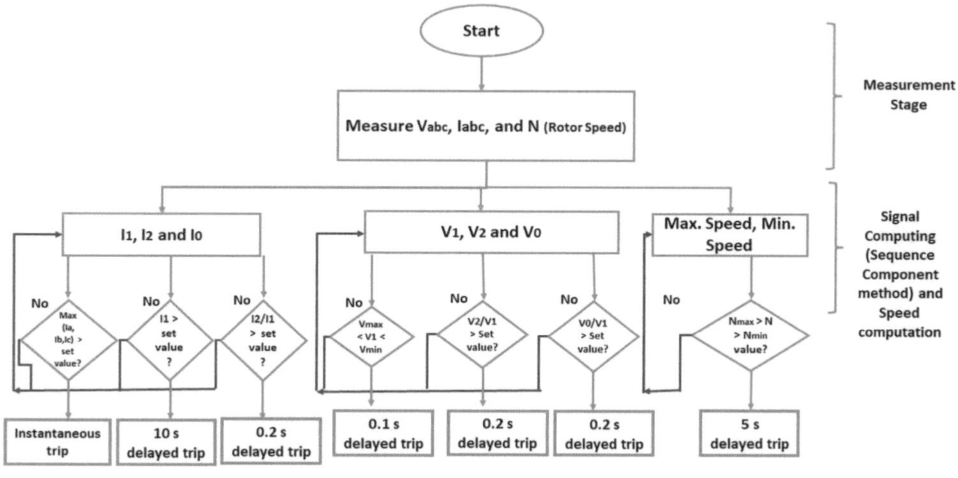

Figure 5.
Protection algorithm for WTG.

sequence, and zero sequence voltage and current are V_1, V_2, V_0 and I_1, I_2, I_0 respectively have been calculated. Based on the different conditions as shown in **Figure 5**, the tripping commands have been sent to the circuit breaker of W1. The next section will describe how the Algorithm detects LG, LL, LLL, LLLG faults for internal and external fault conditions.

5. Different internal and external fault detection by digital relay

5.1 LG faults

LG faults have been applied at F1 to F7 locations as internal and external faults. Considered F1 fault as LG fault and used at 15 s of simulation time for 0.3 s duration which is an internal fault for W1. in this case-1, the voltages are unbalanced significantly which has V0/V1 ratio found 0.985pu which is greater than a set value and as per algorithm the relay issue tripping after 0.001 s which instantaneous. It is important to note that in this case, relay R2 at W2 and R3 at W3 are not affected and remain immune. The tripping coordination of R1 to R3 for this case-1 is shown in **Figure 6**. **Figure 7** shows that positive sequence and zero sequences are present significantly during the fault, and the ratio of V0/V1 exceeds the set value. Similarly, LG fault has been applied at the F6 location. In this case, the fault is at POC, which required the operation of all the relays operated at 0.2 s. **Table 2** shows the other faults and measurement of current and voltage sequence components during the faults at bus 1.

5.2 LL faults

As internal and external faults, LL faults have been applied at F1 to F7 locations. Considered F2 fault as LL fault and used at 15 s of simulation time for 0.3 s duration which is an internal fault for W2. In this case-2, the positive sequence voltages less than set value and as per the algorithm the relay issue tripping after 0.001 s. It is important to note that in this case, relay R1 at W1 and R3 at W3 are not affected

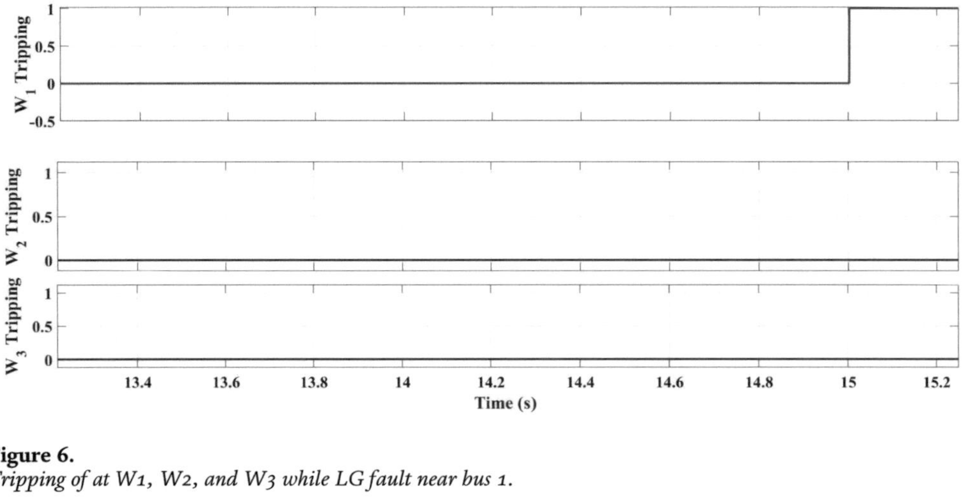

Figure 6.
Tripping of at W1, W2, and W3 while LG fault near bus 1.

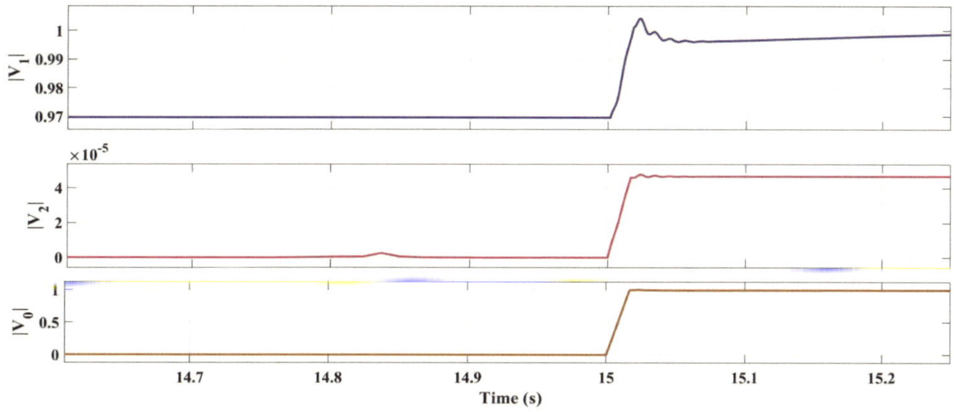

Figure 7.
Positive, negative, and zero sequence voltage variation during internal fault at W1.

Internal fault type	Fault location	Max (Ia,Ib,Ic) (p.u)	I1 (p.u)	I2/I1 (p.u)	V1 (p.u)	V2/V1 (p.u)	V0/V1 (p.u)
LG	Bus 1	R1:0 R2: 0.977 R3:0.979	R1:0 R2:0.983 R3:0.9846	R1:0 R2:0 R3:0	R1:0.998 R2:0.979 R3:0.979	R1:0 R2:0 R3:0	R1:0.985 R2:0.0001 R3:0.0001
LL	Bus 1	R1: 0 R2:0 R3:0	R1: 0 R2:0 R3:0	R1:0.037 R2:0.022 R3:0.022	R1:0.821 R2:0.9015 R3:0.9015	R1:0.458 R2:0.2568 R3:0.2568	R1:0 R2:0 R3:0
LLG	Bus 1	R1:0 R2:0 R3:0	R1:0 R2:0 R3:0	R1:0.037 R2:0.022 R3:0.022	R1:0.0821 R2:0.9015 R3:0.09015	R1:0.458 R2:0.2568 R3:0.2568	R1:0.662 R2:0 R3:0
LLL	Bus 1	R1:0 R2:0 R3:0	R1:0 R2:0 R3:0	R1:0 R2:0 R3:	R1:0.738 R2:0.738 R3:0.738	R1:0 R2:0 R3:0	R1:0 R2:0 R3:0
LLLG	Bus 1	R1:0 R2:0 R3:0	R1:0 R2:0 R3:0	R1:0 R2:0 R3:0	R1:0.665 R2:0.738 R3:0.738	R1:0 R2:0 R3:0	R1:0 R2:0 R3:0

Table 2.
Internal fault on feeder 1, 2, and 3 near SCIG.

and remain immune. Additionally, It provides backup protection after 0.15 s and 0.14 s by R1 and R3 to R2. The V1 and V2/V1 variations are shown in **Table 2** during the LL fault at bus 2.

5.3 LLG faults

As internal and external faults, LLG faults have been applied at F1 to F7 locations. Considered F3 fault as LLG fault and used at 15 s of simulation time for 0.3 s duration which is an internal fault for W3. in this case-3, the positive sequence voltages less than set value and as per algorithm the relay issue tripping after 0.001 s. It is important to note that in this case, relay R1 at W1 and R3 at W3 are not affected and remain immune. Additionally, It provides backup protection after 0.15 s and 0.14 s by R1 and R2 to R3.

5.4 LLLG faults

LLLG faults have been applied at F1 to F7 locations as internal and external faults. Considered F4 fault as LLLG fault and used at 15 s of simulation time for 0.3 s duration which is the internal fault for W1. in this case-4, the positive sequence voltages less than set value and as per algorithm the relay issue tripping after 0.001 s. In this case, relay R2 at W2 and R3 at W3 is also instantaneous as the fault is on the point of Common Coupling (POC). **Table 3** shows that I2/I1 and V1 change significantly during fault at bus 4.

5.5 LLL faults

LLL faults have been applied at F1 to F7 locations as internal and external faults. Considered F5 fault as LLL fault and used at 15 s of simulation time for 0.3 s duration which is the internal fault for W2. In this case-5, the positive sequence voltages less than set value and as per algorithm the relay issue tripping after 0.001 s. It is important to note that in this case, relay R1 at W1 and R3 at W3 is also instantaneous as the fault is on the point of Common Coupling (POC).

5.6 External fault

When LG fault is applied on the F7 location, which is in the grid and considered an external fault to the wind farm, the ratio of negative sequence voltage to positive sequence voltage and negative sequence current to positive sequence current is less than the set value [14]. So, as per the algorithm, relays R1 to R3 issued delayed tripping at 0.21 s after the fault instant. For all the other external faults, the variation of V1, V2/V1, and I2/I1 are shown in **Table 4**.

Fault type	Fault location	Max (Ia,Ib,Ic) (p.u)	I1 (p.u)	I2/I1 (p.u)	V1 (p.u)	V2/V1 (p.u)	V0/V1 (p.u)
LG	Bus 4	R1:0 R2:0 R3:0	R1:0 R2:0 R3:0	R1:0.02 R2:0.02 R3:0.02	R1:0.88 R2:0.88 R3:0.88	R1:0.24 R2:0.24 R3:0.24	R1:0.76 R2:0.76 R3:0.76
LL	Bus 4	R1:0 R2:0 R3:0	R1:0 R2:0 R3:0	R1:0.05 R2:0.05 R3:0.05	R1:0.545 R2:0.545 R3:0.545	R1:1.0 R2:1.0 R3:1.0	R1:0 R2:0 R3:0
LLG	Bus 4	R1:0 R2:0 R3:0	R1:0 R2:0 R3:0	R1:0.046 R2:0.046 R3:0.046	R1:0.473 R2:0.473 R3:0.473	R1:1.0 R2:1.0 R3:1.0	R1:0.994 R2:0.994 R3:0.994
LLL	Bus 4	R1:0 R2:0 R3:0	R1:0 R2:0 R3:0	R1:0 R2:0 R3:0	R1:0.0007 R2:0.0007 R3:0.0007	R1:0 R2:0 R3:0	R1:0 R2:0 R3:0
LLLG	Bus 4	R1:0 R2:0 R3:0	R1:0 R2:0 R3:0	R1:0 R2:0 R3:0	R1:0.0007 R2:0.0007 R3:0.0007	R1:0 R2:0 R3:0	R1:0 R2:0 R3:0

Table 3.
Internal fault on feeder 1, 2, and 3 near POC.

Fault type	Fault location	Max (Ia,Ib,Ic) (p.u)	I1 (p.u)	I2/I1 (p.u)	V1 (p.u)	V2/V1 (p.u)	V0/V1 (p.u)
LG	Bus 5	R1:0 R2:0 R3:0	R1:0 R2:0 R3:0	R1:0.024 R2:0.024 R3:0.024	R1:0.845 R2:0.845 R3:0.845	R1:0.289 R2:0.289 R3:0.289	R1:0 R2:0 R3:0
LL	Bus 5	R1:0 R2:0 R3:0	R1:0 R2:0 R3:0	R1:0.051 R2:0.051 R3:0.051	R1:0.574 R2:0.574 R3:0.574	R1:0.897 R2:0.897 R3:0.897	R1:0 R2:0 R3:0
LLG	Bus 5	R1:0 R2: R3:	R1:0 R2: R3:	R1:0.042 R2: R3:	R1:0.479 R2: R3:	R1:0.877 R2: R3:	R1:0 R2: R3:
LLL	Bus 5	R1:0 R2:0 R3:0	R1:0 R2:0 R3:0	R1:0 R2:0 R3:0	R1:0.001 R2:0.001 R3:0.001	R1:0 R2:0 R3:0	R1:0 R2:0 R3:0
LLLG	Bus 5	R1:0 R2:0 R3:0	R1:0 R2:0 R3:0	R1:0 R2:0 R3:0	R1:0.001 R2:0.001 R3:0.001	R1:0 R2:0 R3:0	R1:0 R2:0 R3:0

Table 4.
External fault of W1, W2, and W3 at grid side.

6. Challenges and possibilities

The main challenge in this protection scheme is that the relays R1 to R3 are affected by fault resistance while power swing occurs near them, and they cannot detect it. It is important to note that power swing blocking and out-of-step tripping functions are available to handle these challenges in the existing system [15].

7. Results and discussions

The results of **Tables 2–4** clearly show the quantity such as instantaneous overload current (Ia, Ib, Ic), I1, I2/I1, V1, V2/V1, and V0/V1 changes during different faults differently. This variation is used in the digital relay to identify internal and external feeder faults. The relays at W1 to W3 remain immune due to correct settings during grid faults. In this protection scheme, instantaneous overload current (Ia, Ib, Ic) provides instantaneous AC Over current, I1 provides AC Overcurrent (positive-sequence), I2/I1 provides Ac Current unbalance, V1 provides AC over and under-voltage (positive sequence), V2/V1 provides AC unbalance voltage (negative sequence), V0/V1 provides C unbalance voltage (zero sequences) protection to W1 to W3 against internal and external feeder faults correctly.

8. Conclusions

F1 from F2 and F3 can be detected using positive sequence voltage because the fault current of F2 and F3 can be seen at W1 via step-up transformer and feeder, due to which current is reduced compared to F1 in case of LG fault. While W2 sees F2 directly and W3 sees F3 directly, that is not affected F1, which is a parallel feeder fault. In the case of external fault, zero sequences cannot be used as they are trapped in the winding. So algorithm used negative sequence current and voltage to positive sequence current and voltage ratio, which is less than the set value in case of external fault. So, R1 to R3 does not operate. The proposed Algorithm based digital relay provides all the different fault detection in a single unit suitable for internal and external fault protection of WTG. The main challenge to this scheme is that fault resistance may mal-operate the scheme in some rare events.

Conflict of interest

The authors declare no conflict of interest.

Abbreviations

WTG	wind turbine generator
IG	Induction generator
WECS	wind energy conversion system
SCIG	squirrel cage induction generators
WRSG	wound rotor synchronous generator
PMSG	permanent magnet synchronous generator
WRSG	wound rotor synchronous generator
WRIG	wound rotor induction generator
KV	Kilo Volt
STATCOM	Static Var Compensator
W1	Wind turbine 1
F1	Fault 1
POC	Point of Common Coupling
R1	Relay 1
LG	Line to ground
LL	Line to line
LLL	Line to line to line
LLG	Line to line to ground
LLLG	Line to line to line to ground
I1	Positive sequence Current
V1	Positive sequence Voltage
I2	Negative sequence Current
V2	Negative sequence Voltage
I0	Zero Sequence Current
V0	Zero Sequence Voltage
pu	Per unit

Appendix A: wind turbine generator modeling

SCIG data			
Parameter	**Symbol**	**Unit**	**Value**
Nominal power	Pn	VA	3.33 MVA
Line-to-line voltage	Vn	Volt	575
frequency	fn	Hz	60
Stator	Rs	pu	0.004843
Stator	Lls	pu	0.1248
Rotor	Rr'	pu	0.004377
Rotor	Llr'	pu	0.1791
Magnetizing inductance	Lm	pu	6.77
Inertia constant	H	s	5.04
Friction factor	F	pu	0.001
Pairs of poles	P		3
Nominal wind turbine mechanical output Power (W)	Pm	W	3 MW
Base wind speed	Nw	m/s	9
Base rotational speed	N	pu	1
Maximum power at base wind speed	Pm(max)	pu	1
Pitch angle controller gain	Kp and Ki	pu	5 and 25
Maximum pitch angle	β_{max}	deg	45
Maximum rate of change of pitch angle	$d\beta/dt$	deg/s	2

Author details

Jigneshkumar P. Desai
Ganpat University, U.V. Patel College of Engineering, Mehsana, India

*Address all correspondence to: jpd.fetr@gmail.com

References

[1] Oussama M, Hamza A. Commande d'une éolienne à base de la MADA pour éliminer le déséquilibre dans les réseaux électriques. 2020. DOI: 10.13140/RG.2.2.25642.03527

[2] Al-Bahadly, editor. Wind Turbines. London, United Kingdom: IntechOpen; 2011. Available from:https://www.intechopen.com/books/115?msclkid=d709f9b6af6911ec94f05b231973e141doi:10.5772/643

[3] Adaramola M. Wind Turbine Technology: Principles and Design. London: CRC Press; 2014

[4] He PW, Ledwich F, Xue G, Yusheng. Small signal stability analysis of power systems with high penetration of wind power. Journal of Modern Power Systems and Clean Energy. 2013;**1**: 241-248

[5] Modeling and Modern Control of Wind Power—IEEE eBooks—IEEE Xplore. Available from: https://ieeexplore.ieee.org/book/8268023?msclkid=ac6ed820af7c11eca53d22aa4d385d42

[6] Multilin GE. GE Consumer Industrial Multilin. W650-Wind Generator Protection System Instruction Manuals. 2006. Available from: http://www.gedigitalenergy.com/app/ViewFiles.aspx?prod=w650&type=3

[7] Schweitzer Engineering Laboratories. SEL-700GW Wind Generator Relay. Available from: http://www.selinc.com/sel-700gw/

[8] Wind Plant Collector Design WG. Wind power plant grounding, overvoltage protection, and insulation coordination. In: Proceedings of 2009 IEEE Power and Energy Society General Meeting. Canada: Calgary; 2009

[9] Wind Plant Collector Design WG. Wind power plant substation and collector system redundancy, reliability, and economics. In: Proceedings of 2009 IEEE Power and Energy Society General Meeting. Canada: Calgary; 2009

[10] Haslam SJ, Crossley PA, Jenkins N. Design and field testing of a source based protection relay for wind farms. IEEE Transactions on Power Delivery. 1999; **14**(3):818-823

[11] Richard Gagnon (Hydro-Quebec) group. Wind Farm (IG) – MATLAB Simulink. 2021-2022. Available from: https://www.mathworks.com/help/physmod/sps/ug/wind-farm-ig.html?msclkid=6969af7daf7b11ecb30af44d7d733271

[12] Zheng TY, Kim YH, Crossley PA, Kang YC. Protection algorithm for a wind turbine generator in a large wind farm. IEEE Trondheim PowerTech. 2011;**2011**:1-6

[13] Desai JP, Makwana VH. Phasor measurement unit incorporated adaptive out-of-step protection of synchronous generator. Journal of Modern Power Systems and Clean Energy. 2021;**9**(5): 1032-1042

[14] Desai JP, Makwana VH. Modeling and implementation of percentage bias differential relay with dual-slope characteristic. IEEE Texas Power and Energy Conference (TPEC). 2021;**2021**: 1-6. DOI: 10.1109/TPEC51183.2021.9384987

[15] Desai JP, Makwana VH. A novel out of step relaying algorithm based on wavelet transform and a deep learning machine model. Prot Control Mod Power Syst. 2021;**6**:40

Chapter 3

Low Specific Power Wind Turbines for Reduced Levelized Cost of Energy

Balaraman Kannan and Bastin Jeyaraj

Abstract

Wind turbines with Low Specific Power (LSP) are envisaged as one of the modern-day manifestations to reduce the variability in wind generation, lower the cost of energy, increase the penetration to larger areas and better utilize the transmission system. In this regard, this chapter analyzes the characteristics of a LSP turbine synthesized close to a target Specific Power of 100 W/m^2 (LSP-105) based on ground-based measurements at varying site conditions representing various IEC wind classes. The overall analysis suggests that, under reasonable scenarios, low-specific power turbines could play a significant role in the future wind energy fleet, with their impact being particularly noticeable in low wind areas of the world. The analysis reveals that LSP turbines would provide a higher capacity utilization factor (CUF), even in low wind sites, and may reduce the Levelized Cost of Energy (LCOE) to an extent of 60%. On the other hand, the grid utilization pattern is found to be improving with LSP wind turbines in the medium and high wind sites. The results further suggest that reducing the cut-off wind speed could be one of the successful strategies to optimize the cost of LSP turbine in low wind sites.

Keywords: capacity utilization factor, grid utilization, LCOE, power curve, specific power, wind class, wind turbine

1. Introduction

Although renewable energy is now a considerable element of our energy mix [1], it is predicted that it will play an even greater role in the future transition toward clean energy and sustainable power supply. Wind and solar power, two of the most important sources of the clean energy transition, suffer from intermittency and variability, which, if not addressed, could slow the pace of the climate-neutral transition and raise total energy system costs [2]. Furthermore, with increasing penetration, the variability will increase the portion of underutilized grid infrastructure, necessitating the need for grid flexibility and ancillary support. From the standpoint of market value, this may also result in lower marginal costs, which will have an impact on the project's economics [3]. During high wind periods, the marginal value of wind is expected to reduce due to the higher amount of wind energy flowing into the grid [4], and during low wind hours, the generation tends to be low to meet the demand.

In general, considering different wind regimes, wind turbines are designed to withstand a class of wind speeds as specified by International Electro-technical Commission (IEC). IEC Class I is for the high windy sites, those with an annual average wind speed of 10 m/s. Class II is designed for locations with an average wind speed of 8.5 meters per second at hub height. Class III is for even lower windy sites, with an average wind speed of no more than 7.5 m/s. Class IV was also described in older IEC Standards, which is for very low-wind sites with an average wind speed of 6 m/s, but it has been superseded by Class S. This Special class "S" turbines are with design values chosen by the designer based on site-specific conditions. It is worth noting that while the majority of class I and class II wind sites are exhausted and presently large areas suitable for class III and IV are available for development. Due to their extremely low CUF and higher LCOE, wind turbines now available in the market are uneconomical to operate in class III and IV locations. With LCOE being the prominent factor in deciding the share of the renewable mix, solar PV generation has surpassed wind generation in many countries. When these considerations are taken into account, the solution that is foreseen is the design of wind turbines with low specific power (LSP) in order to increase the deployment of wind power with reduced variability, lower LCOE, and suitability for low wind sites.

Modern-day wind turbines, especially the dominant, three-bladed, upwind turbine configuration, have undergone significant design improvements toward increased energy generation and reduced cost of energy [5]. Wind turbines have also become physically larger in several dimensions, including rotor diameter, hub height, and nameplate/rated capacity. Wind turbine design has also become tailor-made to the market environment in which the technology is going to be deployed, based on wind regimes, grid access, etc., and low specific power (LSP) turbines are considered as one of such manifestations.

As swept area increases with the square of blade length, increasing the blade length of a turbine will increase the power extraction and, with fixed generator capacity (rated power), reduce the Specific Power (SP) rating of the turbine. On the other hand, theoretically, decreasing the generator capacity while maintaining a constant swept area will also result in the desired designed specific power.

Reduced Specific Power turbines provide a number of advantages that have resulted in their widespread use in several wind markets, including India, China, and the United States, among others. Due to the larger size of the rotor, more energy can be captured when the wind moves past the blades of the wind turbine. Hence, for a given turbine generator capacity, the generator runs closer to or at its rated capacity for an increased percentage of the time duration in such reduced specific power wind turbine.

Hence, reduced specific power wind turbines naturally result in higher energy generation for their installed capacity, resulting in a higher capacity utilization factor (CUF). With the highly sophisticated control systems of modern wind turbines, this higher capacity utilization factor can often be achieved with a relatively limited impact on the overall turbine cost. In such cases, LSP wind turbines provide a direct path toward a lower levelized cost of energy (LCOE) by providing a higher generation per investment.

Further, as reduced specific power turbines have lower-rated wind speeds, resulting in evenly distributed power production over a wide range of wind speeds, the variability of wind generation is well mitigated with such turbines. With significant penetration of RE generation into the grid, such variability management will be extremely beneficial to grid managers, and the overall energy mix can be well managed with reduced storage requirements.

Further, the energy generation profiles resulting from reduced specific power turbines have also been found to increase the wholesale market value of wind energy [6]. Turbines with reduced specific power and taller towers can be conceptually correlated with higher electricity prices in some markets by producing less during high wind hours and producing more during low wind hours [7, 8].

With such a background, this chapter analyses the LSP turbine synthesized for a target SP of close to 100 W/m^2 (for the study, a wind turbine with 105 W/m^2 is considered). Based on ground-based measurements, these LSP wind turbines are compared with other prevalent wind turbines in the Indian market with a view toward evaluating the opportunities to continue the specific power reduction in the future.

This chapter analyses the improvement in %CUF with these turbines for the IEC site classifications (i.e., high wind, medium wind, low wind, and also one of the coastal site conditions) and induces thoughts on how grid utilization will be influenced by the low-specific power turbine compared to the present-day wind turbines. In order to reduce the cost of wind turbines, the chapter further analyses the opportunity of reducing the cost of wind turbines by reducing cut-off wind speeds (varying cut-off wind speeds to 20 m/s, 18 m/s, 15 m/s, and 13 m/s) in LSP turbines as it allows the turbine blades to be lighter [9].

This chapter is expected to be useful to various stakeholders in the sector by encouraging further research in this area, as LSP wind turbines are expected to play a vital role in the wind generation fleet going forward, particularly as wind penetration increases in lower wind speed regions.

2. Low specific power (LSP) wind turbine

This section explains the key properties of the LSP wind turbine, as well as the characteristics of other wind turbine types currently on the market, in greater detail. The influence of the CUF on different sites with the wind turbines under consideration is also investigated.

2.1 Wind turbine characteristics

The study covers available wind turbines with specific power (SP) ranging from 379 W/m^2 to 173 W/m^2, which represent multi-MW scale turbine types prevalently installed in the Indian Market. These wind turbine types considered in this study, when correlated with its period of deployment, clearly show that there is a definite, reduced specific power trend in India, the United States, China, and Brazil. However, this stands in contrast to most of the European market, where the average specific power is found to be high, although it is difficult to generalize.

Turbine name (units)	Rated power (kW)	Rotor diameter (m)	Specific power (W/m^2)
SP-379	2000	82	379
SP-345	2100	88	345
SP-295	2000	93	295
SP-271	2000	97	271
SP-245	3000	125	245
SP-210	3465	145	210
SP-200	2000	113	200
SP-180	2100	122	180
SP-173	3300	156	173

Table 1.
Wind turbine models considered in this study.

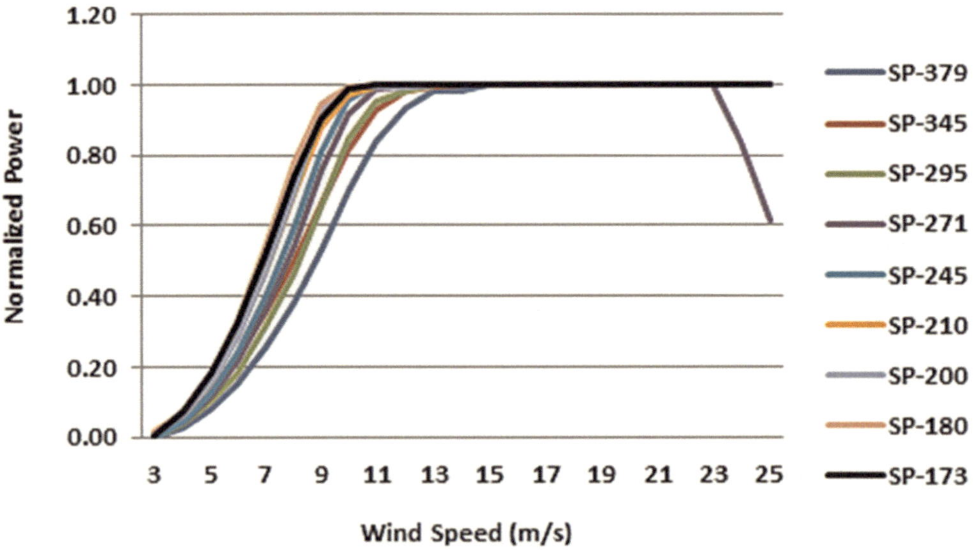

Figure 1.
Power curves considered for the study.

Table 1 shows the wind turbines considered in the comparison study against the LSP wind turbine model. The respective power curves (normalized) are shown in **Figure 1**. It is clearly evident from **Figure 1** that the power curve moves toward the left and is able to generate more energy at lower wind speeds, while the specific power is decreasing. To eliminate the influence of hub height variation in the comparison study, all wind turbines in the study are considered with the same hub height of 120 m.

2.2 Power curve of LSP wind turbine

The power curve of the LSP wind turbine is derived/synthesized in a unique way from one of the latest wind turbine models (SP-173) in the Indian market, not by increasing the blade length, but by reducing the generator capacity and keeping

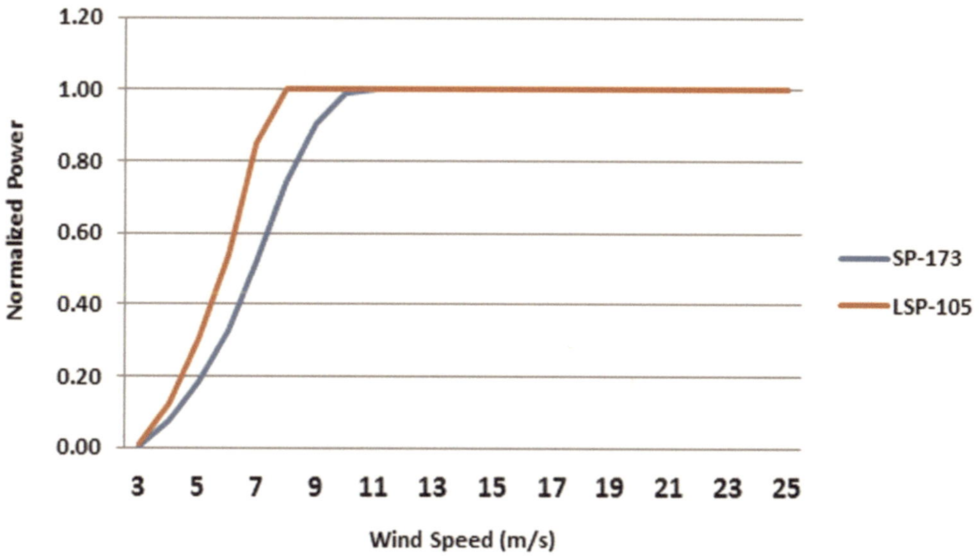

Figure 2.
Comparison of SP-173 and LSP-105. (derived from SP-173 by reducing the rated capacity to 2000 kW instead of 3300 kW).

the blade length constant. Considering the logistical and transportation-related constraints imposed by the longer blades [10], especially in complex terrain conditions, such a reverse approach seems to be justifiable. **Figure 2** shows the normalized power curves of SP-173 and LSP-105 (derived from SP-173 by reducing the rated capacity to 2000 kW from 3300 kW).

2.3 Wind sites

In order to understand the impact of Low Specific Power wind turbines on varying wind climate, four different wind sites have been chosen for the study. The sites have been chosen to represent high (High W), medium (Med W), low (Low W), and low-coastal wind (Low coast W) regimes, considering the future wind farm development possibilities. The sites are defined by 120 m hub height wind speed data derived from one continuous year of met. Mast-based standard measurement. **Table 2** depicts the site details, wherein **Figure 3** shows the wind speed distribution of the said four sites as below:

Site name (units)	Average wind speed @ 120 m (m/s)	Weibull "A" (m/s)	Weibull "k"
High W	10.4	11.69	2.29
Med W	8.5	9.60	2.62
Low W	6.9	7.80	2.66
Low coast W	6.3	7.02	2.82

Table 2.
Details of sites considered for the analysis.

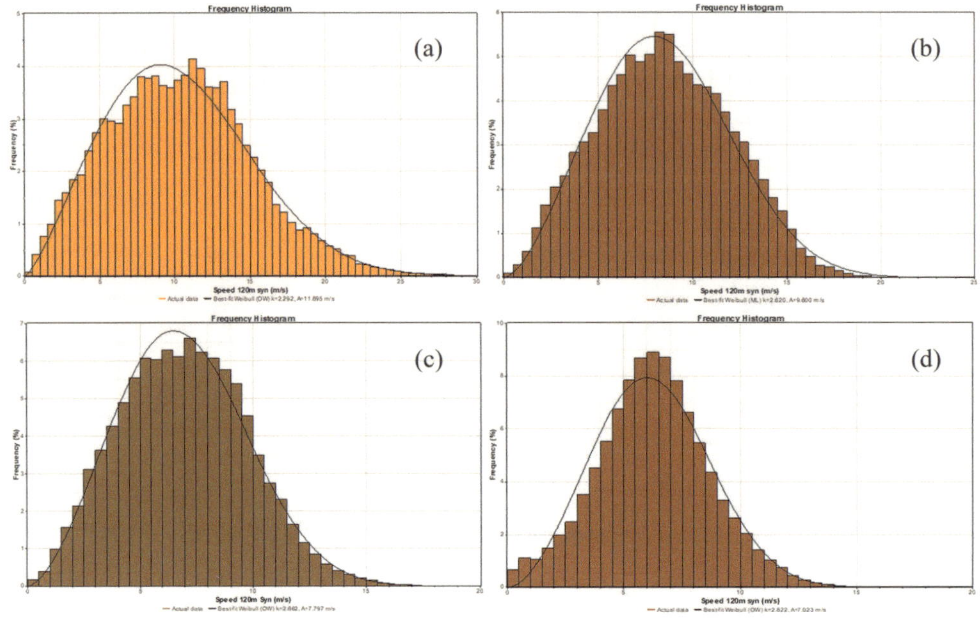

Figure 3.
Wind speed distribution of the sites considered in the study. (a) Represents the histogram of high W site, wherein (b), (c), and (d) represents the histograms for med W, low W and low coast W sites.

2.4 Comparative analysis

The generation for the entire year is estimated for each of the considered wind sites and wind turbines using the latest tools and % CUF are plotted in **Figure 4**. The trend seems instructive and shows the industry's ability to achieve higher capacity

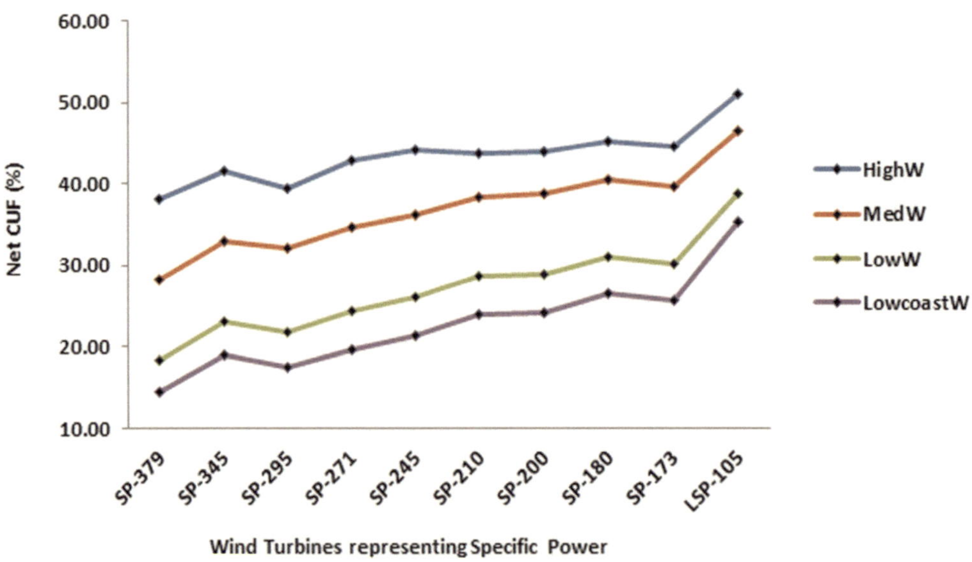

Figure 4.
Turbines representing specific power vs. net capacity utilization factor (%).

utilization factors by lowering the Specific Power of the turbines in the low and medium wind sites. Although the %CUF varies with respect to the wind potential, as obvious, there is a strong positive correlation found between the reductions in Specific Power with the increase in Capacity Utilization Factor (%) estimations. However, the increase in generation in the high wind sites is considerably lower (in the order of around 30%) compared to low wind sites (an increase of 110–145%). Moreover, the futuristic Low Specific Power wind turbine (LSP-105) simulated in the study is expected to produce appreciable generation in low wind sites resulting in lower LCOE. This is an encouraging phenomenon in particular for a situation when the wind penetration is increasing into the lower wind-speed regions.

Hence, it is evident that reducing Specific Power is undoubtedly enhancing the % CUF in all type of site conditions considered in this study, but highly significant in low wind sites. Of course, a higher capacity utilization factor is a mid-point, and it is not necessary that such wind turbine will have the economic superiority, as there is a cost to achieve it – a larger rotor assembly needs to be built and maintained for a given wind turbine capacity. As it is a trade-off between the increment in %CUF (indirectly, Energy yield) and cost for maintaining the larger rotor assembly, Levelized Cost of Energy (LCOE) may be the better metric to understand the realistic benefits (in terms of revenue) achieved from reducing Specific Power.

Interestingly, till date, the industry is able to maintain in such a way that the cost increment toward reducing Specific Power has not been enough to outweigh the LCOE benefit derived from the corresponding increase in generation. The empirical data from the United States shows that the historical trend toward reduced Specific Power turbines in the United States produces a trend with higher capacity factors and lowers LCOE [11].

In order to understand such trade-offs between cost and generation in the reducing specific power scenario, at varying wind regimes, LCOE was computed with respective cost and capacity factor differentials. The model inputs align with the traditional scaling theory - lower SP turbines will have a higher initial cost. For the financial estimation, standardized terms and numbers were assumed, which include, Debt: Equity of 70:30, Interest rate of 8%, Return on Equity (RoE) of 12% with the lifetime of the plant as 25 years. With regard to O&M expense, the same was assumed to be 1.5% of Capital Expense (CAPEX) with 3.84% escalation every year.

Figure 5 present the LCOE trend with the wind turbines with different SP for varying wind regimes. It is clearly evident from **Figure 5** that LCOE trend seems to be marginal variation for High and Medium wind sites but a definite LCOE reduction is seen in Low wind sites with the lower SP wind turbines. In case of the Low coast W site, the LCOE shows almost a 60% reduction with the LSP-105 wind turbine. It is noted that the trend shown is portrayed under the *traditional scenario* aligning with the scaling theory (lower SP turbines will have higher initial cost). In case, a *favorable lower SP scenario* is assumed, wherein cost/MW of both higher SP and lower SP turbines are same, then LCOE reduction in High and Medium wind sites may also be assured with reduced SP. We cannot ignore such *favorable lower SP scenarios* as Low SP turbines, being the present trend and gaining more market share, can have reduced cost through supply chain optimization and increased purchase volume [11].

In addition to the %CUF and LCOE, when looking into the market value of wind also, research results reveal that LSP turbines appear to be advantageous and generally provide greater wholesale market value than higher Specific Power turbines by

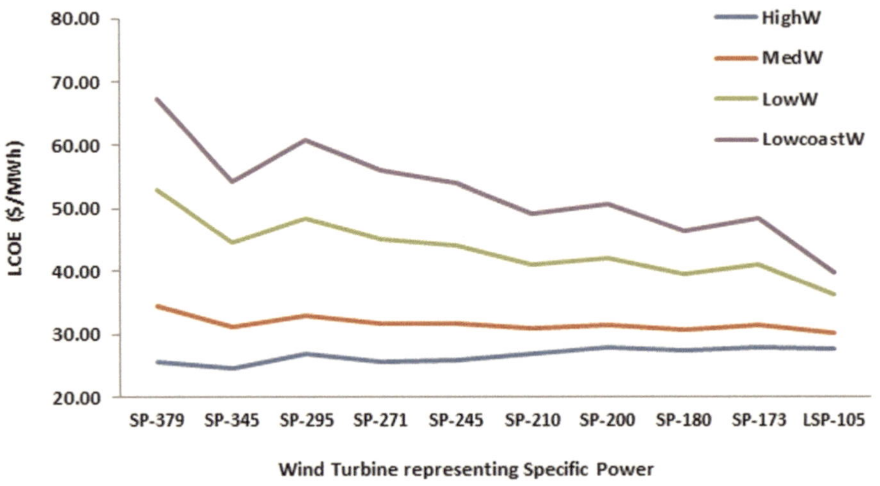

Figure 5.
LCOE trend with respect to specific power for different wind conditions.

shifting generation to lower wind hours. Recent European research [7, 12, 13] suggests that, in the future, the enhancement in market value provided by LSP turbines could become an increasingly significant selling point, presuming that wind penetration continues to increase.

3. Impact of low specific power wind turbine on grid infrastructure

One of the major problems of having high wind penetration in the grid is higher variability. On the one hand, this will call for under-utilized grid infrastructure (during low wind times). On the other hand, it will necessitate more grid flexibility and ancillary services/storage to compensate for the fluctuation. With more wind penetration into the grid expected in the future, it will be interesting to see how a LSP wind turbine will impact grid utilization versus conventional wind turbines. One such analysis is carried out and is represented in **Figure 6**. The hourly power production from all the wind turbines considered in the study is plotted against the number of hours (in percentage). For the purpose of comparison, all the turbines are normalized to 1000 kW. **Figure 6** clearly depicts that the grid utilization pattern generally increases with decreasing specific power. In particular, the futuristic LSP-105 is expected to utilize the allotted grid capacity (by generating at rated power) for a significantly longer time. At the high wind site, it is expected that the LSP-105 can generate at rated capacity for more than 50% of the time, even after accounting for the realistic loss factors. Despite the fact that the percentage time decreases with respect to medium and low wind sites, as shown in **Figure 6b** and **d**, we can see an improvement in grid utilization with lower specific power in all sites, with an outstanding performance from LSP-105.

With LSP turbines having higher capacity utilization factors, which lead to more consistent energy generation, the need for grid flexibility is expected to be less with a reduction in ancillary support. The European energy modeling study [9] justifies the same as the share of low Specific Power technology is seen to be higher in places where transmission constraints prevail. The study further emphasizes the

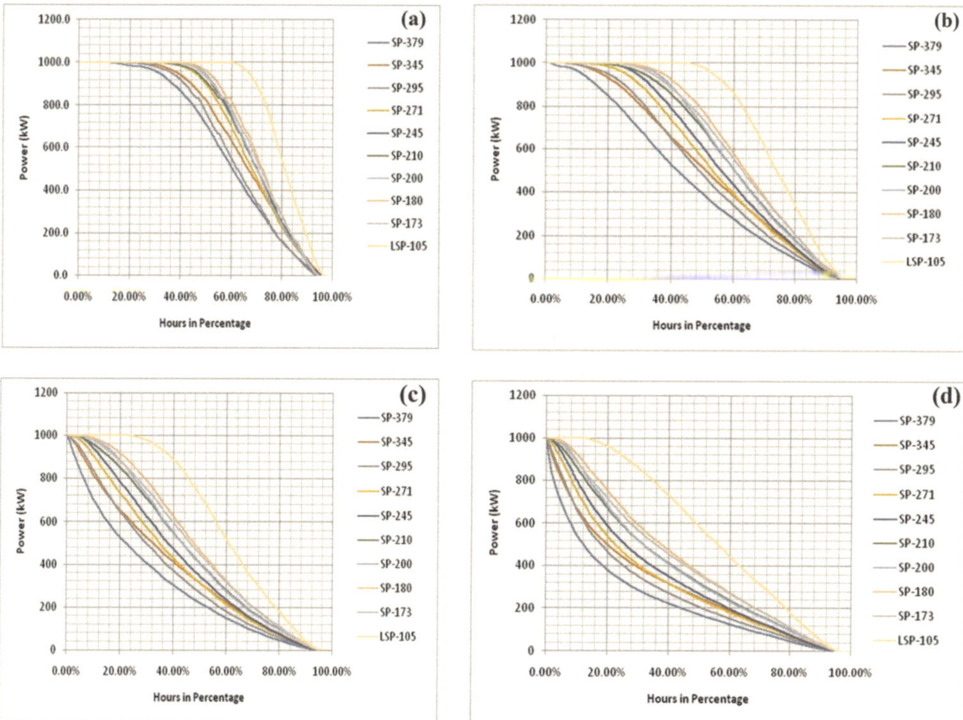

Figure 6.
Graphs to depict grid utilization pattern at high (a), medium (b), low (c) and low-coastal (d) sites.

introduction of low Specific Power technology into the Northern European energy system leads to a decrease in transmission investment, solar PV investment, and offshore wind investment.

4. LSP wind turbine with reduced cut-off wind speed

The square-cube law states when the diameter of a wind turbine's rotor increases, theoretical energy output increases by the square of the rotor diameter, but the volume and mass of material required to scale the rotor increases as the cube of the rotor diameter [14]. As on date, the wind industry has been able to maintain the scaling process economically by streamlining its processes, operations, material selection, etc. [15, 16]. Consequently, at some size, the cost of a LSP turbine may increase faster than the resulting energy output, making the further reduction of Specific Power uneconomical [17]. In such a scenario, reducing the cut-off wind speed of the wind turbine may be an area for consideration to reduce the cost. This low cut-off wind speed is particularly important since it does not need to be able to operate during high wind conditions leading to lighter turbine blades and reduced overall cost.

A recent European study used such unique LSP feature/Wind technology combining lower Specific Power (100 W/m^2) and low cut-off wind speed of 13 m/s [18] and the impact that both of these specifications has been found significant [9], wherein the comparative study reveals that the reduction of cut-off wind speed from 25 m/s to 13 m/s could have a blade mass that is up to 33% lighter than a conventional turbine of the same rotor diameter and that could lead to cost reduction. Furthermore, the mass of the blades is likely to have an impact on the cost of the rest of the turbine, meaning there could be savings in other parts such as nacelle, tower, and foundation as well.

Inspired by European studies [9, 18], a wide range of cut-off wind speeds were analyzed. The change in energy generation possibility at different wind sites is evaluated by reducing the cut-off wind speed of our Low Specific Power wind turbine (LSP-105) from 25 m/s to 20 m/s (LSP-105-20), 18 m/s (LSP-105-18), 15 m/s (LSP-105-15) and 13 m/s (LSP-105-13) as shown in **Table 3**. It can be seen from **Figure 7** that energy generation reduces on different scales, with respect to the wind sites, as cut-off wind speed reduces. It is seen that, in the low wind regions (Low W & Low coast W), the reduction in the generation is found to be very minimal and even the LSP-105-13 shows a similar generation profile compared to the LSP wind turbine with a 25 m/s cut off wind speed—a 2% reduction for the Low W site and a 0.6% reduction for the Low coast W site. This would have a major economic effect on the low wind sites, which have hitherto proved to be uneconomical.

It is noted that in the high and medium wind sites, it is not so economical to reduce the cut-off speed wherein there is a considerable reduction in generation (as

% Energy difference with respect to LSP-105 (25 m/s Cut-off wind speed)				
Sites	**LSP-105-20**	**LSP-105-18**	**LSP-105-15**	**LSP-105-13**
High W	3.58%	7.66%	19.65%	34.78%
Med W	0.01%	0.11%	2.56%	11.25%
Low W	0.00%	0.00%	0.22%	2.03%
Low coast W	0.00%	0.02%	0.12%	0.60%

Table 3.
Reduction in energy generation (%) while reducing the cut-off wind speed of the low SP wind turbine.

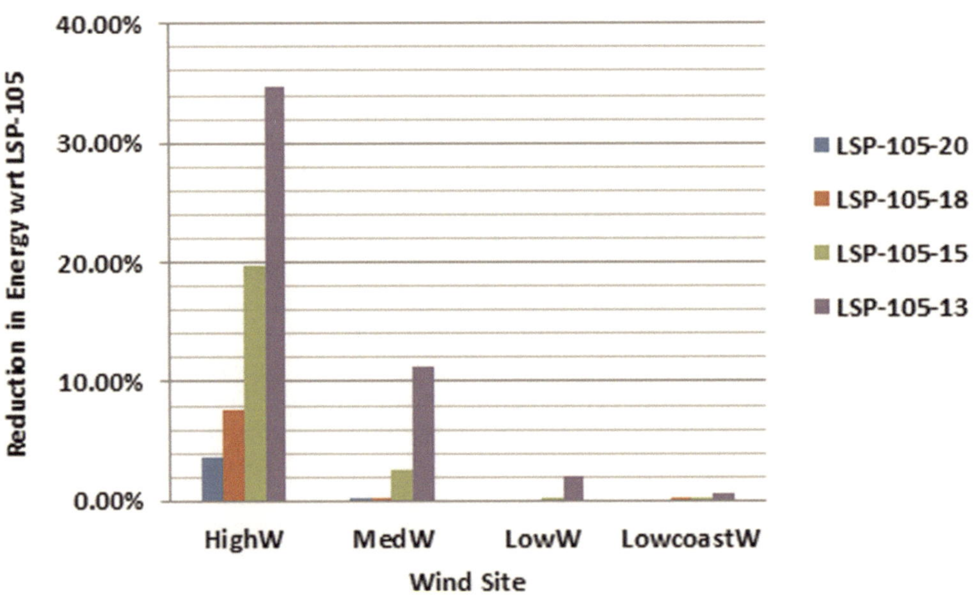

Figure 7.
Reduction in energy generation (%) while reducing the cut-off wind speed of the low SP wind turbine.

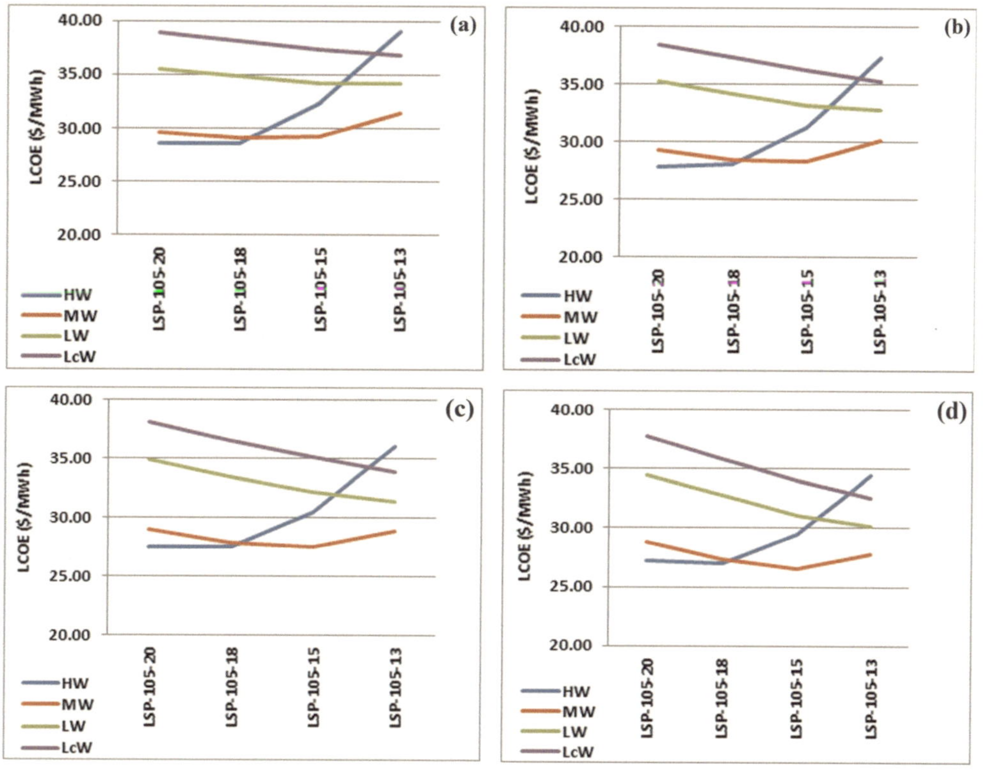

Figure 8.
LCOE sensitivity analysis for the reduction in cut-off wind speed of the low SP wind turbine – (a) shows 2% reduction in capital cost and (b), (c) and (d) shows 3%, 4% and 5% reduction respectively.

compared to a cut-off wind speed of 25 m/s, there is a reduction to the extent of 35% in terms of energy). The LCOE estimation also supports the said statement, as shown in **Figure 8**. The figure depicts a sensitivity analysis to understand the impact of reducing cut-off wind speed on LCOE, wherein the sensitivity analysis was carried out by reducing the capital cost of the LSP-105 by 2%, 3%, 4%, and 5% with respect to the reduction in cut-off wind speeds. Based on the graph, it is evident that even if the capital cost is categorically reduced to 5% for every cut-off wind speed reduction (viz., 20 m/s, 18 m/s, 15 m/s, and 13 m/s), the LCOE seems to be on an increasing trend in high and medium wind sites, mainly because of the anticipated energy loss at the higher wind speeds.

The results suggest that though reducing the cut-off wind speed will lead to significant energy loss in high wind sites and may impact the revenue and energy generation balance, in low wind sites it is economical to reduce cut-off wind speeds, as the expected reduction in blade mass and related savings will definitely outweigh the drop in energy generation. This is an encouraging takeaway in the present scenario and may lead to a conducive environment for more wind penetration into low wind sites.

5. Conclusion

The design of modern wind turbines has seen major changes over the course of history. For example, one of the most noticeable manifestations is the steady decline in the specific power of wind turbines over time. With an eye toward analyzing the prospects of further specific power reduction in the future, this chapter discusses the technological advantages provided by the low specific power (SP) turbine synthesized close to a target SP of 100 W/m^2, which was determined by ground-based measurements.

In accordance with the findings, low-specific power wind turbines can improve the capacity utilization factor, lower the cost of electricity, increase the value of wind, and better utilize the transmission system in all wind circumstances, albeit varying degrees. However, while the continuation of this trend toward lower specific power may not be sustainable, this analysis suggests that, under reasonable scenarios, low-specific power turbines could play a significant role in the future wind energy fleet, with their impact being particularly noticeable in low-wind areas of the world. Research into this area is predicted to be critical in the future, particularly in medium-to-low wind regimes, as these LSP wind turbines may be useful in identifying new potential sites and facilitating increased wind penetration into the grid.

Acknowledgements

Contributions from my colleagues Mr. Senthilkumar, Dr. Boopathi, Mr. Haribhaskar & others at the National Institute of Wind Energy, and our colleagues in the Ministry of New and Renewable Energy are duly acknowledged.

Appendix A: Basic definitions

A.1 Capacity utilization factor

Capacity Utilization Factor (CUF) is defined as the ratio of actual energy generation (kWh) to the maximum possible energy generation from a wind turbine/wind farm in a year.

$$\text{CUF} = \frac{Actual\ Energy\ Generation}{Rated\ Capacity\ x\ 365\ x\ 24.} \tag{A.1.1}$$

CUF is a metric often used to evaluate the technical performance of a wind turbine/wind farm. It is a measure of “how well the plant is utilized”.

A.2 Levelized Cost of Energy (LCOE)

Levelized Cost of Energy (LCOE) is the cost of generating electricity over its lifetime. It is an economic assessment and defines the minimum price at which energy must be sold for a project to break even.

A.3 Specific Power

Specific Power (SP) is defined as the ratio between the rated power of the turbine and its swept area [19], and is expressed in units of Watts per square meter (W/m^2) as shown below:

$$\text{Specific Power (SP)} = \frac{\text{Rated } Power}{\text{Swept Area}} \tag{A.3.1}$$

The specific power is required to be lower in order to extract more output from lower wind speeds. The following equation can explain the reason behind it by showing that power is proportional to the rotor area and wind speed [20].

$$P = \frac{1}{2}\rho A V^3 \tag{A.3.2}$$

Where, P–Power in the wind (W), ρ – Air Density (kg/m^3), A–Rotor swept area (m^2) and V–Wind Speed (m/s)

A.4 Wind class of a wind turbine

Wind class of a wind turbine helps to choose a suitable turbine for a particular site. The design of wind turbine and site conditions should complement each other for the successful operation of the wind turbine at the particular site, throughout its design life. As per the International Electro-technical Commission (IEC) standard IEC 61400-1, three wind classes (Class I, II & III) are categorized to represent high, medium, and low wind regimes. The classification is governed by the average annual wind speed (measured at the turbine's hub height), the speed of extreme gusts that could occur over 50 years, and how much turbulence is there at the wind site.

Generally, Class I turbines are designed to cope up with high average wind speeds in the range of 10 m/s. A Class II turbine is designed for windier sites up to 8.5 m/s average wind speed, whereas a Class III turbine is designed for a low wind site with the annual average wind speed up to 7.5 m/s. Each of the mentioned Wind Class further has subclasses to design turbines matching up with the wind turbulence at the site.

A.5 Power curve of wind turbine

Power curve of a wind turbine shows the relationship between the output power of the turbine and wind speed, provides a convenient way to model the performance of wind turbines. A typical power curve for a pitch regulated wind turbine is shown below (**Figure A1**):

A.6 Wind speed frequency distribution

Wind speed frequency distribution refers to the probability density function of wind speed and shows how well the wind speed values are distributed over the time period (maybe in a year).

The Weibull distribution is a two-parameter function (A & k) commonly used to fit the wind speed frequency distribution.

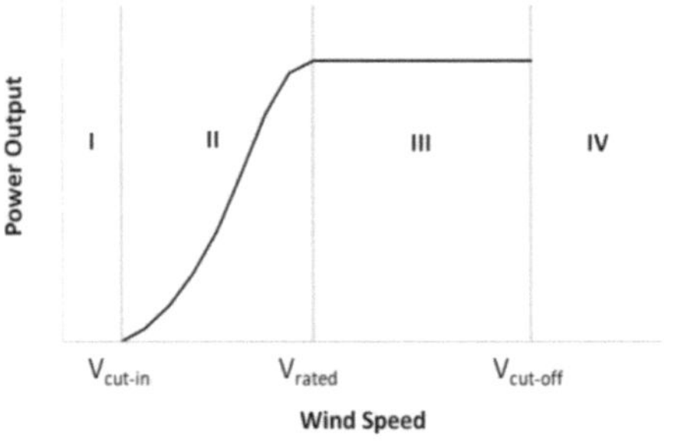

Figure A1.
Power curve of pitch-regulated blades.As shown in the figure, the power output in region-I is zero, as the wind speeds are less than the threshold minimum, known as the cut-in speed. In region-II, between the cut-in and the rated speed, the power production increases proportionally to the wind speed. In the region-III, a constant output (rated) is produced until the cut-off speed is attained through regulating the system. Beyond this speed (region IV) the turbine is shut down to protect its components from high winds; hence, it produces zero power in this region.

Weibull "A" (m/s) is known as the Weibull scale parameter; a measure for the characteristic wind speed of the distribution and it is proportional to the mean wind speed.

"k" is the Weibull shape parameter. It specifies the shape of a Weibull distribution and normally falls between 1 and 3. A small value for "k" signifies highly variable winds, while larger "k" describes relatively constant winds.

Author details

Balaraman Kannan* and Bastin Jeyaraj
National Institute of Wind Energy, Chennai, India

*Address all correspondence to: balaraman@rediffmail.com

References

[1] GWEC, Global Wind Report. 2021. Available from: https://gwec.net/wp-content/uploads/2021/03/GWEC-Global-Wind-Report-2021.pdf

[2] Winter, Wilhelm. European Wind Integration Study, ENTSO-E. 2010. Available from: http://www.wind-integration.eu/downloads/library/EWIS_Final_Report.pdf

[3] López Prol J, Steininger KW, Zilberman D. The cannibalization effect of wind and solar in the California wholesale electricity market. Energy Economics. 2020;**85**:104552. DOI: 10.1016/j.eneco.2019.104552

[4] Mills A, Wiser R. Changes in the economic value of wind energy and flexible resources at increasing penetration levels in the Rocky Mountain Power Area. Wind Energy. 2014;**17**(11):1711-1726

[5] Wiser R, Yang Z, Hand M, et al. Wind energy. In: Edenhofer O, Pichs-Madruga R, Sokona Y, et al., editors. IPCC Special Report on Renewable Energy Sources and Climate Change Mitigation. Cambridge; New York: Cambridge University Press; 2011. pp. 535-607

[6] Molly JP. Rated power of wind turbines: What is best. DEWI Magazine. 2011;**38**:49-57

[7] Dalla Riva A, Hethey J, Vitinxa A. Impacts of Wind Turbine Technology on The System Value of Wind In Europe, NREL/TP-6a20-70337. Golden, CO: National Renewable Energy Laboratory; 2017

[8] May N. The impact of wind power support schemes on technology choices. Energy Economics. 2017;**65**: 343-354

[9] Philip S et al. Competitiveness of a low specific power, low cut-out wind speed wind turbine in North and Central Europe towards 2050. Applied Energy. 2022;**306**(Part B):118043. DOI: 10.1016/j.apenergy.2021.118043

[10] DNV GL Supersized wind turbine blade study: R&D pathways for supersized wind turbine blades. Report, DNV GL, Seattle, WA, 2019

[11] Bolinger M, Lantz E, Wiser R, Hoen B, Rand J, Hammond R. Opportunities for and challenges to further reductions in the "specific power" rating of wind turbines installed in the United States. Wind Engineering. 2021;**45**(2):351-368. DOI: 10.1177/0309524X19901012

[12] Hirth L, Muller S. System-friendly wind power. Energy Economics. 2016; **56**:51-63

[13] Johansson V, Thorson L, Goop J, et al. Value of wind power—Implications from specific power. Energy. 2017;**126**: 352-360

[14] Burton T, Sharpe D, Jenkins N, et al. Wind Energy: Handbook. Chichester; New York: John Wiley; 2001

[15] Razdan P and Garrett P, Life cycle assessment of electricity production from an onshore V110–2.0 wind plant. Report, Vestas Wind Systems A/S, Aarhus, 2015

[16] Razdan P and Garrett P, Life cycle assessment of electricity production from an onshore V120–2.0 wind plant. Report, Vestas Wind Systems A/S, Aarhus, 2018

[17] Sieros G, Chaviaropoulos P, Sørensen J, et al. Upscaling wind

turbines: Theoretical and practical aspects and the impact on the cost of energy. Wind Energy. 2010;**15**:3-17

[18] Aa Madsen H et al. Initial performance and load analysis of the Low Wind turbine in comparison with a conventional turbine. Journal of Physics: Conference Series. 2020;**1618**:032011

[19] R. Wiser, M. Bolinger, G. Barbose, N. Darghouth, B. Hoen, A. Mills, J. Rand, D. Millstein, S. Jeong, K. Porter, N. Disanti, F. Oteri, 2018 Wind Technologies Market Report (2018) 103.

[20] G. Masters, Renewable and efficient electric power systems, 2004.

Chapter 4

Toward Self-Reliant Wind Farms

Anubhav Jain

Abstract

Large-scale integration of renewable energy generators, inverter-based resources and network interconnections into the grid brings forth a massive penetration of power electronic converters. This results in a highly dynamic environment that poses a risk to stability of system voltage and frequency and can ultimately trigger wide-area blackouts. Since conventional synchronous generation is being phased out, alternate sources must be included to provide support through ancillary services in future power networks. In a completely decarbonized system, they must also take the lead in ensuring stability and security by participating in blackout defense and network restoration. Offshore wind power plants are deemed suitable candidates due to their capability of providing large amounts of power with fast startup times and advanced control functionalities. However, a change in control philosophy to *grid forming* is required to enable a more active participation from the next-generation wind turbines. Such changes also have the potential to minimize dependence on auxiliary diesel gensets for a greener carbon footprint. This chapter aims to give insight into the forthcoming challenges and highlight potential solutions to make wind farms more self-reliant resulting in wind energy as cornerstone of the future electricity supply.

Keywords: wind, power electronics, converters, grid forming, greenstart, islanding, transient, stability

1. Introduction

It is evident that the undeniable rise of global warming owing to global greenhouse gas emissions from worldwide energy consumption that is not showing any signs of slowing down must be curbed to avoid its irreversible impact. Fossil fuels accounted for nearly 70% of the growth in energy demand in 2018 despite solar and wind growing at a double-digit pace since renewables were not able to catch up, with the power sector accounting for nearly two-thirds of emission growth [1]. Thus, green energy transition is of paramount importance, and the highest levels of ambition and effort on a global scale are needed to achieve the 1.5°C Paris climate goal, as highlighted in **Figure 1**. It is clear that the energy system of the carbon-neutral world of the future will have electricity as its backbone being responsible for almost half of the increase in total energy demand in 2018. However, a threefold expansion of power generation is required for electricity to assert itself as the *fuel of the future*, with its total share exceeding 60% by 2050 compared with 20% today [2].

The integration of renewable energy sources (RESs) on a large scale into power grids all around the world is currently the most efficient, cleanest and cost-effective way of

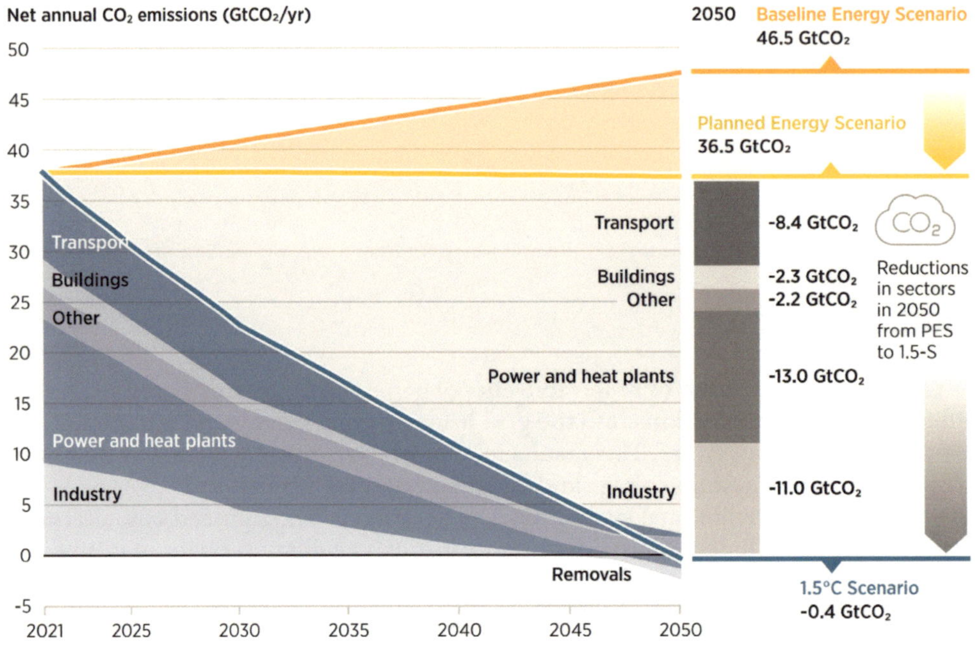

Figure 1.
Annual net CO_2 emissions (in Gt/yr) from 2021 to 2050: it is clear that current planned policies will yield only stabilization of global emissions by 2050 but a 27% baseline rise is likely if not fully implemented; reproduced from [2].

electrifying the world. Out of 170 countries in the world that have set up ambitious targets for decarbonization, 30 are already set to achieve net zero in the coming decades with strategic action plans [2]. A recent example of a significant milestone in a country's energy system can be seen in Denmark, where 50% of the electricity consumption in 2019 was supplied by wind and solar—with the former contributing a staggering 47% [3]. Overall strong renewables growth is expected beyond 2022 when the global installed capacity of coal-fired plants is set to peak before starting to decline in the following years and be overtaken by solar and wind energy in 2025 [4].

Since conventional thermal generation is being replaced by RES distributed across different time zones and climates, located far from consumers, cross-border interconnections over long distances have an undeniable role to play in the unified electric network of the future. They allow cost-effective grid expansion without significantly upgrading the current transmission grid infrastructure, thus ensuring efficient, flexible and resilient flow of clean energy. Although high-voltage alternating current (HVAC) connections are currently more common, its requirement of special reactive power compensation to prevent capacity drop-off becomes costly at longer distances. This makes high-voltage direct current (HVDC) connections a more economic alternative for efficient long-range bulk power transmission between countries, islands and offshore resources, and additionally their controllability due to voltage source converters (VSC) allows advanced functionalities to enhance stable grid operation [5].

1.1 The changing power grid

The aforementioned steps to electrify the globe, namely large-scale integration of renewables and dense interconnections via high-voltage corridors, are the key to achieving the climate neutrality target. However, such a transformation in the grid infrastructure introduces a massive amount of power electronic converters (PECs) that is essential to manage the variability of wind and solar energy sites. Additionally, the widely scattered distribution of resources necessitates larger power transits, which must be coped with expanding the transmission network either through capacity boosting using FACTS[1] or new HVDC corridors that both rely on PEC. Such a change in operational philosophy is paramount for efficient grid usage [6].

This paradigm shift in generation, transmission and demand naturally results in future power grids being very different from the current one, mainly owing to PEC introducing control interactions with faster time constants although needed for faster decision-making and advanced control functionalities. This creates a highly dynamic environment and poses a risk to power system stability, which has been investigated in detail in the MIGRATE[2] project [7]. Firstly, the PEC interface leads to inertial decoupling of rotating machines such as wind turbine generators (WTG) leading to a reduction in total inertia, which causes *frequency stability* issues due to to higher RoCoF[3] and dynamic frequency nadirs or peaks during power imbalance. Secondly, reduced fault-current contribution due to limited overloading capability of semiconductors makes *fault detection* harder in a converter-dominated environment. Morevoer, overburdened reactive power reserves due to increasing distance between load centres and generation, coupled with limited voltage control capabilities in the transmission grid, can lead to local/regional *voltage stability* issues and a reduced *transient stability* margin, especially during system contingencies.

The declining strength of the network and increasing threat to stability make it challenging to contain voltage and frequency excursions due to faults exposing a greater proportion of PEC-interfaced units to sudden under-voltage trips, which ultimately can trigger wide-area blackouts if large generation such as offshore wind farms[4] is involved, as has already been seen, for example, in South Australia (2016) and around London in the United Kingdom (2019) [8, 9].

1.2 Offshore wind as a cornerstone

The massive penetration of PEC in the grid due to the prevalence of renewable generation and inverter-based resources (IBR)[5] has increased the risk to power system stability and reliability, which translates to more frequent blackouts, especially in areas with high volume of RES [10]. Thermal generation plants that are conventionally responsible for maintaining power system stability and security are now being phased out in favor of renewables and non-traditional technologies due to societal decarbonization aims, rising fuel costs coupled with aging assets and decreasing load factors. Since this increases the cost of ancillary services and of warming-up the generators (cold start) to provide blackstart services, maintaining the status quo is not an option. Thus, considerable changes are required in developing technological

[1] Flexible AC transmission system.

[2] Massive integration of power electronic devices @ www.h2020-migrate.eu.

[3] Rate-of-change of frequency.

[4] Interchangeably referred to as offshore wind power plants.

[5] Encompassing FACTS, batteries, HVDC links and PEC-regulated loads such as electric vehicle battery chargers and variable speed motor drives.

capabilities and opening up new markets that facilitate non-traditional technologies[6] to support the system, adding more resilience against dependence on a single technology and alleviating reliance on specific transmission routes [11, 12].

Offshore wind is one of the fastest growing RESs in the world, and its rise has been possible thanks to technological innovations and strong policy support despite higher capital and operational costs to cope with the rough sea conditions. Contrary to space constraints for onshore wind, higher capacity factors and full load hours due to steady wind conditions together present a good business case for offshore wind, and coupled with economies of scale, its LCoE[7] is expected to drop to about 6–7 = €c/kWh by 2025, becoming competitive with onshore wind prices, which is the cheapest generation source in majority of places in the world [13]. Moreover, wind power has been shown to have CO_2 emissions about four times lower than solar and with offshore wind turbines and farms getting enormous in size, as highlighted in **Figure 2**, their carbon footprint could beat even the original large-scale zero-carbon source nuclear power [15].

Thus, offshore wind power has a significant role to play as an electricity generation source in the future power system. However, only decarbonization of the grid is not sufficient to meet our climate goals since sectors such as heavy industry and transport[8] are *difficult-to-electrify*. This highlights the need for alternate energy vectors that can be obtained from renewable sourced electricity, referred to by the umbrella term *Power-to-X* (P2X). Recently, offshore wind has gained attention to generate hydrogen for sector coupling as P2X can reduce curtailment needs since excess wind output can be transformed into hydrogen as energy storage also, thus enabling flexible demand and conferring grid benefits. Thus, *green hydrogen* has the potential to transfer the benefits of renewables beyond the electricity sector by facilitating decarbonization of all sectors of the economy, where currently no climate-neutral alternatives exist [16].

2. Looking into the future

As discussed above, large offshore wind power plants (WPPs) are deemed suitable candidates to take up the responsibility of maintaining power system stability and security, potentially even participating in early state network restoration. Since the fast-growing capacity of the overall site and the individual turbines is pushing offshore WPPs further away from the shore and into deeper waters, as indicated in **Figure 3**, HVDC transmission is more suited to export the power to the onshore grid. Although more expensive, the fully controllable VSC interface allows HVDC to provide various dynamic grid support services that enhance system stability and resilience [5].

However, large offshore WPPs today consist of upto 100s of WTGs connected in a large inter-array network of upto 70 km of subsea cables, with long HVAC or HVDC transmission corridor that transports the bulk power onshore, requiring either special reactive power compensation or large converter substations both offshore and onshore to manage the power flow efficiently. This makes the offshore WPP an aggregated unit with a converter-dominated environment and a very rich resonance

[6] Like interconnectors, sites with trip-to-houseload operation and aggregated units such as wind and solar, supported by energy storage systems.

[7] Levelised cost of electricity.

[8] For example, refining and metallurgical industry, long-distance trucking and shipping.

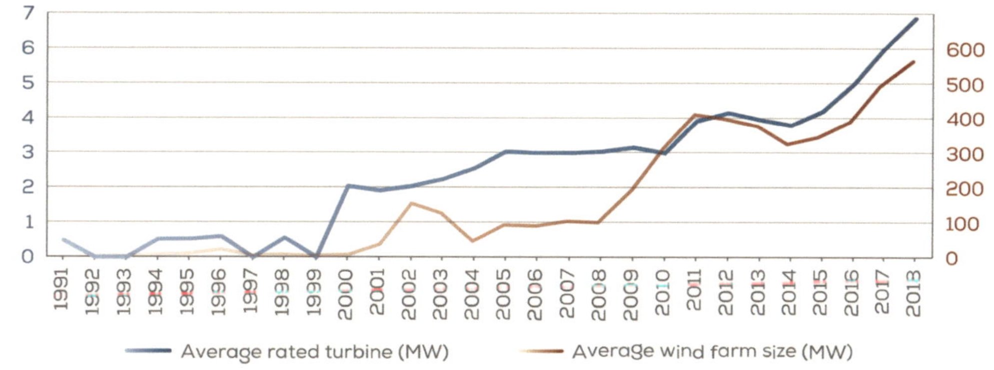

Figure 2.
Evolution of yearly average newly installed capacity of offshore wind turbines and farms; reproduced from [14]. Today the world's largest offshore wind farm is 1.2 GW Hornsea-1 (UK) and offshore wind turbines are already reaching ratings upto 15 MW, as of 2021.

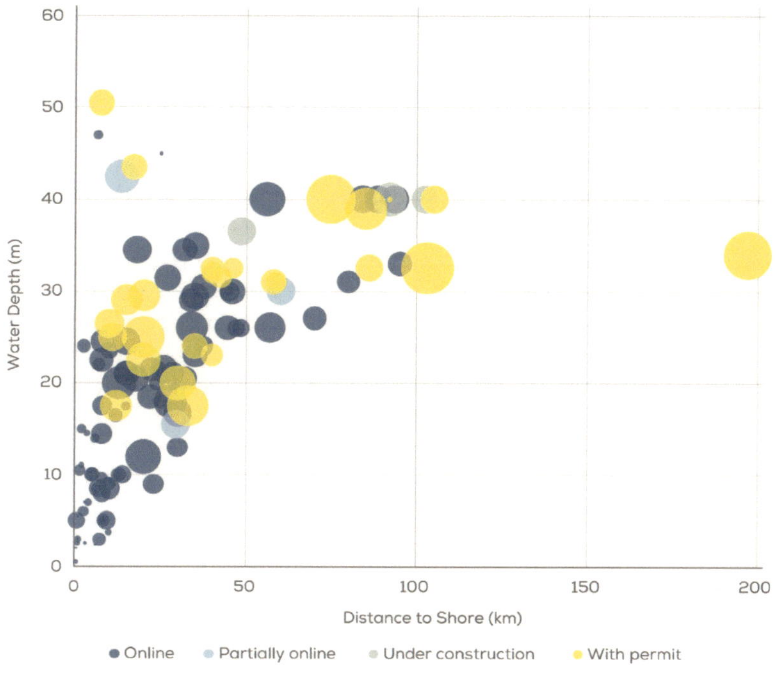

Figure 3.
Average water depth and distance to shore of bottom-fixed offshore wind farms: the overall site capacity is indicated by bubble size; reproduced from [14].

spectrum that must be first operated in a stable and robust manner before providing onshore grid services [17].

2.1 Toward next-generation wind farms

Traditionally, the first-generation grid-connected PEC-interfaced *grid-feeding* sources only supplied set-point based real and reactive power with basic survivability over a certain voltage and frequency range, while the second-generation devices now

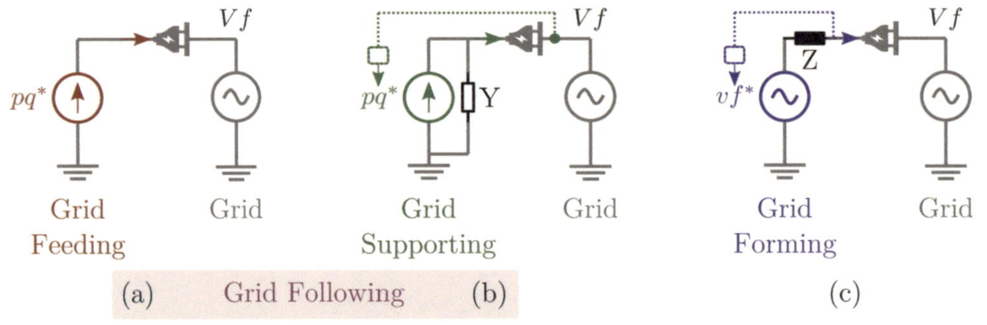

Figure 4.
Working philosophy of (a) grid feeding, (b) grid supporting and (c) grid forming units. Together (a) and (b) are referred to as grid following [18].

are required to provide more support to the grid, especially in areas of high RES energy penetration, hence classified as *grid-supporting* units [6], shown in **Figure 4**. Today wind turbines and wind farms already contribute to system stability through provision of ancillary services such as fast power reduction as frequency response to create a spinning reserve margin for primary frequency regulation, steady-state and dynamic reactive power control for voltage support, and fault ride through characteristics that involve reactive current injection for improving voltage stability during faults. In countries of high RES penetration, more active participation of wind power in voltage and frequency stability is required. Thus, local vocal control in weak grid scenario and fast frequency response is being increasingly demanded by the new grid codes. Furthermore, latest requirements include power oscillation damping and synthetic inertia provision that mimics the exchange of kinetic energy from a synchronous rotating mass by injecting active power in proportion to calculated RoCoF [19].

However, while synthetic inertia is an important service to compensate for reduced inertia in the power system, it requires measurement-based activation unlike conventional inertia that is based on inherent physical characteristics of synchronous machines [20]. This makes it an insufficient replacement for inertia as true synchronous inertia-like response can be achieved only by *grid-forming* units [21]. Today wind turbines are *grid-following (GFL)* in that they rely on an external grid voltage[9] to which the control latches using a PLL[10] for stable operation, effectively making the WTG behave as a current source injecting controlled power. Such wind turbines are currently exempt from network restoration services as not only can they not create their own voltage and lack controlled islanding capabilities but also connection in early stages of blackstart process can result in a recurrence of blackout due to the grid not being strong enough for large wind farms [17].

Now as the penetration of PEC increases, assumptions valid for stronger, traditional grids may no longer hold resulting in improper PLL behavior causing a negative impact of the controller power sharing and system stability [22]. This necessitates the next generation of converter-based units, called *grid-forming (GFM)* and shown in **Figure 4**, to be capable of proactively supporting the grid in all states, especially emergency and blackout without having to rely on services from synchronous generators. Thus, such GFM units must be able to take the lead in creating system voltage to *control instead of just supporting* its amplitude and frequency, prevent adverse control

9 Either the main onshore grid or the offshore grid formed by the HVDC converter.

10 Phase locked loop.

interactions, counter harmonics and unbalances and support system survival while contributing to short circuit power and system inertia—limited by the boundaries of energy storage capacity and available power rating [6]. These are key to ensure flexible, efficient and reliable operation of future decentralized converter-rich grids.

2.2 Some potential advantages

Contrary to conventional WTGs, a GFM wind turbine behaves as a voltage source, thus not only allowing outward energization without having to wait for completion of network reconstruction, but also potentially participating in sectionalizing strategy for defense against blackouts by ensuring continuity of power supply in a regional island or at the very least switching to trip-to-houseload operation that reduces restoration time compared with cold startup and facilitates bottom-up grid recovery [19].

Moreover, GFM-WTGs can potentially minimize dependence on the offshore auxiliary diesel generator that is associated with capital and operational costs[11], not to mention the high emissions—especially when the diesel generator is operating at full-load during unscheduled outages that can last upto 4–6 months. Since a GFM-WTG can produce produce power to keep itself warm avoiding the risk to its health[12] as long as the wind blows, replacing the diesel genset with few of these can yield significant economic and reliability benefits over the project's lifetime [17]. The CO_2 displacement can also contribute to reducing carbon-footprint taxes and ensure a smoother/faster granting of permits in the future.

Additionally, GFM-WTGs will also come in handy in the future to supply charging stations offshore essential for maritime vessel electrification, thus displacing a significant amount of marine fuel with green electricity and ultimately playing a key role in achieving our climate goals. Thus, blackstart and islanding capabilities unlocked by GFM-WTGs are an essential feature of self-reliant WPPs that not only make them more actively participate in advanced voltage and frequency control but also enable them to take up the responsibility of ensuring stable and robust grid operation without relying on synchronous generation, thus accelerating net-zero transition.

3. Wind-powered virtual synchronous generators

With the displacement of conventional generation, the most cost-effective way of ensuring reliable and secure operation of the power system while continuing with the decarbonization strategies is to make converters integrated into the grid behave like synchronous machines as it is impractical to now change the entire philosophy of operation and control of the entire power system without incurring significant losses to economy and livelihood all over the world. Thus, as mentioned before, GFM converter units are key to ensure flexible, efficient and reliable operation of future decentralized converter-rich grids. This concept is, however, not new having existed as different names in different converter applications. A simple representation of a

[11] Diesel genset on the offshore platform not only occupies extremely costly space but also requires annual refueling, special fire protection, personnel safety protocols and maintenance. In addition, a backup diesel generator is present to combat startup issues.

[12] due to moisture damage, icing up of electronics and equipment, bearing deformation, standstill marks and vibrations due to unfavorable yaw-axis orientation

GFM converter is shown in **Figure 4**, acting as a controlled AC voltage source behind an impendace. In principle, this is to some extent functionally similar to electronic oscillators used, for example, in clock generators in microelectronics, and has been researched upon so far in the power system context mainly for microgrids and to a limited extent in FACTS applications where it is traditionally called *voltage injection*.

Today GFM is a hot topic in power system research, especially in the context of converter-rich networks that are expected in the future due to high integration of RES and IBRs. The simplest way to mimic system-level functionalities of grid-connected synchronous generators such as self-regulation capability and communication-less power sharing is by employing traditional droop-based power controllers [23]. Further complexities can be added to fully utilize the controllability of PEC interface and emulate inertia and damping characteristics of synchronous machines—of course limited by the the stored energy capacity and available power rating that is connected at the converter backend. In this scope, the *Virtual Synchronous Machine* (VSM)-based concept was introduced using a detailed implementation of synchronous machine dynamics in its power control loop [24]. This *power-based* synchronization inspired from the swing equation that acts as a self-synchronizing block presents a more stable solution than the traditionally used *voltage-based* PLL, which requires enhancements to ensure stability under unbalanced and distorted voltage conditions such as voltage sag, weak grids or off-grid operation [25].

Since a converter can be made to behave as needed by modifying its controls, different schemes of the VSM family have been developed to emulate synchronous generator characteristics with varying degree of details, as reviewed in [26], resulting in a range of dynamic and transient stability performance needs such as independently adjustable inertia, damping and steady-state droop or highly non-linear behavior during grid faults and connection-disconnection processes [25]. Another synchronous machine-inspired control with different implementations and enhancements is the *Synchronverter*, a detailed review of which is given in [27]. Finally, non-linear GFM control strategies relying on the duality between PECs and synchronous machines have been recently developed, such as *Machine-Matching* and *Virtual Oscillator Control*, which have demonstrated robust steady-state droop-like behavior with a faster and better damped response during transients, albeit for low-power (microgrid) applications [28].

3.1 From grid following to grid forming…

Today the experience of power system operators with GFM-PEC is currently limited to battery facilities in South Australia where the Dalrymple Battery Energy Storage System (BESS) has successfully demonstrated some of the most immediately sought-after benefits of GFM converter-based resources such as virtual inertia. This was seen during the system separation event on November 16, 2019, when it provided almost instantaneous power injection proportional to RoCoF due to a slip/difference between its internal *virtual rotor* frequency and the grid frequency exactly mimicking the mechanism of inertial response from a synchronous spinning mass. Contrary to FFR, this does not require any measurement or frequency detection to start responding. The high-power GFM control additionally allows the BESS to behave closely to a synchronous generator during both steady-state and transient conditions, enabling advanced performance in stand-alone operation, when paralleling with other voltage and/or current sources or when grid-connected [29]. In addition, a secondary control housing the main automation and functional logic allows the provision of

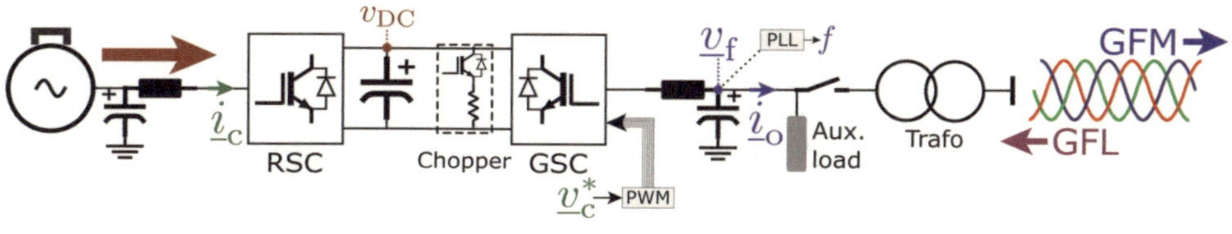

Figure 5.
Commonly used detailed electrical model for WTG to study electro-magnetic transients in simulation. Average models for the converter (i.e. approximating its behavior as a voltage or current source mainly considering only the control scheme) can also be used when switching transients are not of concern based on study needs.

reliability and flexibility services such as very low SCR[13] operation, seamless islanding transition and live-live grid resynchronization, support to non-synchronous system strength via short-term fault current injection[14], controlled islanded operation[15], blackstart capability with soft-start for limiting transformer inrush, and fast active power injection as part of SIPS[16] [29].

Advanced system support GFM functionalities can also be obtained from solar PV and wind energy. However, not all players are equally predestinated for GFM as the cost of development for each differ. While making a GFM-BESS is relatively straightforward since the backend is simply a voltage source, solar PV and wind require maximum power point tracking control for its backend resource capture. This adds some complexities as the backend RES control must now be integrated into the DC link controller. In this regard, making a GFM WTG is likely to require highest efforts, since the mechanical rotating mass puts limitations on the power and energy buffer that is needed to provide transient performance, for example, during a phase-jump event on the grid side.

Thus changes are needed mainly in control (software) and some in hardware to transform a GFL-WTG for operation as a GFM unit. For reference, a typical WTG electrical model used in simulations to study electro-magnetic transients can be seen in **Figure 5**, which consists of Grid Side Converter (GSC) and Rotor Side Converter (RSC) with their respective controls in different levels of detail depending on the study needs, along with the simplified generator electro-mechanical model and the turbine controller, for example, in [30]. In conventional GFL-WTG today, voltage is provided by an external grid, and the WTG connects with the aim to normally supply maximum power extracted from wind. This is achieved by controlling the generator speed to operate at optimal tip-speed ratio for each wind speed[17]. For this, the RSC uses standard vector-based (or field-oriented) torque control to extract electrical power from the generator based on a reference (torque/power) obtained from the turbine controller, which ensures maximum power point tracking. Additionally, pitch control is present in the turbine controller to limit the power captured that is essential to avoid over-speeding of the rotor at high/above-rated wind speeds or for intentional de-rated operation (using set-point control) [31]. The GSC then is tasked with

---

13 Short circuit ratio; very low means ≪1.5.

14 Overloading capability of 2 pu for 2 s.

15 Including wind farm power dispatch/curtailment for reducing unserved energy and distributed energy resources curtailment to avoid conditions due to uncontrolled local generation such as rooftop solar PV.

16 System Integrity Protection Scheme, which is a sectionalizing strategy to protect against complete area blackouts.

17 That varies, and hence, this is called maximum power point tracking.

controlling the DC link voltage by transporting all the power extracted to the grid. However, this requires a stiff grid point which can absorb the WTG power output without excessive voltage/frequency rise.

The primary difference between a GFM and GFL WTG's operation is that the former generates its own voltage, and so there is no need for an external grid voltage. This can be achieved in a relatively straightforward manner by implementing GFM control in the GSC. However, since power flow is now set by the load (when in islanded mode) or by set-point control (in grid-connected mode or parallel sharing), RSC must extract equal electrical power from the generator to regulate the DC link voltage. Thus, the torque or power reference now (needed by RSC) comes from a DC link controller rather than the turbine controller, as stated previously for GFL-WTGs. The turbine controller's main task now is to regulate the speed at rated and prevent over-speeding by using the pitch controller when necessary [32]. Alternatively, the generator speed can be controlled for sub-optimal tip-speed ratio which requires a speed controller that receives reference from DC link control to feed the generator torque control [33]. It is likely that pitch control would be operated more in GFM operation and so along with alternate revenue streams to make up for the wasted wind power, the impact on mechanical loading and in turn the lifetime of the turbine must also be investigated since GFM controls can result in a different power ripple spectra than traditional GFL WTGs potentially leading to higher levels of vibrations at frequencies close to the natural resonant modes of the generator shaft, rotor and tower [34].

3.2 But it is not so easy ...

Recently, for the first time worldwide, Scottish Power Renewables in collaboration with Siemens-Gamesa Renewable Energy has successfully demonstrated the ability of onshore GFM-WTGs to operate in island condition supplying local loads while supporting conventional GFL-WTGs and ultimately energizing the upstream grid transmission network [35]. While GFM control allows WTG to provide frequency stability services, notably phase-step power injection in response to phase jumps in the grid and inertial response proportional to RoCoF to arrest frequency events autonomously and immediately, there are certain limitations that must be overcome for robust operation [34]. Since an individual WTG can find it difficult or impossible even with a high inertia setting to extract the inertial response from the background power ramps due to wind speed fluctuations, farm-level aggregation must be taken into account. Additionally, at low/zero power, only a small power/energy response is possible from the WTG DC link capacitance as the rotor does not have sufficient energy yet. Although an extra energy storage device[18] can allow a more guaranteed response over a wider range of operating conditions but adding significant additional cost due to high energy required for more extreme events (more than 1 Hz/s). In absence of this, there is a risk of large reduction in rotor speed drawing the WTG into recovery (resulting in a second power output dip), or worse below cut-out speed if wind is low enough. This is the opposite of what is desired and can cause further grid instabilities and even lead to blackouts due to system separation if many WTGs are involved [34]. While dynamic inertia, curtailment and deliberate sub-optimal

[18] Integrated within the DC bus of the WTG or connected as a separate unit through converter interface.

operation are technical solutions to be considered, challenges of grid code compliance must be overcome and alternate revenue streams be opened.

3.3 Non-existent markets

Recently, UK's energy regulator OFGEM has approved the first ever technical specification GC-0137 of GFM control from PEC integrated into the grid, proposed by National Grid ESO [36]. Although non-mandatory, it marks a step as signficant as RES integration itself in the net-zero transition because of providing essential clarity for describing synchronous coupling with power grid in a technology neutral manner, which will enable any connecting power module utilizing PEC technology (e.g. wind, solar, HVDC) to offer grid stability services more actively. Despite a long way yet to go with testing and coordination coming next, such a specification already breaks the circular problem faced by manufacturers and system operators, fed by lack of widely available functionalities from IBRs today due to unclear specifications or demand leading to operational constraints making it even less attractive for them to develop resulting in shrinking market volumes for OEMs[19].

The full potential of GFM power generators is however unlocked through the provision of blackstart and islanding capabilities along with voltage and frequency control that are essential for ensuring stability, reliability and security in future converter-rich grids without relying on synchronous generators, while compensating for the cost of developing such functionalities in RES such as wind and solar. The Dalrymple BESS in Australia current relies on only a few revenue streams compared with its technical capabilities, as mentioned before, namely inertial response for frequency stability, islanding to reduce unserved energy, frequency control ancillary services and energy arbitrage [29]. However, more services are possible such as blackstart, short-term fault current provision, voltage regulation and pre-emptive response for SIPS, but these are not yet monitized due to lack of any mechanism to do so under current market and regulatory frameworks [29]. While GFM-related capabilities are relatively more straightforward to implement in individual PEC-interfaced units such as BESS, solar-PV or WTGs, there are many challenges to ensure robust and reliable operation of GFM units aggregated into parks such as large offshore WPPs, despite the numerous advantages available from them such as reduced LCoE, increased energy production due to steadier wind conditions at sea, no inland space and noise constraints and higher reliability by combining electrical resource and maintenance facilities, justifying the cost of grid connection.

3.4 Green-starting wind farms

Since a large offshore WPP is considered as an aggregated unit consisting of many active components such as WTG converters, offshore HVDC converter in case of HVDC-connected WPPs that can have adverse control interactions in certain operating conditions and are a source of harmonics which can trigger resonances due to the presence of long high-voltage inter-array and export cables, transformers and filters, especially in the offshore network. Furthermore the offshore grid has low damping in the network provided mainly by auxiliary load that is limited to 1% for WTG and 0.1% for offshore substation, which can create situations in certain scenarios such as

[19] Original equipment manufacturers.

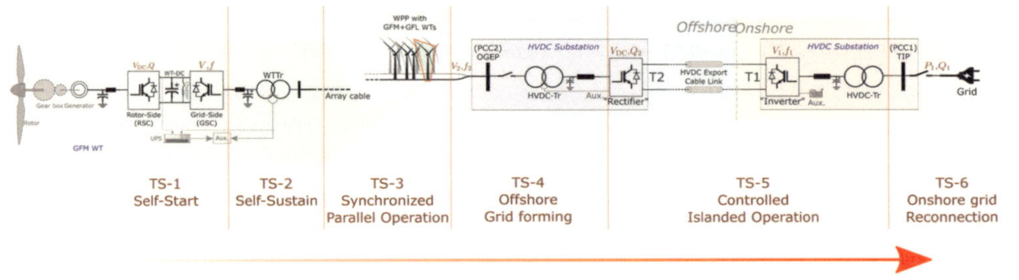

Figure 6.
Target states in the greenstart energization sequence of an HVDC-connected offshore WPP; reproduced from [18].

energization where transient and harmonic stability can be a challenge to ensure. Thus, stable and robust operation of the offshore network and export link must be ensured before the WPP can actively participate in not only supporting the grid with voltage and frequency ancillary services but also provide essential stability and reliability services through GFM control functionalities, like the ones mentioned before.

The energization and stable operation of a large offshore WPP upto the transmission interface point where it connects to the onshore grid, as shown in **Figure 6**, have recently been referred to as *greenstart* to distinguish from the commonly used blackstart of the power grid on a larger scale since it is of more concern to the wind farm developer. However, to better understand the range of technical challenges associated with it, the entire sequence can be divided into different *target states* just like traditional power system restoration being comprised of different stages, namely preparation and defensive actions, system build-up by blackstart units and transmission backbone energization followed by load restoration and meshing for resilience. These target states as highlighted in **Figure 6** start with initial energization of WTG3 auxiliary load by a backup supply (TS-1) followed by houseload operation when the rotor is oriented to the wind (TS-2). Then multiple GFM and GFL WTGs must synchronize for operating in parallel (TS-3) to emulate a voltage source strong enough to energize the offshore network (TS-4) while ensuring stable and robust islanded operation of the HVDC link (TS-5) before finally connecting to the onshore grid for block load pickup or re-synchronization (TS-6) [17]. The challenges associated with each stage are discussed below.

3.4.1 Self-start and sustain

At the individual WTG level, an industrial grade UPS[20] as auxiliary power supply is currently used to keep energized the central control units for braking, yaw and pitch, dehumidifiers and heating units, grease lubrication system, fire protection, relays, hub computers, distribution boards and the SCADA[21] interfaces and positioning and warning lighting system[22]—all of which are essential to ensure safe and reliable operation of the WTG offshore. However, a UPS can provide idling mode energy sufficient only for a grid outage upto few days after which the WTG enters shutdown, which is not good for its health due to potential vulnerability to damage

[20] Uninterruptible power supply.

[21] Supervisory control and data acquisition.

[22] To avoid collisions with ships and aeroplanes.

from moisture, icing up of electronics, bearing deformation, standstill marks and vibrations due to unfavorable yaw-axis orientation—all of which impact the lifetime and efficiency, contributing to high O&M[23] costs.

Since a GFM-WTG can produce power to sustain its own houseload as long as wind blows, not only relieving any dependence on external supplies but also re-charging the UPS, it can more importantly provide outward energizing power to inter-array cables, transformers and filters, for supplying auxiliary loads of other GFL-WTGs in the network and even the offshore substation [35]. However, additional energy storage is necessary to support the WTG in dealing with the demanding *power* (MW) transients during energization (even if soft-start is considered) and network configuration changes (such as connection-disconnection of WTGs) by avoiding severe mechanical stree, but also ensuring enough *energy* capacity (MW h) in the system for the entire duration to avoid speed recovery related insecurities, especially at the start when the rotor has insufficient energy. While research is happening in integrating batteries, supercapacitors and flywheels as high-power-density sources, high-energy-density alternatives to diesel such as hydrogen are gaining momentum for long-term energy management [17]. The not-so-insignificant costs of such storage systems are expeceted to be compensated by future upcoming markets that utilize enhanced services from GFM converters such as blackstart, islanding and voltage/frequency control.

3.4.2 Multi-unit grid forming

As mentioned before, since an aggregated unit such as an offshore WPP consists of many (order of upto 100) WTGs, their synchronized parallel operation is essential to allow any energization capabilities of enhanced stability services onshore. Consequently, many questions must be answered to ensure cost-effective self-reliant operation. Firstly, since GFM-WTGs add to the capital cost of the project, it is essential to fine-tune the number of GFM-WTGs required. This must take into account numerous factors, especially the application under consideration, which could be auxiliary power needs for which reliability comparisons and carbon emissions must be taken into account, or for providing onshore services for which relevant markets and regulatory frameworks must exist to provide a sound business case. From a technical point of view, the most important is the choice of GFM control to be implemented in the WTG-GSC and while a single GFM unit is relatively straightforward to operate maintaining stability in grid-connected and islanded modes, the challenge is to optimally tune the parameters for operating many units in parallel that maintain synchronism in the face of large network transients and configuration changes.

Recent studies have shown that while different GFM control strategies are able to deal with the transients such as energization in a controlled manner maintaining stability of voltage and frequency at the offshore terminal, there transient behavior exhibits differences and some are prone to more oscillations than others such as VSM-based control due to reduced system damping owing to lower control bandwidths that push the system closer to instability, while Direct Power Control–based strategy exhibits more stiff control over the voltage and frequency resulting in superior performance [37].

[23] Operational and maintenance.

Thus, for the entire offshore WPP to behave as a strong enough GFM source without any loss of synchronism between the multiple parallel units, extensive system-level studies are required for an optimal tuning of all the different levels of control loops, which ensures transient stability across different operational scenarios and can help reduce costs too. For example, while it is suggested that a ratio of 3:1 (GFM:GFL) is safe to use, especially for a mix of turbines from different manufacturers, an optimized set of control parameters can allow a single GFM-WTG to support upto 20 GFL-WTGs providing robust operation at least in small load steps [35]. This is however valid only for onshore WPPs since offshore WPPs tend to be much larger in capacity with longer and higher cross section of inter-array cables thus lead to more demanding transient and dynamic requirements for GFM WTGs.

That said, ensuring stability for high-power converters can be quite a challenge since the higher rating that puts a limit on the switching frequency of the semiconductor devices due to loss considerations and so the controller bandwidths allowed are lesser, ultimately translating into lower stability margins [37]. Furthermore to complicate matters, improving one oscillation mode can trigger another and the tuning strategy used for a single converter unit might not be applicable directly to multiple units operating in parallel [38]. This necessitates case-specific enhancements for active damping including but not limited to virtual impedance, cross feed-forward compensations and lead lag controllers. It is important to note here that solutions such as master-slave approach cannot be used for numerous assets spread across several kilometers as in large offshore networks because high-bandwidth communication links needed to ensure reliable, robust, low-power and secure operation not only become increasingly costly but also the additional delays are undesirable as they can trigger control instability [23].

3.4.3 Stable, robust and safe islanding

In order for a large offshore WPP to supply essential GFM services to the onshore grid[24] which can help compensate for the extra developmental cost of GFM-WTG technology and any additional energy storage needs, the offshore network of inter-array cables, transformers, filters and WTG converters must be controlled in a stable and robust manner with the ability to deal with contingencies and maintain high security and reliability. This is however quite challenging to achieve since the offshore network is a converter-dominated environment which makes its dynamics very different from traditional onshore grids now, which are also expected to exhibit similar characteristics in the decarbonized future.

The large share of high-voltage cables, transformers and filters offshore provides a resonance-rich spectrum which is not static due to various configurations of cables and WTGs in service, especially during contingencies. These resonances are prone to be excited by the harmonic injection of converters, especially when resonant frequencies corresponding to longer cable lengths (and thus larger capacitance) are in range of controller bandwidths, making *harmonic stability* critical to assess. Additionally, reduced online generation and loading in the early stages of energization lead to lesser system damping resulting in sharp resonant points which can be triggered by slight changes in the network configuration. This must be avoided as sustained over-

---

[24] Like inertia, voltage and frequency control and support, blackstart and islanding capabilities.

voltages cause accelerated aging, insulation degradation and component failure due to dielectric and thermal stress on the equipment [39].

A completely new regime of challenges is introduced due to unexpected interactions between controllers and filters of nearby converters present in the system since cross-coupling between electro-mechanical dynamics and electro-magnetic transients due to the wide-ranging control timescales of PEC can lead to negative damping in the control output admittance. This makes it far more complicated to tune parameters and ensure stability and robustness of operation in different scenarios, especially with changing network configurations during large load steps, WTG connection/disconnection and energization sequence involving long cable switchings. Impedance and eigenvalue-based methods are commonly used for system-level stability analysis and while reduced order models can reveal great insight, detailed models are becoming increasingly important to get a more holistic view since assumptions valid for small-signal models and traditional strong grids do not hold for large transients and the offshore weak grid case.

Furthermore, since the offshore network formed by GFM-WTGs represents a relatively weak grid compared with today's offshore HVDC-VSC-based grid that is backed a strong onshore grid, *transient stability* of PLL-connected GFL-WTGs proves to be a challenge, which can be attributed to the well-known problems of an un-enhanced PLL in weak grid operation [22], especially during large reactive power steps when large cable switchings are involved. This further affects the choice GFM control strategy and its tuning since instabilities can be highly sensitive to certain parameters making it difficult to maintain synchronism and cause maloperation of protection, posing a risk of disconnection of WTGs triggering a re-blackout, especially in the early stages of energization or during low-power operation when less generation and load are connected [18].

Last but not the least, resilience to faults in the offshore grid and HVDC transmission is essential to allow robust operation and reduce the risk of a re-blackout. Although the Dalrymple BESS in Australia can provide short-term overload current for clearing faults, the normal protection settings based on high fault current in-feeds (over-current and earth fault) are insufficient to protect the network in island or blackstart mode since the injection from the WTGs is too low to trigger the relay pick-up, especially at low numbers of WTGs. Thus, a special set of settings along with voltage protection is required to protect the network over the full range of planned operating scenarios. However, several fault scenarios may still not be picked up and a re-design of the protection scheme may be needed [40]. GFM-WTGs can potentially help by actively limiting the current causing the fault to automatically extinguish [41].

4. Summary

The extensive integration of renewable energy and HVDC that is yet to grow in the coming years is already changing the dynamics of the grid and pushing it to its limits, which necessitates the advent of a new class of converter-based power modules, namely grid forming. It is foreseen that such modifications will not only enable them to participate in supporting grid stability but also allow them to contribute more actively in providing essential services to ensure reliable and secure operation of future decarbonized electric network. Offshore wind power plants are deemed to play a key role in achieving this reality, but there are many obstacles yet to overcome. However, a dialog between system operators, developers and manufacturers to facilitate the development of required technology and market for a faster uptake of the responsibilities conventionally targeted to large thermal power plants by renewable sources and aggregated non-traditional technologies is already gaining momentum.

Recent studies and demonstrations have shown that grid forming wind turbines can not only provide blackstart and islanding capabilities but also support conventional grid following wind turbines while maintaining complete control over voltage and frequency with a desired response to transient events that contributes to stability of the grid. However, there are limits to the capabilities both at the individual turbine level and the aggregated system/farm level. In addition to the need of additional energy storage and new revenue streams along with potential re-design of the protection scheme, adverse control interactions in the resonance-rich offshore network make harmonic and transient stability critical to assess for ensuring reliable and secure operation. Thus, there is still a long way to go to make self-reliant wind farms a reality, but their potential to yield significant operational cost benefits while also reducing the carbon footprint over the project's lifetime makes them an unavoidable player in helping meet our climate goals while ensuring high reliability and resilience of electricity supply with the most cost-effective and efficient usage of grid infrastructure.

Acknowledgements

The author would like to thank the InnoDC project that has received funding from the European Union's Horizon 2020 research and innovation programme under the Marie Skłodowska-Curie grant agreement number 765585.

Conflict of interest

The authors declare no conflict of interest.

Thanks

The author would like to thank his Ph.D. study supervisors, Prof. Dr. Nicolaos A. Cutululis and Dr. Jayachandra N. Sakamuri, discussions with whom contributed significantly to this work. The author would also like to thank the Department of Wind Energy at the Technical University of Denmark (DTU) and Vattenfall Vindkraft A/S (Denmark) for hosting the author's Ph.D. study and start of his career in the renewable energy industry.

Abbreviations

BESS	battery energy storage system
GFL	grid following
GFM	grid forming
HVAC	high-voltage alternating current
HVDC	high-voltage direct current
IBR	inverter-based resources

PEC	power electronics converter
RES	renewable energy sources
VSM	virtual synchronous machine
VSC	voltage source converters
WPP	wind power plant
WTG	wind turbine generator

Author details

Anubhav Jain
Vattenfall Vindkraft A/S, Kolding, Denmark

*Address all correspondence to: zefirebznt@gmail.com

References

[1] International Energy Agency (IEA). Global Energy & CO_2 Status Report (GECO). 2019

[2] International Renewable Energy Agency (IRENA). World Energy Transitions Outlook: 1.5°C Pathway. 2021

[3] International Energy Agency (IEA). Electricity Information: Denmark. 2020

[4] International Energy Agency (IEA). Renewables 2020: Analysis and forecast to 2025. 2020

[5] Michael PB, Johnson BK. The ABCs of HVDC transmission technologies. IEEE Power and Energy Magazine. 2007;**5**(2): 32-44

[6] ENTSO-E. High Penetration of Power Electronic Interfaced Power Sources and the Potential Contribution of Grid Forming Converters. 2019

[7] Massive Integration of Power Electronic devices (MIGRATE). D1.1 Report on Systemic Issues. Project deliverable. 2016

[8] Australian Energy Market Operator. Black system, South Australia, 28 September 2016—Final report. 2017

[9] National Grid. Technical report on the events of 9 August 2019. 2019

[10] The National HVDC Centre. Maximising HVDC Support for GB Black Start and System Restoration HVDC-BS-001. 2019

[11] Elia. Study on the Review of the Black Start Ancillary Services. 2018

[12] National Grid. Black Start from Non—Traditional Generation Technologies. 2019

[13] International Renewable Energy Agency (IRENA). Renewable Power Generation Costs in 2018. 2019

[14] Wind Europe. Offshore wind in Europe: Key trends and statistics 2018. 2019

[15] Helman C. How Green Is Wind Power, Really? 2021

[16] Stori V. Offshore Wind to Green Hydrogen: Insights from Europe. Report. Clean Energy States Alliance (CESA); 2021

[17] Jain A, Das K, Göksu Ö, Cutululis NA. Control solutions for blackstart capability and islanding operation of offshore wind power plants. In: 17th International Wind Integration Workshop; Stockholm. Darmstadt Germany: Energynautics GmbH; 2018

[18] Jain A. Green & black-starting HVDC-connected offshore wind power plants: Grid forming control, energization transients and islanding capabilities [PhD thesis]. Denmark: Department of Wind Energy, Technical University of Denmark (DTU); 2022. DOI: 10.11581/10.11581/DTU:00000103

[19] Jain A, Sakamuri JN, Das K, Göksu Ö, Cutululis NA. Functional Requirements for Blackstart and Power System Restoration from Wind Power Plants. In: 2nd International Conference on Large-Scale Grid Integration of Renewable Energy in India; New Delhi. Energynautics GmbH; 2019

[20] Eriksson R, Modig N, Elkington K. Synthetic inertia versus fast frequency response: A definition. IET Renewable Power Generation. 2018;**12**(5):507-514

[21] Fang J, Li H, Tang Y, Blaabjerg F. On the inertia of future more-electronics

power systems. IEEE Journal of Emerging and Selected Topics in Power Electronics. 2019;**7**(4):2130-2146

[22] Ray I, Tolbert LM. The case against phase-locked loops in weak AC grids. In: 2019 IEEE Electrical Power and Energy Conference, EPEC 2019. 2019. pp. 1-5

[23] Rocabert J, Luna A, Blaabjerg F, Rodriguez P. Control of power converters in AC microgrids. IEEE Transactions on Power Electronics. 2012; **27**(11):4734-4749

[24] Beck HP, Hesse R. Virtual synchronous machine. In: 9th International Conference on Electrical Power Quality and Utilisation (EPQU); Barcelona. 2007

[25] Wang X, Taul MG, Wu H. Grid-synchronization stability of converter-based resources—an overview. IEEE Open Journal of Industry Applications. 2020;**1**:115-134. DOI: 10.1109/OJIA.2020.3020392

[26] D'Arco S, Suul JA. Virtual synchronous machines—Classification of implementations and analysis of equivalence to droop controllers for microgrids. In: Processing of IEEE Grenoble Conference Grenoble, France: IEEE (Institute of Electrical and Electronics Engineers); 2013

[27] Vasudevan KR, Ramachandaramurthy VK, Babu TS, Pouryekta A. Synchronverter: A comprehensive review of modifications, stability assessment, applications and future perspectives. IEEE Access. 2020; **8**:131565-131589

[28] Unruh P, Nuschke M, Strauß P, Welck F. Overview on grid-forming inverter control methods. MDPI Energies. 2020;**13**(10). DOI: 10.3390/en13102589

[29] Cherevatskiy S, Sproul S, Zabihi S, Korte R, Klingenberg H, Buchholz B, et al. Grid forming energy storage system addresses challenges of grids with high penetration of renewables (a case study). In: CIGRE 2020 Paris Session 48, Paris, France: CIGRE; 2020. Available from: https://e-cigre.org/publication/SESSION2020_C2-C6-322

[30] Clark K, Miller NW, Sanchez-Gasca JJ. Modeling of GE Wind Turbine-Generators for Grid Studies. Report. General Electric International, Inc; 2010

[31] Thongam JS, Ouhrouche M. MPPT control methods in wind energy conversion systems. In: Carriveau R, editor. Fundamental and Advanced Topics in Wind Power. Rijeka: IntechOpen; 2011. Chapter 15

[32] Shan M, Shan W, Welck F, Duckwitz D. Design and laboratory test of black-start control mode for wind turbines. Wind Energy. 2020;**23**(3):763-778

[33] Yuan X, Wang F, Boroyevich D, Li Y, Burgos R. DC-link voltage control of a full power converter for wind generator operating in weak-grid systems. IEEE Transactions on Power Electronics. 2009;**24**(9):2178-2192

[34] Roscoe A, Brogan P, Elliott D, Knueppel T, Gutierrez I, Campion J-CP, et al. Practical experience of operating a grid forming wind park and its response to system events. In: 18th Wind Integration Workshop, Dublin, Ireland: Energynautics GmbH Robert-Bosch-Straße Darmstadt Germany; 2019

[35] Roscoe A, Brogan P, Elliott D, Knueppel T, Gutierrez I, Crolla P, et al. Practical experience of providing enhanced grid forming services from an onshore wind park. In: 18th Wind Integration Workshop, Dublin, Ireland:

Energynautics GmbH Robert-Bosch-Straße Darmstadt Germany; 2019

[36] National Grid ESO. GC0137: Minimum Specification Required for Provision of GB Grid Forming (GBGF) Capability. 2021. Available from: https://www.nationalgrideso.com/industry-information/codes/grid-code-old/modifications/gc0137-minimum-specification-required

[37] Jain A, Sakamuri JN, Cutululis NA. Grid-forming control strategies for black start by offshore wind power plants. Wind Energy Science. 2020;**5**(4):1297-1313

[38] Sun C, Joos G, Bouffard F. Identification of low-frequency oscillation mode and improved damping design for virtual synchronous machines in microgrid. IET Generation, Transmission and Distribution. 2019;**13**(14):2993-3001

[39] Cigré, WG B3.36. Special Considerations for AC Collector Systems and Substations Associated With Hvdc-Connected Wind Power Plants. CIGRE Technical Brochure. 612. 2015

[40] Gutierrez I, Crolla P, Roscoe A, Brogan P, Elliott D, Knueppel T, et al. Operator considerations for the implementation of testing enhanced grid forming services on an onshore wind park. In: 19th Wind Integration Workshop, Dublin, Ireland: Energynautics GmbH Robert-Bosch-Straße Darmstadt Germany; 2020

[41] Arasteh A, Jain A, Göksu O, Zeni L, Cutululis NA. Fault ride through capability of grid forming wind turbines: A comparison of control schemes. In: The 9th Renewable Power Generation Conference (RPG Dublin Online 2021), Dublin, Ireland: IET (Institution of Engineering and Technology). Vol. 2021. 2021. pp. 37-41

Chapter 5

Offshore Wind Farm Grid Connection with Diode Rectifier Unit HVDC and Phase Shifting Transformer

Lijun Cai

Abstract

This chapter describes a new method for operating the offshore wind farm (OWF) with diode rectifier unit (DRU)-high-voltage direct current (HVDC) (offshore side), where a medium voltage (MV) submarine cable is in parallel operation with DRU-HVDC link. In order to avoid uncontrolled current flow through the MV submarine cable, a phase shifting transformer (PST) is applied on the onshore side of the MV submarine cable. The application of PST is to ensure the smooth blackstart and stable operation of the OWF and DRU-HVDC link. Both static and dynamic behaviors of the proposed method are presented in this chapter and the simulation results validate the proposed method.

Keywords: Offshore wind farm, phase shifting transformer, blackstart, diode rectifier unit (DRU), HVDC

1. Introduction

Since the wind energy is renewable and environmental natural resource, the utilization of wind power plant increased quickly. In the future, the development of wind power utilization will focus on large offshore wind farms (OWFs) [1–4].

Especially, many planned OWFs become larger and more distant from the onshore grid. Conventional HVAC transmission is not flexible and limited due to large charging currents of the submarine cables. With the development of power electronics, OWFs with voltage source converter (VSC)-high-voltage direct current (HVDC) grid connection become more popular. The VSC-HVDC technology is based on insulated gate bipolar transistors (IGBTs), and it offers significant advantages over the thyristor-based line-commutated converter-HVDC (LCC-HVDC) technology. VSC-HVDC converters can be used to supply weak grids, offer blackstart capability, and can provide decoupled active and reactive power controls [1–4].

Due to the continuous increase of the OWF capacity, the capacity of offshore converter stations should also be increased. This could result in a larger dimension and heavier offshore converter stations. In order to reduce the costs of offshore

VSC-HVDC converter station and offshore platform, the diode rectifier unit (DRU) concept for offshore converter stations is proposed in [5–7], as shown in **Figure 1**.

1.1 Advantages of DRU

Compared to the VSC-HVDC and LCC-HVDC converter stations, the main advantages of DRU unit are summarized as follows [5–7]:

- Simpler structure.
- Works robust since diodes do not need protection against the steep rise of current when fired and are not susceptible to failure events within the recovery time [5].
- Compared to IGBTs, diodes have lower switching and conducting losses due to a lower on-stage voltage and switching frequency [5–7].
- Easy transport and installation: according to [7, 8], replacing the VSC offshore converter station by a DRU, the total top side volume could be reduced by 80% and weight could be reduced by 65%. The installation time could be reduced by 20%.
- According to [7, 8], by applying DRU, the transmission loss could be reduced by 20% and transmission capacity could be increased by 30%. The total costs could be reduced by 30%.
- Higher reliability, modular design, and reduced operation and maintenance costs [5–8].

1.2 Technical challenges and possible solutions for DRU-HVDC link

Compared to the well-controlled VSC-HVDC, the DRU is a passive device without any controllability, and therefore it cannot provide the AC reference voltage for the OWF [8, 9]. Furthermore, in order to meet the power quality and reactive power requirements on the point of common coupling (PCC), filters and reactive compensations should be equipped on the DRU [10].

In order to deal with the above-mentioned challenges, different solutions were proposed:

1. First solution: since the DRU has not the blackstart capability, an additional 33 kV or 66 kV medium voltage (MV) AC cable is suggested in [5] for the startup of OWFs, as shown in **Figure 1** [5]. However, during the startup of OWFs, there

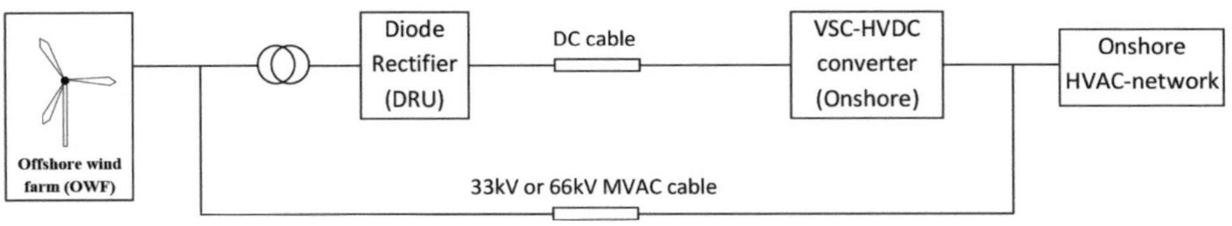

Figure 1.
OWF with DRU-HVDC.

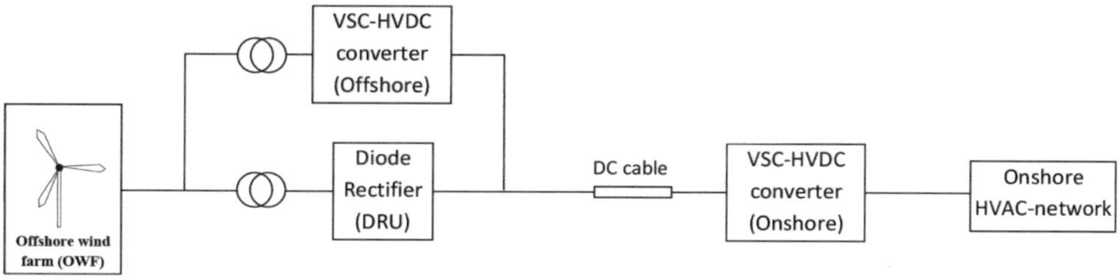

Figure 2.
Hybrid topology on the offshore side.

could be a period where the DRU-HVDC and medium voltage AC (MVAC) cables are in parallel operation. Large current on the MVAC cable could occur because of the uncontrolled DRU operation. After the startup of the OWFs, the MVAC cable will be disconnected and part of the offshore wind turbine will take over the control of the OWF [5, 9], e.g. grid forming controls (because the DRU is a passive rectifier without any control capability). Therefore, modifications on the wind turbine controllers (from grid following to grid forming) are necessary. Furthermore, the coordinated control of the wind turbines in OWF is necessary [9].

2. Second solution: in [6], a hybrid topology with parallel operation of DRU and VSC is suggested for controlling the voltage and frequency of the offshore grid, as shown in **Figure 2**. However, considering the dimension and costs of the offshore VSC converter station, the cost-effectiveness could be significantly reduced [11]. Therefore, the first solution is concentrated in this chapter.

1.3 Main objective of this chapter

The main objective of this chapter is to ensure the stable operation and improvement of the dynamic behavior of OWF with DRU-HVDC grid connection. Therefore, a new method is applied, where the phase shifting transformer (PST) is used on the onshore side of the MVAC cable [12].

This chapter is organized as follows: following the introduction, the proposed system structure, startup procedure, wind turbine and HVDC models are introduced in Section 5. In Section 6, both static and dynamic behaviors are analyzed. Finally, brief conclusions are deduced.

2. System structure and operation

2.1 Proposed system structure

The proposed method is shown in **Figure 3**, where the PST is applied on the onshore side of MVAC cable. The main objective of PST is for the smooth startup and stable operation of the OWF [12–14].

Similar to the method proposed in [5], the MVAC cable is used for the startup of OWF. Due to the application of PST, the MVAC cable and DRU-HVDC link could be operated in parallel (no additional switching procedure necessary).

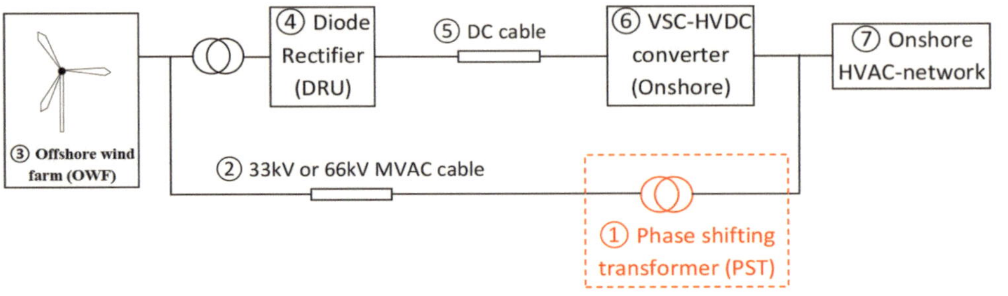

Figure 3.
Proposed system structure.

2.2 System startup

The startup procedure of the OWF with DRU grid connection is summarized as follows (the numbers are illustrated in **Figure 3**):

1. Start the onshore VSC-HVDC converter (⑥) (which is connected to the onshore power system (⑦) directly).

2. Start the PST (①).

3. Charging the MVAC cable (②).

4. Charging one of the MVAC cable strings in the OWF (③).

5. Start further wind turbines on the charged OWF string (depend on the capacity of the MVAC cable (②)).

6. Use the OWF park controller (③) [15] to regulate the reactive power flow on the MVAC cable (②) (equals to the natural charging reactive power of the MVAC cable (②)). The total apparent power (active power generated by the OWF and the MVAC cable (②) charging reactive power) should be less than the capacity of the MVAC cable (②).

7. Start the DRU (④).

8. Charging the DC cable (⑤).

9. Start the rest wind turbines of the OWF (③).

10. Use the PST (①) to regulate the active power transfer on the MVAC cable (②) to a minimum value ($\approx 0MW$). This enables the parallel operation of MVAC cable and the DRU-HVDC grid connection.

11. Use the OWF (③) park controller to regulate the reactive power flow on the MVAC cable (②) (equals to the natural charging reactive power of the MVAC cable (②)).

2.3 Advantages of the proposed method

The proposed method could have following advantages:

1. The PST will regulate the active power flow and the OWF park controller (③) [15] will regulate the reactive power flow on the MVAC cable (②). Overloading is avoided for the parallel operation of MVAC cable (②) and DRU-HVDC link (④,⑤,⑥).

2. MVAC cable (②) will not be switched off after the startup of OWF (③) and therefore no additional switching is necessary.

3. No modification of the wind turbine controller and onshore VSC-HVDC converter controller is necessary.

3. System models

The wind turbine operated in the OWF could be the doubly fed induction generator (DFIG) or full converter (FC). Both wind turbine models are briefly summarized in this section.

The onshore VSC-HVDC converter applies the conventional DC voltage and AC voltage (or reactive power compensation) controller. The DRU model is also introduced in this section.

3.1 DFIG wind turbine model

The basic configuration of a DFIG wind turbine is shown in **Figure 4** [1–3].

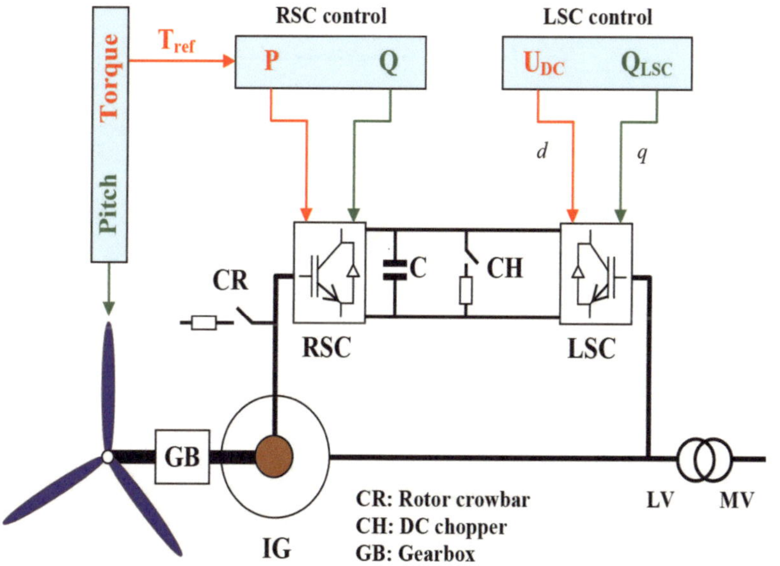

Figure 4.
Structure of DFIG.

The stator of the induction machine is connected directly to the grid, and the rotor is connected to the grid through two converters: line-side converter (LSC) and rotor-side converter (RSC). The LSC and RSC consist of two three-phase pulse-width modulated (PWM) converters, and they have a common DC bus. Conventionally, both converters are of three-phase two-level type having three legs, each of which consists of two IGBTs and two anti-parallel diodes as illustrated in **Figure 5**. Voltage control in these converters is done by the PWM using a carrier frequency in the order of kHz [1–3]. In this chapter, the general DFIG controllers introduced in [1–3] are applied.

3.2 FC wind turbine model

The typical configuration of the FC wind turbine is given in **Figure 6**. Usually, the generator could be a multi-pole synchronous generator designed for low speed, and this allows for gearless design. The generator can either be electrically excited or permanent magnet synchronous generator. For allowing variable speed operation, the synchronous generator is connected to the grid through two full power converters

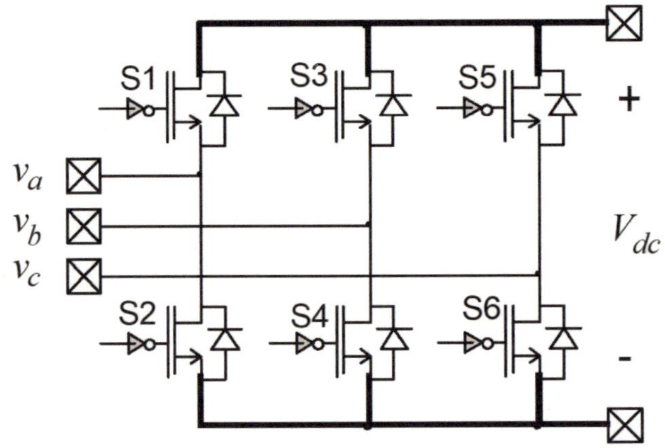

Figure 5.
Two-level VSC circuit.

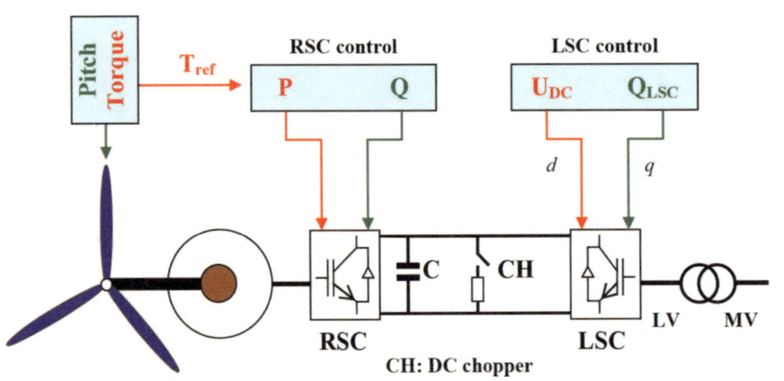

Figure 6.
Structure of FC.

(LSC and RSC), where they convert the variable frequency output power of the generator to AC power with grid frequency [1–3].

3.3 DFIG and FC controller

The controls of both DFIG and FC wind turbines are achieved by controlling LSC and the RSC utilizing vector control techniques [1–3].

Vector control allows decoupled control of both active and reactive power. RSC is used to control the active and reactive powers delivered to the grid. LSC is used to maintain the DC bus voltage regardless of the magnitude and direction of the rotor power. The reactive power controllability of the LSC is always applied to reinforce fault-ride-through (FRT) capability and provides grid voltage/reactive power control [1–3].

3.4 Onshore VSC-HVDC converter model

In this chapter, the onshore VSC-HVDC converter applies the modular multilevel converter (MMC) topology, and it is illustrated in **Figure 7** [15]. Each converter phase consists of upper and low multi-valve units. Each multi-valve unit has a modular structure with series-connected sub-modules (SMs). Each SM contains a capacitor and two IGBTs/diodes as illustrated in **Figure 8** [15]. This chapter is concentrated on the operation of OWF and DRU, and the half-bridge SMs are applied [15].

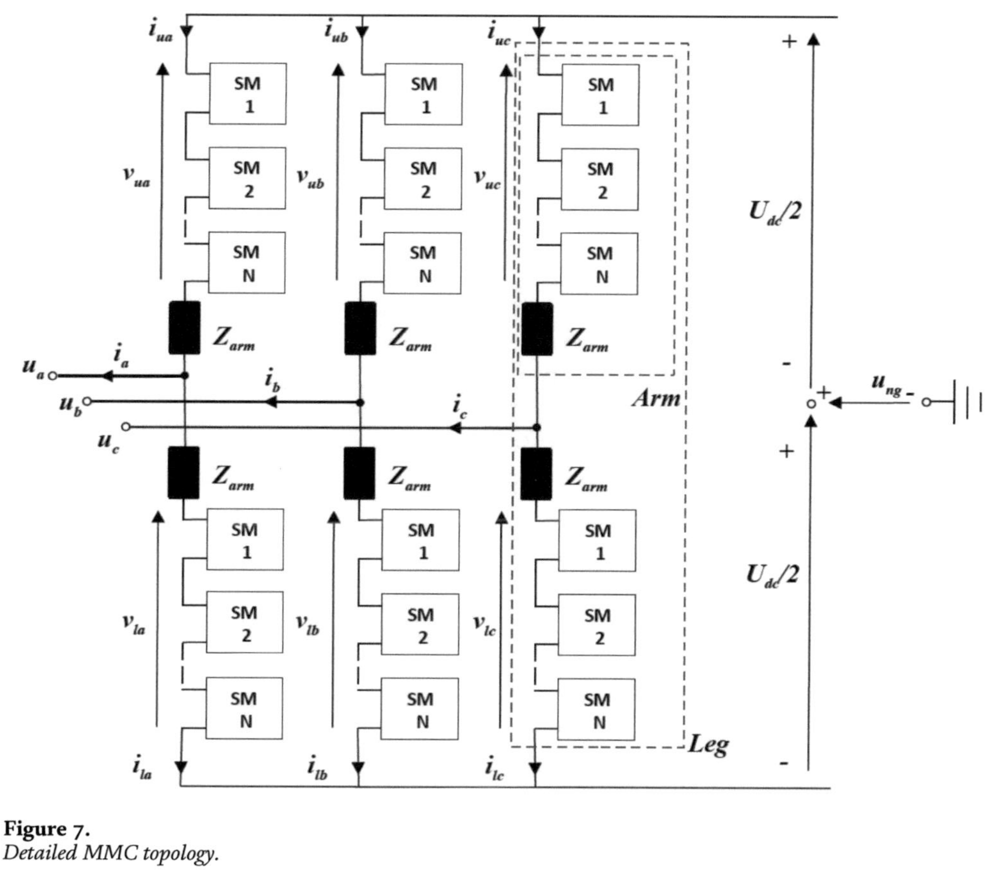

Figure 7.
Detailed MMC topology.

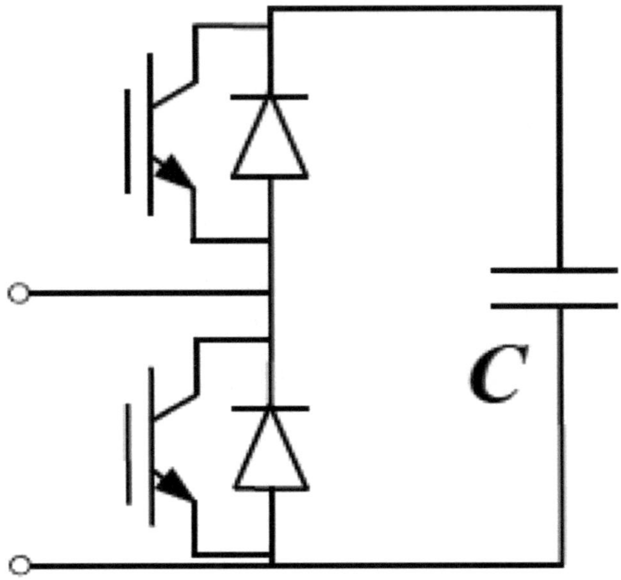

Figure 8.
MMC sub-module.

3.5 Onshore VSC-HVDC controller

The onshore MMC injects the active power transmitted by the offshore DRU to the onshore AC grid while maintaining the DC voltage at desirable level. In addition, it supports the onshore AC grid voltage in steady state operation and during faults. It uses a vector control [1–3, 15].

The frame of the onshore VSC-HVDC is shown in **Figure 9** [15].

3.6 DRU model

As described in [5], the DRU combines a transformer with a diode rectifier and DC smoothing reactors in a common tank filled with synthetic ester. In this chapter, the 6-pulse DRU is considered [16].

All the system is modeled in PowerFactory [15]. The main system components, e.g. offshore DRU, DC cable, onshore VSC-HVDC, 33-kV MVAC submarine cable and PST, are shown in **Figure 10**, where the OWF is operating with nominal power.

4. Simulation results

The system illustrated in **Figure 3** is simulated in this section. Both static and dynamic behaviors of the proposed method are considered.

4.1 System parameter

4.1.1 33-kV cable parameter

The ABB 33-kV submarine cable is applied in this simulation and the parameters are given in **Table 1** [15].

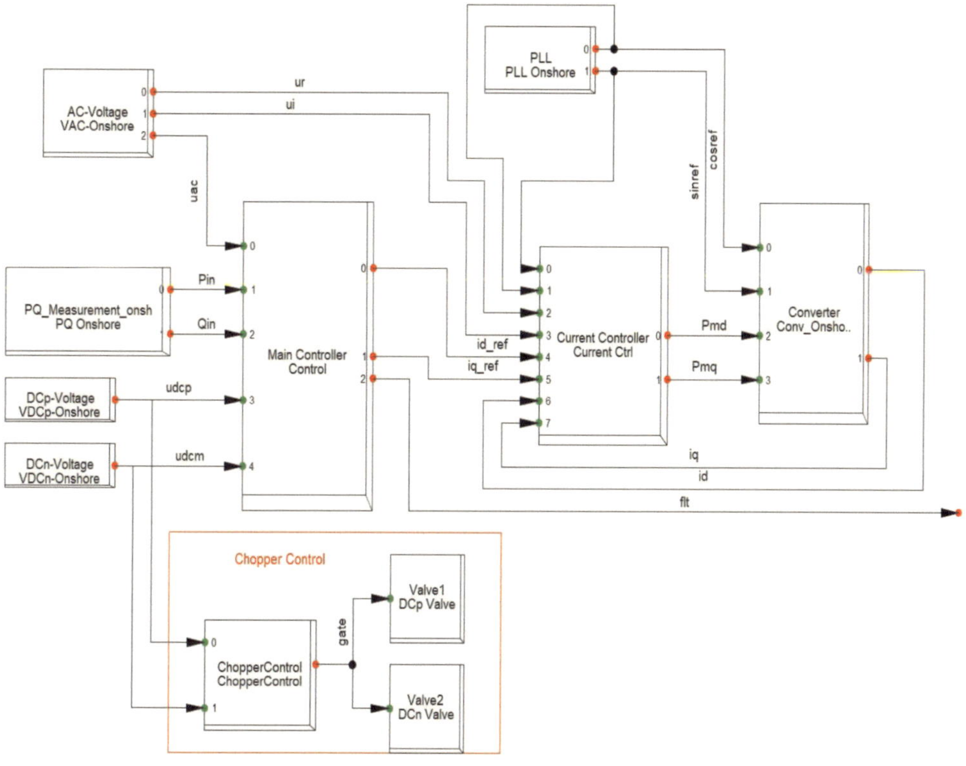

Figure 9.
Control frame of the onshore VSC-HVDC converter.

4.1.2 PST parameter

A standard transformer model is modified to enable the PST function. The parameters are given in **Table 2** [15]. For simulating the energization, saturation is also considered.

4.1.3 Wind turbine settings

The OWF consists of 20 strings, and 10 FC wind turbines ($S_N = 5.6$ MVA, $P_N = 5$ MW) are equipped on each string. One string structure is given in **Figure 11a**. The total OWF capacity is 1 GW [15].

4.1.4 Static operation

Firstly, the static operation of the proposed method is considered. OWFs operating with 20% and 100% of nominal power are selected for the demonstration of the proposed approach.

4.1.4.1 OWF operating with 20% of the nominal power

Figure 11a shows the power flow of one string in the OWF, where the active power of each wind turbine is 20% of its nominal power (1 MW). The wind turbines

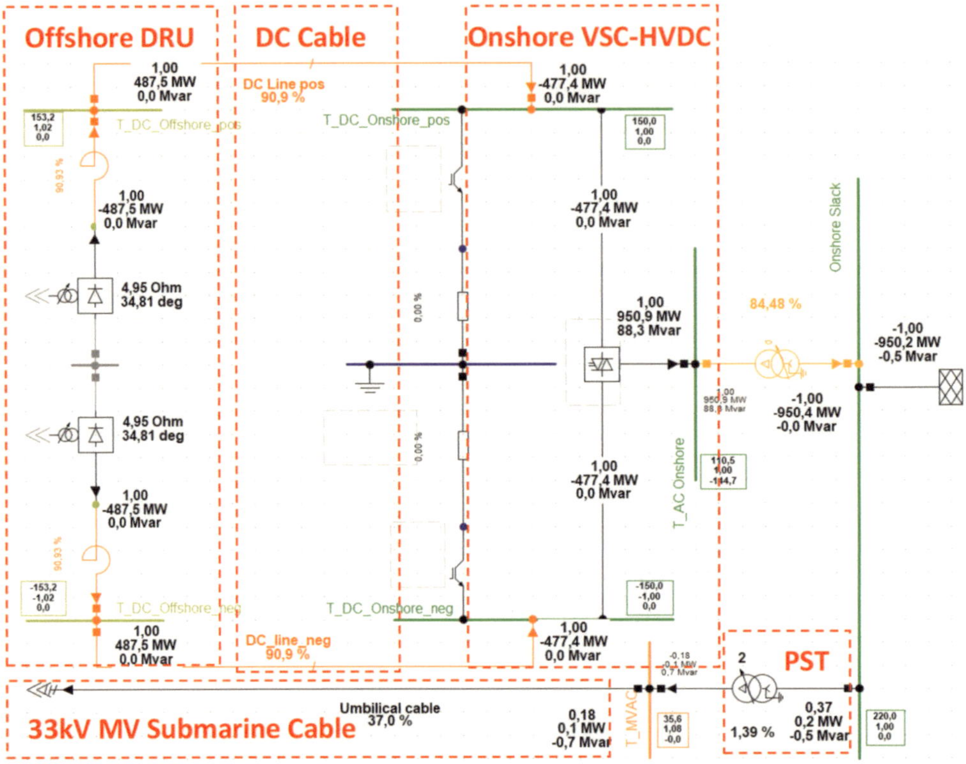

Figure 10.
DRU and VSC-HVDC with nominal power output of OWF.

33-kV submarine cable parameter	Value
Rated current (kA)	0.437
Rated voltage (kV)	33
AC resistance (ohm/km)	0.0754
Reactance (mH/km)	0.36
Capacitance (μF/km)	0.23
Cable length (km)	120

Table 1.
33-kV submarine cable parameter.

on the string are connected by 33-kV submarine cable and the cable length (between two wind turbines) is 1.5 km.

Power control for the 33-kV cable between onshore and offshore:

1. Active power control: the active power is controlled by the PST and it is 0 MW on the offshore side. The total active power generated by the OWF is transferred by the DRU-HVDC link, as shown in **Figure 11b**.

2. Reactive power control: the reactive power is controlled by the OWF park controller (offshore grid station) [15] to ensure the cable is

System parameter	Value
Rated power (MVA)	60
Rated voltage HV (kV)	220
Rated voltage MV (kV)	33
Vector group	YN/D
Short-circuit voltage (%)	20
Copper loss (kW)	180
Knee flux (p.u.)	1.1
Linear reactance (p.u.)	200.6431
Saturated reactance (p.u.)	0.2
Saturation exponent	15

Table 2.
PST parameter.

operating on its natural charging reactive power (9.4 MVar), as illustrated in **Figure 11b**.

As illustrated in the power flow result, all node voltages are in nominal range.

4.1.4.2 OWF operating on nominal power output

Figure 12a shows the power flow of one string in the OWF, where each wind turbine generates nominal power (5 MW).

As illustrated in **Figure 12b**, the offshore grid station can control the reactive power on the 33-kV onshore-offshore cable with its charging reactive power (9.4 MVar, shown in **Figure 12b**). Similar to the results in Section 6.1.4.1, the total active power generated by the OWF is transferred by the DRU-HVDC link.

It is clear that all the node voltages are in the nominal range.

4.1.4.3 Startup of OWF and HVDC system

Since the dynamic behaviors during the startup of the PST and 33-kV onshore-offshore cable were discussed in [13], this chapter is concentrating on the startup of OWF wind turbines and DRU.

4.1.4.4 Startup of OWF

The wind turbines of the OWF are started with 20% of the rated power (1 MW). At 0.5 s, the first string of the OWF is switched on. From 1 s to 5.5 s, the wind turbines on this string are started sequentially.

At 6 s, the second string of the OWF is switched on and from 6 s to 10 s, and the wind turbines on the second string are started.

In order to demonstrate the capability of OWF park controller (offshore grid station) [15], the reactive power of the 33-kV cable between onshore and offshore is set to 14 MVar, as shown in **Figure 13**.

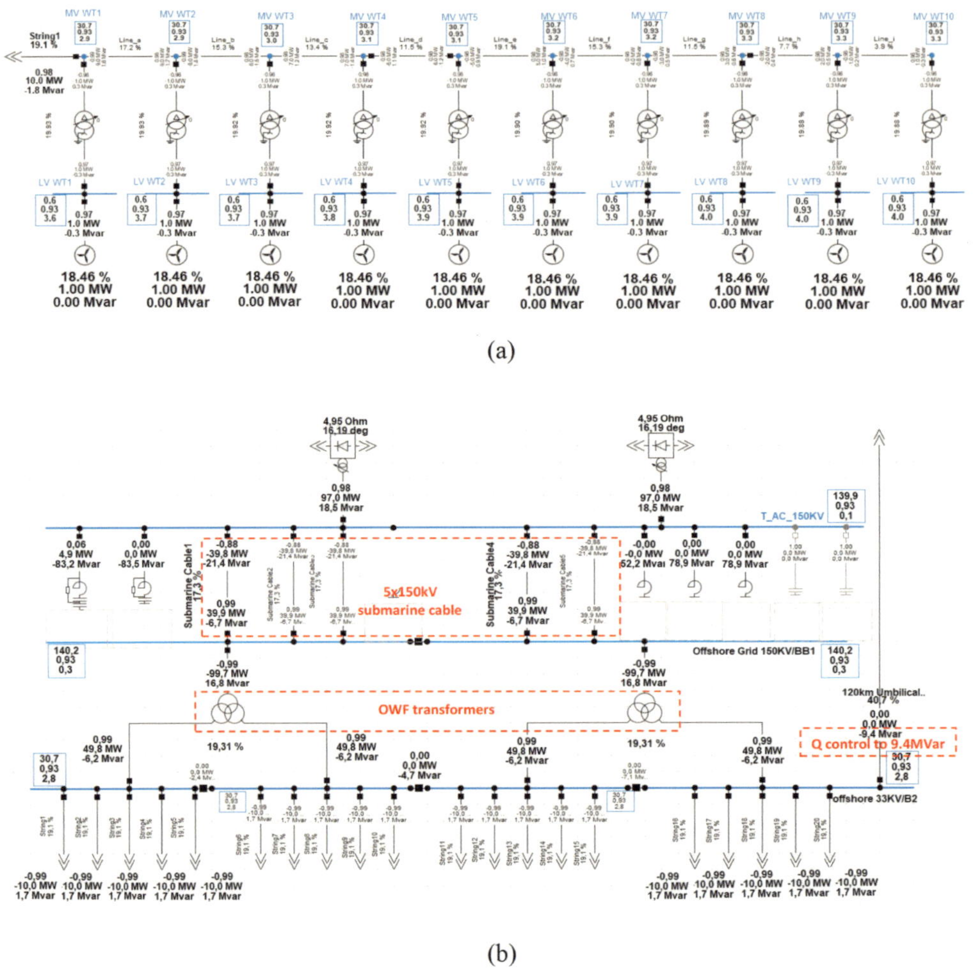

Figure 11.
OWF operates with 20% of nominal power. (a) One string of OWF and (b) OWF with DRU.

After starting the first string (1–5.5 s), since the wind turbines on the first string are working on 20% of their nominal power, the total reactive power capability of the first string (10 wind turbines) is not enough to maintain the regulated reactive power on the 33-kV cable (14 MVar).

After the startup of the second string, the reactive power of the 33-kV cable can be controlled effectively.

During the startup of the OWF, it is obvious that there are oscillations on the reactive power. This is mainly due to the parameter of the onshore-offshore 33-kV cable and the saturation parameters of the PST.

4.1.4.5 Startup of OWF transformer and 150-kV submarine cable

After starting the OWF wind turbine, the OWF transformers (illustrated in **Figure 11b**) are started at 12 s. Then two 150-kV submarine cables (illustrated in **Figure 11b**) are switched on at 15.5 s.

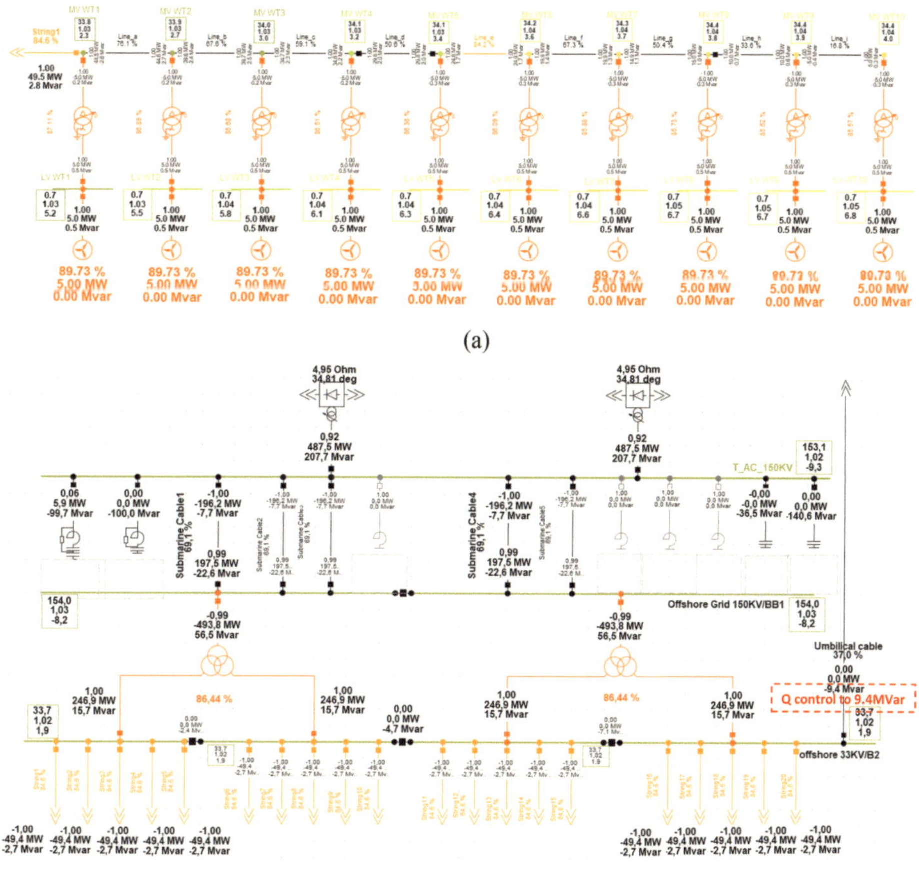

(a)

(b)

Figure 12.
OWF operates with nominal power. (a) One string of OWF and (b) OWF with DRU.

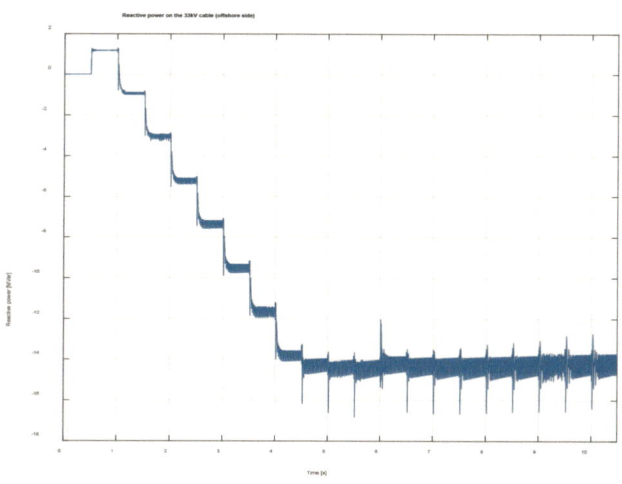

Figure 13.
Reactive power on the 33-kV cable (offshore side).

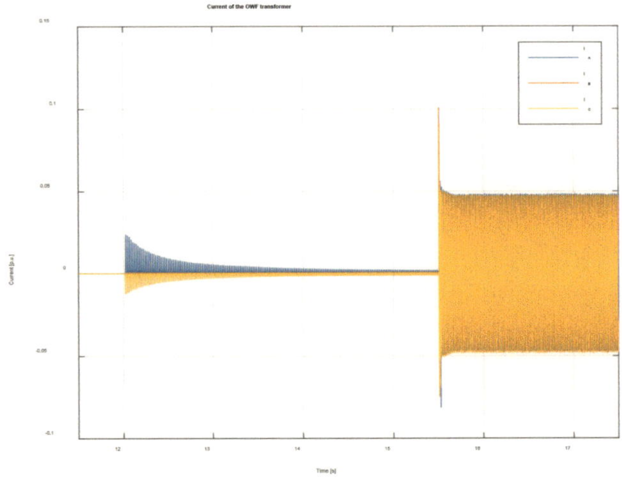

Figure 14.
Current of the OWF transformer.

Figures 14 and **15** show the current and the voltage of the OWF transformer. The transformer inrush current is limited effectively by means of the controllable switch of the circuit breaker (switching on at voltage maximum) [15] and the voltage is also in the nominal range.

4.1.4.6 Startup of DRU

After starting the two 150-kV submarine cables, the DRU is switched on at 18.5 s. **Figure 16** shows the current and the voltage of on the DRU. Both the voltage and current are in the nominal range. Since the AC filters on the DRU are deactivated, the

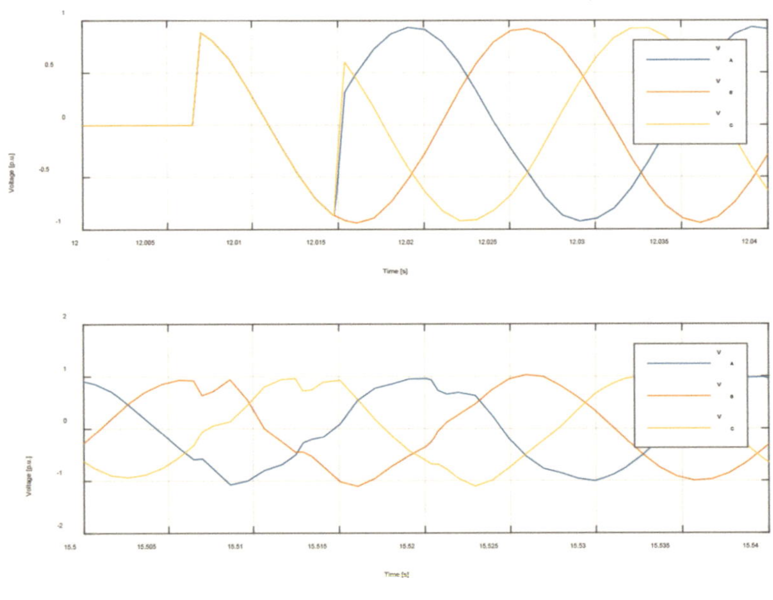

Figure 15.
Voltage of the OWF transformer.

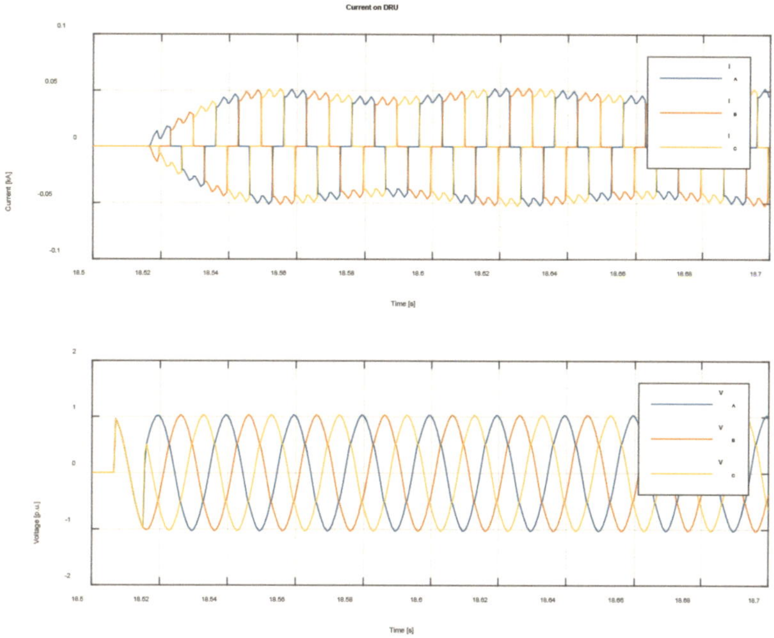

Figure 16.
Current and voltage on the DRU.

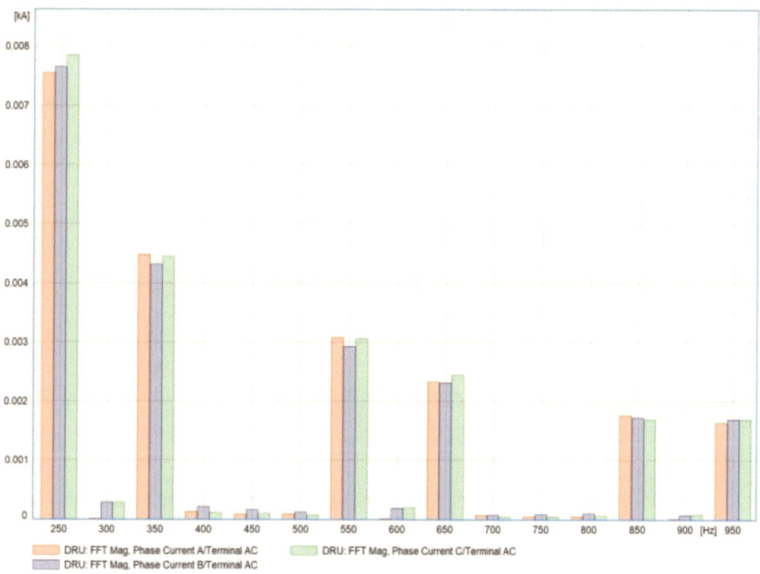

Figure 17.
FFT analysis of the DRU current.

DRU current contains 5th, 7th, 11th, and 13th harmonics as the fast fourier transformation (FFT) analysis shown in **Figure 17**.

After starting DRU, the whole OWF and DRU-HVDC link are in normal operation.

5. Conclusion

The stable and economical planning of OWF with HVDC grid connection is quite important for the future power systems. This chapter proposed a new method, where the application of PST ensures the blackstart and stable operation of the OWF with DRU-HVDC link. Applying the proposed method, no modification of the wind turbine controllers and onshore VSC-HVDC controllers are necessary. Moreover, since the PST is located onshore, it reduces the maintenance, operation, and installation costs of the PST.

Abbreviations

DFIG	doubly fed induction generator
DRU	diode rectifier unit
FC	full converter
HVDC	high-voltage direct current
IGBT	insulated gate bipolar transistor
LCC	line-commutated converter
LSC	line-side converter
MMC	modular multilevel converter
MV	medium voltage
OWF	offshore wind farm
PCC	point of common coupling
PST	phase shifting transformer
PWM	pulse-width modulation
RSC	rotor-side converter
SM	sub-module
VSC	voltage source converter

Author details

Lijun Cai
Institute of Electrical Power Engineering, University of Rostock, Germany

*Address all correspondence to: lijun.cai@uni-rostock.de

References

[1] Karaagac U, Mahseredjian J, Saad H, Jensen S, Cai LJ. Examination of fault ride-through methods for off-shore wind farms connected to the grid through VSC-based HVDC transmission. In: Paper Published in 11th International Workshop on Large-Scale Integration of Wind Power into Power Systems; November 13-15, 2012; Lisbon Portugal

[2] Cai LJ, Yin H, Lan YL, Lan T, Wu X, Eckel HG, et al. Sub-synchronous harmonic impedance analysis of doubly-fed induction generator wind turbine. IFAC-PapersOnLine. 2019;**52**(4):165-169

[3] Cai LJ, Fan Z. Doubly-fed induction generator offshore wind farm with VSC-HVDC grid connection: start-up procedure. In: International Conference on Renewable Power Generation RPG. London, UK: IET; 2015

[4] Bozhko SV et al. Control of offshore DFIG-based wind farm grid with line-commutated HVDC connection. IEEE Transactions on Energy Conversion. 2007;**22**(1):71-78

[5] Hammer T, Seman S, Menke P, Hacker F, Szangolies B, Meth J, et al. Diode-Rectifier HVDC link to Onshore Power Systems: Dynamic Performance of Wind Turbine Generators and Reliability of Liquid Immersed HVDC Diode Rectifier Units, CIGRE 2016, 21, rue d'Artois, F-75008 PARIS. Paris: CIGRE;

[6] Hoffmann M, Hemdan N, Kurrat M. AC Fault Analysis of DRU-VSC Hybrid HVDC Topology for Offshore Wind Farm Integration. VDE/IEEE Power and Energy Student Summit. London: IEEE; 2018

[7] 2nd Generation DC Grid Access for Large Scale Offshore Wind Farms. Siemens AG; 2015. Available from: https://rave-offshore.de/files/downloads/konferenz/konferenz-2015/Session7_2015/7.2_Menke

[8] Yu L, Li R, Xu L. Operation of offshore wind farms connected with DRU-HVDC transmission systems with special consideration of faults. Global Energy Interconnection. 2018:608-617

[9] Neumann C, Eckel HG, Achenbach S. A fault handling current control strategy for offshore wind power plants with diode rectifier HVDC transmission. In: Paper Published in 17th International Workshop on Large-scale Integration of Wind Power into Power Systems; October 17-19, 2018. Stockholm, Sweden: IEEE;

[10] Blasco-Gimenez R, Aparicio N, Ano-Villalba S, Bernal-Perez S. LCC-HVDC connection of offshore wind farms with reduced filter banks. IEEE Transactions on Industrial Electronics. 2013;**60**(6): 2372-2380. DOI: 10.1109/TIE.2012.2227906

[11] Ramachandran R, Poullain S, Benchaib A, Bacha S, Francois B. On the black start of offshore wind power plants with diode rectifier based HVDC transmission. In: 2019 21st European Conference on Power Electronics and Applications (EPE '19 ECCE Europe). Europe: ECCE; 2019. pp. 1-10. DOI: 10.23919/EPE.2019.8914779

[12] Patent P26336DE: Verbindungzwischen Offshore-Energiesystemen und Landstromnetzensowie Schwarzstartverfahrenfür Offshore-Energiesysteme

[13] Cai LJ, Meng X, Latif Q, Eckel HG, Weber H. Application of phase shifting transformer (pst) for blackstart and stable operation of offshore wind farm

with diode-rectifier unit HVDC link. In: Paper published in 19th Wind Integration Workshop; November 11-13, 2020. Berlin, Germany: IEEE;

[14] Cai LJ, Meng X, Zhang H, Eckel HG, Weber H. Dynamic behavior of phase shifting transformer (PST) for blackstart and stable operation of offshore wind farm with diode-rectifier unit HVDC link. In: Paper Published in 20th Wind Integration Workshop, September 29-30, 2021. Berlin, Germany: IEEE;

[15] Available from: https://www.digsilent.de/en/cable-analysis.html

[16] Kundur P. Power System Stability and Control. New York, USA: McGraw-Hill Education Ltd. ISBN-10: 007035958X

Chapter 6

Optimizing Berthing of Crew Transfer Vessels against Floating Wind Turbines: A Comparative Study of Various Floater Geometries

Laurent Barthélemy

Abstract

Securing the return on investment for commercial floating wind farms by a proper estimate of the operation and maintenance (O and M) downtime is a key issue to triggering final investment decisions. That is why crew transfer vessel (CTV) weather stand-by issues should be assessed together with new floating wind floater concepts, to boost their cost attractivity. However, such issues as the numerical investigation of the landing manoeuvre of a service ship against a floater reveal complex to calculate. Based on similarities with seakeeping, we investigate various floater geometries. To estimate the weather limitations associated with each configuration. Most recent works find that calculation compares with 5% accuracy to an experiment from a test tank at a model scale. Method description: (A) Vessel seakeeping: (1) assess vessel responses (amplitude and phase angles) and (2) compare them with vessel responses of available publications, as a benchmark. (B) Vessel berthing: (1) model both vessel and floater, (2) account for the wave masking effect of existing floater designs, and (3) compare the ratio of wave vertical force over wave horizontal force and the grip coefficient at the interface between the vessel fender and the floater boat landing. Findings: The wave masking effect calculation for a square floater is cross-checked favorably with an existing demonstrator.

Keywords: operation and maintenance, crew transfer vessel, floating wind farm, significant wave height, wave period

1. Introduction

The development of floating wind farms implies the issue of offshore O and M workers safety. It is therefore of upmost importance to know the constraints and acceptable conditions for berthing a CTV.

For berthing with the "bump and jump" method, a CTV comes and pushes its fender against the boat landing ladder. The fender studied here is the stiff fender [1].

IntechOpen

The present work relies on the results of a study performed by HSVA [2] and endeavors to meet the results obtained in its CTV model tank test. However, our approach here is different from HSVA numerical berthing calculations, which are more sophisticated.

2. Method description

2.1 General

The calculation is based on a simplified linear diffraction-radiation model applied in the frequential domain [1].

The studied ship is a CATamaran CTV (CAT CTV) [2, 3]: 27 m long, 8.2 m wide, twin hulls 3.2 m wide. It is modelled with Wigley hulls (**Figure 1**) [1].

The software used are GMSH for meshing [4] and NEMOH for hydrodynamics [5].

2.2 Loads of a unidirectional wave on CTV (seakeeping)

The Wigley hull, due to the wave excitation, moves in a vertical plane as follows (**Figure 2**) [1]:

1 rotation against her floatation centre	1 degree of freedom	θ (pitch)
1 translation against her original position O	2 degrees of freedom	τx (surge) τz (heave)

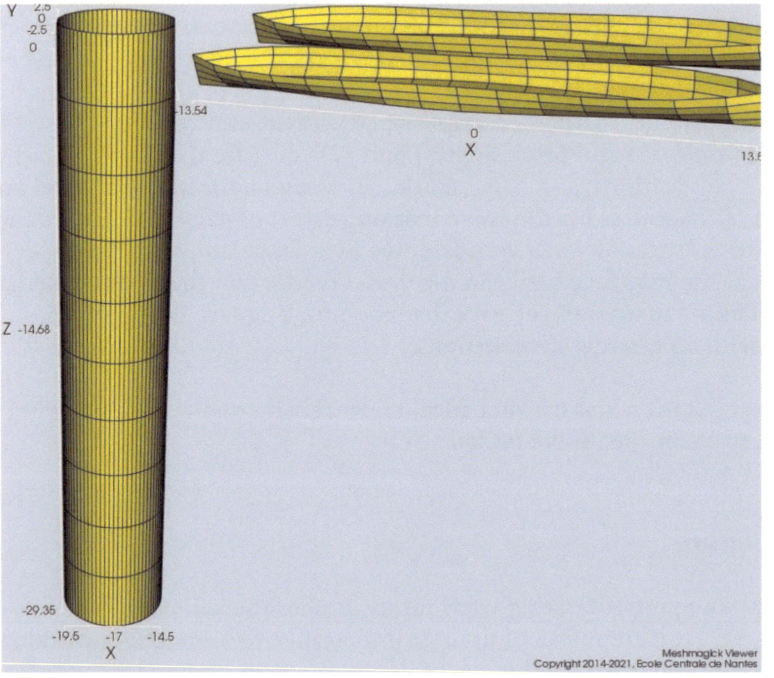

Figure 1.
CTV berthing against monopile (3D view).

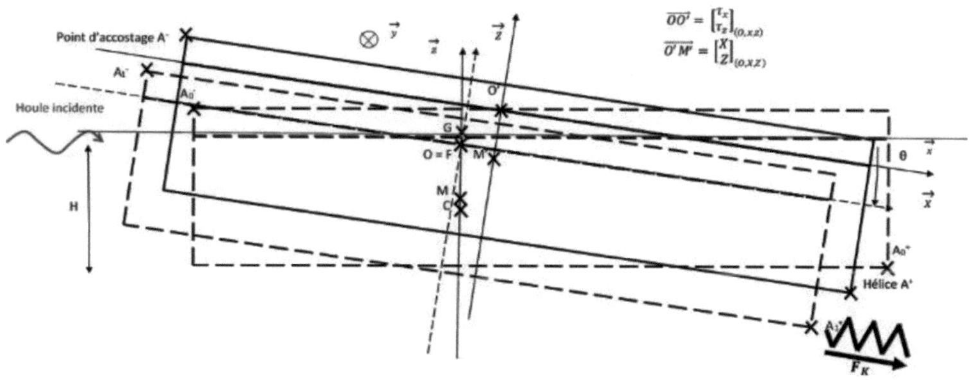

Figure 2.
CTV sea keeping without berthing (elevation).

The equations of dynamics are:

$$\begin{aligned}&(\mathrm{I}+\mathrm{I_a})\ddot{\mathrm{X}}+\mathrm{B}\dot{\mathrm{X}}+\mathrm{KX}=\mathrm{F_{excit}}\\&\Rightarrow \mathrm{I}\ddot{\mathrm{X}}=\mathrm{F_{excit}}-\mathrm{I_a}\ddot{\mathrm{X}}-\mathrm{B}\dot{\mathrm{X}}-\mathrm{KX}\Rightarrow I\ddot{\mathrm{X}}=\sum \mathrm{F_{ext}}\end{aligned} \tag{1}$$

$\mathrm{I}=\begin{bmatrix}\mathrm{I_{11}} & 0 & \mathrm{I_{15}}\\ 0 & \mathrm{I_{33}} & 0\\ \mathrm{I_{15}} & 0 & \mathrm{I_{55}}\end{bmatrix}=\begin{bmatrix}\mathrm{m} & 0 & \mathrm{mZ_G}\\ 0 & \mathrm{m} & 0\\ \mathrm{mZ_G} & 0 & \mathrm{I_G}+\mathrm{mZ_G^2}\end{bmatrix}$	$\mathrm{F_{excit}}$: vector of wave loads $\mathrm{I_a}$: matrix of added inertia (both calculated by NEMOH [5])
$\mathrm{B}=\mathrm{B}_R+\mathrm{B}_V,\mathrm{B}_V=\begin{bmatrix}\Lambda\mathrm{H} & 0 & -\Lambda\frac{\mathrm{H}^2}{2}\\ 0 & \mathrm{b_3} & 0\\ -\Lambda\frac{\mathrm{H}^2}{2} & 0 & \Lambda\frac{\mathrm{H}^3}{3}\end{bmatrix}$	$\mathrm{b_3}\overset{\mathrm{def}}{=}2\rho\mathrm{BC}\sqrt{\mathrm{gH}}\times\mathrm{b_c}$ $\lambda\overset{\mathrm{def}}{=}\Lambda\omega$ with : $\Lambda\overset{\mathrm{def}}{=}4\rho\mathrm{CC_d}(\mathrm{x_{max}}+\mathrm{H}\theta_{\mathrm{max}})/(3\pi)$ $\mathrm{B_R}$ = radiation damping matrix (calculated with NEMOH [5]) $\mathrm{B_V}$ = linearised viscous damping matrix
$\mathrm{K}=\begin{bmatrix}0 & 0 & 0\\ 0 & \mathrm{PBCG} & +\mathrm{P}\\ 0 & -\mathrm{P} & -\mathrm{mgCG}+\mathrm{mg}\frac{\mathrm{B}^2}{12\mathrm{H}}\end{bmatrix}$	P = propeller thrust. If: $\mathrm{P}=(\mathrm{m}+\mathrm{m_a})\mathrm{a}\sqrt{\mathrm{g/h}}\omega$ then $\lim\limits_{k\to 0}\mathrm{x_m/a}=0$

2.3 Loads of a unidirectional wave on CTV (berthing)

The friction coefficient without sliding is, about the principle of action and reaction (**Figure 3**) [1]:

$$\mathrm{f}=\frac{\text{Tangential force at vertical wall}}{\text{Normal force at vertical wall}}=\frac{\mathrm{T}}{\mathrm{N}}\Longrightarrow \mathrm{f}=\left(-\sum \mathrm{F_{ext\ z}}\right)/\left(-\sum \mathrm{F_{ext\ x}}\right) \tag{2}$$

Assumptions:

1. A thrust P is added to N, in order never to reach N < 0:

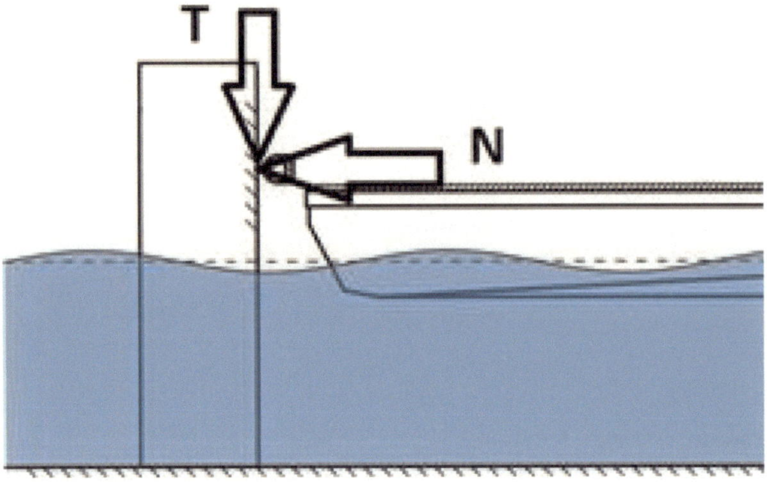

Figure 3.
Coulomb's friction law.

$$\vec{P} = -\left\|\vec{P}\right\|\vec{x} = m\omega^2\mathcal{G}\vec{x}(\text{avec } P < 0 \text{ et } \mathcal{G} < 0) \tag{3}$$

2. For a low friction berthing:

$$\Sigma F_{ext\ z} = I_{33}\ddot{Z} \tag{4}$$

The equations at the berthing point A = A- become therefore:

$$\sum F_{ext\ z} = I_{33}\ddot{Z} \text{ and } \sum F_{ext\ x} + P = I_{11}\ddot{X} + I_{15}\ddot{\theta} \tag{5}$$

$$\Rightarrow f = \frac{-I_{33}\ddot{Z}}{-I_{11}\ddot{X} - I_{15}\ddot{\theta} + P} \Longrightarrow f = \frac{-\ddot{Z}}{-\ddot{X} - Z_G\ddot{\theta} + \mathcal{G}\omega^2} \text{ where } \mathcal{G} = P/\left[m\left(1 + C'_{M1}\right)\omega^2\right] \tag{6}$$

We define t_T and t_N respectively as the time phase corrections required to get the calculated loads T(t) and N(t) in phase with HSVA test results T_{HSVA} (t) and N_{HSVA} (t) [2].

$$\begin{aligned} Z &= Z_m \text{COS}[\omega(t - t_T) + \varphi_z] \\ X &= X_m \text{COS}[\omega(t - t_N) + \varphi_x] \\ \theta &= \theta_m \text{COS}[\omega(t - t_N) + \varphi_\theta] \end{aligned} \tag{7}$$

We also define the following notations:

$$\begin{aligned} &T \overset{\text{def}}{=} \tan(\omega t/2) \\ &A \overset{\text{def}}{=} Z_m \cos(-\omega t_T + \varphi_z), \\ &B \overset{\text{def}}{=} Z_m \sin(-\omega t_T + \varphi_z), \\ &C \overset{\text{def}}{=} X_m \cos(-\omega t_N + \varphi_x) + Z_G\theta_m \cos(-\omega t_N + \varphi_\theta), \\ &D \overset{\text{def}}{=} X_m \sin(-\omega t_N + \varphi_x) + Z_G\theta_m \sin(-\omega t_N + \varphi_\theta) \\ &\Rightarrow f = \left[-AT^2 - 2BT + A\right]/\left[-(C - \mathcal{G})T^2 - 2DT + (C + \mathcal{G})\right] \end{aligned} \tag{8}$$

In order for the function f(t) to get relative extremes, the numerator of the quotient in Eq. (8) must have a positive discriminant δ′:

$$\delta' > 0 \Leftrightarrow |P| > m\omega^2\sqrt{(AD - BC)^2)}/\sqrt{A^2 + B^2} \tag{9}$$

Moreover, the denominator in Eq. (8) must never be null. Physically that means that the CTV propeller thrust P should be great enough in order never to get $N < 0$. Mathematically that implies both that its discriminant Δ' must be negative and that $(\mathcal{G} - C) < 0$:

$$\Delta' < 0 \text{ and } (\mathcal{G} - C) < 0 \Leftrightarrow |P| > m\omega^2\sqrt{D^2 + C^2} \tag{10}$$

If the conditions (3) and (4) are met, then, over a wave period, the friction coefficient will reach its extremes at the instants t_+ and t_-, which correspond to the following values T_+ and T_-:

$$T_\pm = \left[A\mathcal{G} \pm \sqrt{(A^2 + B^2)\mathcal{G}^2 - (AD - BC)^2}\right]/[AD - BC + B\mathcal{G}] \tag{11}$$

Then the maximum friction coefficient over a wave period is:

$$|f_{max}(T)| = \max\left[|f(T_+)|, |f(T_-)|\right] \tag{12}$$

We must therefore choose. Physically, that is the CTV surge over which the CTV captain has the time to adjust the propeller thrust P, in order for the fender never to lose contact with the boat landing. Nevertheless, the CTV is limited by her maximum thrust P_{max}. We infer that:

$$|P| = \min\left(m|\mathcal{G}|\omega^2, |P_{max}|\right) \tag{13}$$

Or, in other words:

$$\mathcal{G} = -\min\left(L, |P_{max}|/\left[m\omega^2\right]\right) \tag{14}$$

The selected criterion for CTV boarding at the berthing point is that the friction coefficient must never exceed the grip factor:

$$|f_{max}(T)| < f_{grip} \text{ with } f_{grip} = 0,8 \text{ (rubber – very wet soil [6])} \tag{15}$$

3. CTV against monopile

For benchmarking purpose, the first calculation models the "bump and jump" against a monopile (**Figures 1** and **4**). The water depth is 29 m.

The studied monopile has a 5 m diameter [2].

It may be noted that the CAT CTV is wider than the monopile: therefore the former is not masked from the waves by the latter.

Figure 5 compares for 2 m significant wave height (Hs) the calculated ratio of wave vertical force over wave horizontal force (T/N) with the grip coefficient of rubber against very wet soil (f_{grip}) [6].

For comparison reference [7] estimates the berthing limit to be for a ratio wavelength over boat length of 1.85 (λ/B): both results meet with 5% accuracy.

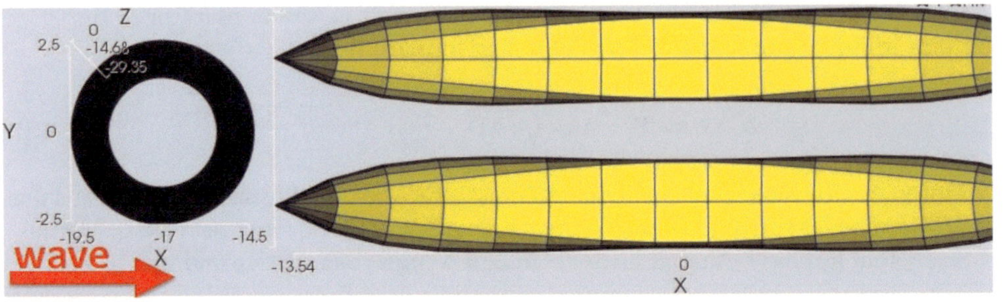

Figure 4.
CTV berthing against monopile (plane view).

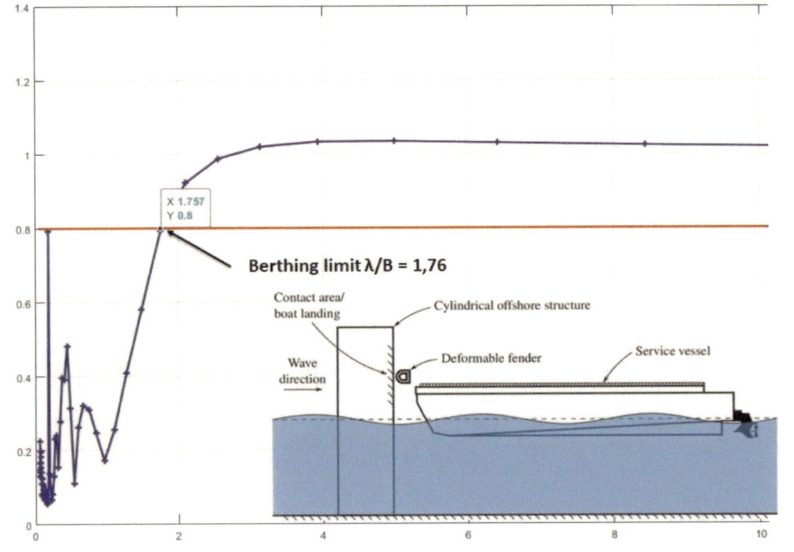

Figure 5.
Curves T/N and f_{grip} versus wavelength over boat length for 2 m Hs.

4. CTV against 13 m diameter cylindrical floater

The second calculation models the "bump and jump" against a planned [8] cylindrical wind turbine floater (**Figures 6** and 7). The water depth is 70 m [8].

The studied cylinder has [8]: a 13 m diameter, 14 m draft, 2000 T displacement.

This time, the floater masks the CAT CTV from the incidental waves, therefore the horizontal incident wave loads are masked, while the vertical incident wave loads are only the ones passing below the floater keel.

Figure 8 compares for 2 m Hs the calculated ratio T/N with f_{grip} versus λ/B.

This time berthing may take place for 2 m Hs whatever the wavelength.

5. CTV against 41 m diameter cylindrical floater

The third calculation models the "bump and jump" against a cylindrical wind turbine floater (**Figures 9** and **10**). The water depth is 23 m.

The studied cylinder has: a 41 m diameter, 7 m draft, 9300 T displacement.

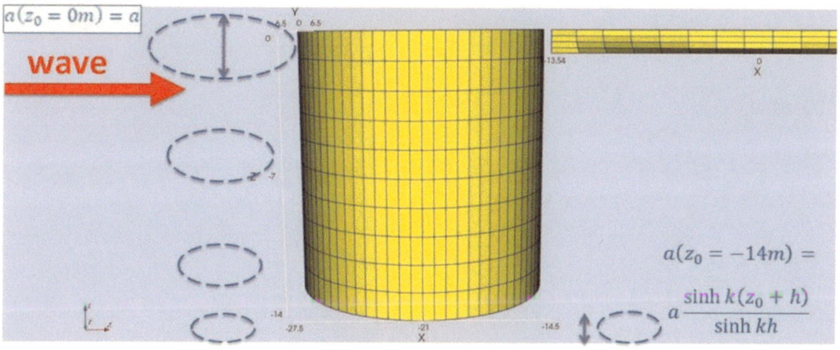

Figure 6.
CTV berthing against 13 m diameter cylindrical floater (3D view).

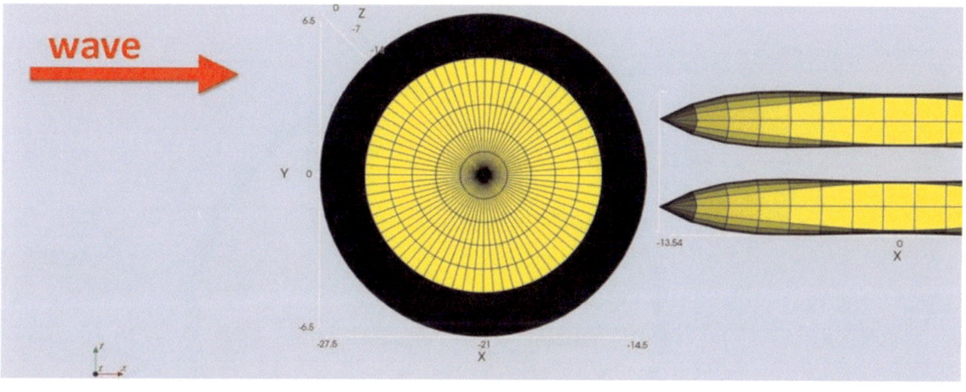

Figure 7.
CTV berthing against 13 m diameter cylindrical floater (plane view).

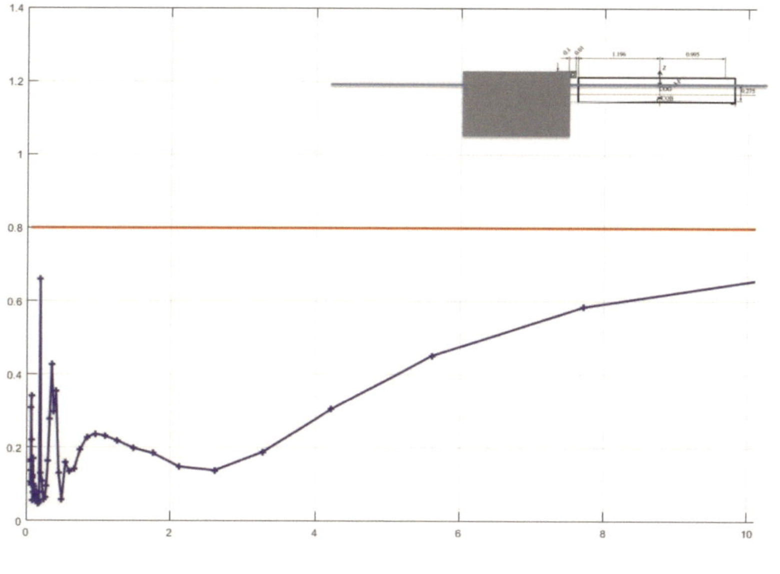

Figure 8.
Curves T/N and f_{grip} versus λ/B for 2 m Hs.

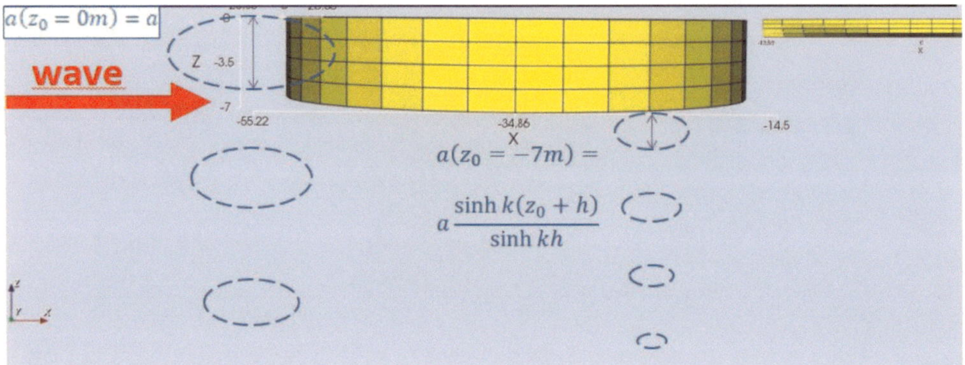

Figure 9.
CTV berthing against 41 m diameter cylindrical floater (3D view).

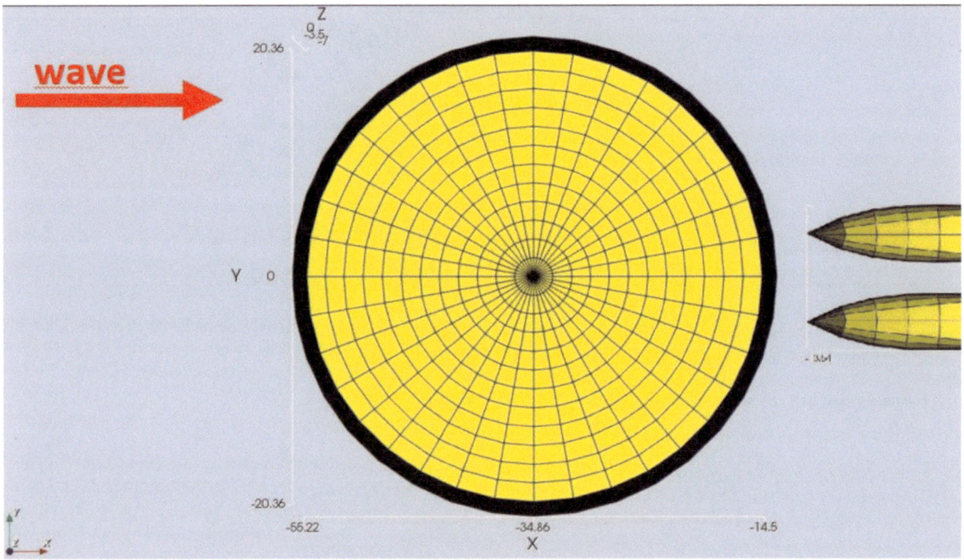

Figure 10.
CTV berthing against 13 m diameter cylindrical floater (plane view).

One more time, the floater masks the CAT CTV from the incidental waves, therefore the horizontal incident wave loads are masked, while the vertical incident wave loads are only the ones passing below the floater keel.

Figure 11 compares for 2 m Hs the calculated ratio T/N with f_{grip} versus λ/B.

This time berthing may take place for 2 m Hs whatever the wavelength.

6. CTV against 36 m side parallelepipedal floater

The fourth calculation models the "bump and jump" against a parallelepipedal wind turbine floater (**Figures 12** and **13**). The water depth is 23 m.

The studied cylinder has: a 36 m sides, 7 m draft, 9300 T displacement (same draft and displacement as in Section 5).

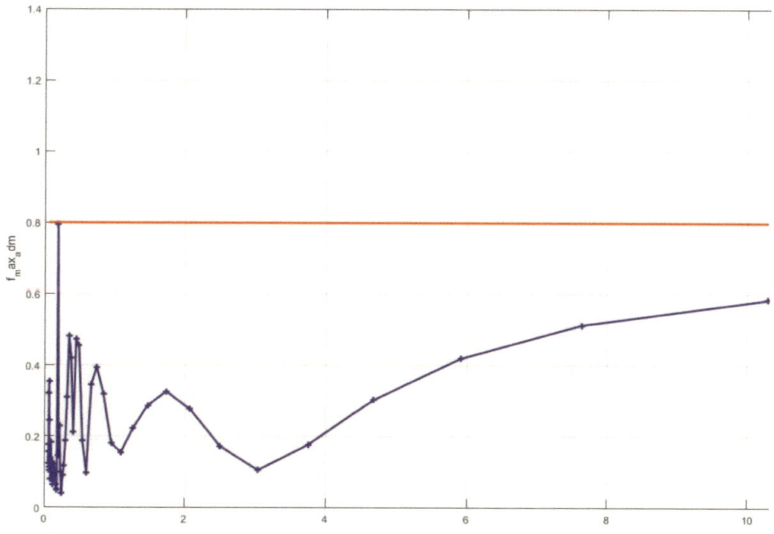

Figure 11.
Curves T/N and f_{grip} versus λ/B for 2 m Hs.

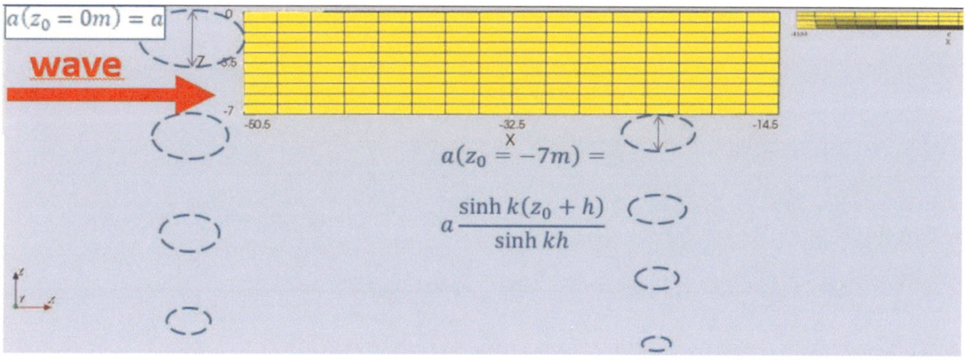

Figure 12.
CTV berthing against 36 m side parallelepipedal floater (3D view).

One more time, the floater masks the CAT CTV from the incidental waves, therefore the horizontal incident wave loads are masked, while the vertical incident wave loads are only the ones passing below the floater keel.

Figure 14 compares for 2 m Hs the calculated ratio T/N with f_{grip} versus λ/B.

This time berthing may take place for 2 m Hs whatever the wavelength.

7. CTV against FLOATGEN floater

The fifth calculation models the "bump and jump" against an existing [9] square hollow floater (**Figures 15** and **16**). The water depth is 23 m [10].

The studied square has [9, 10]: 36 m side, 7 m draft, 6000 T displacement (same draft as in Sections 5 and 6).

Once again, the floater masks the CAT CTV from the incidental waves, therefore the horizontal incident wave loads are masked, while the vertical incident wave loads are only the ones passing below the floater keel.

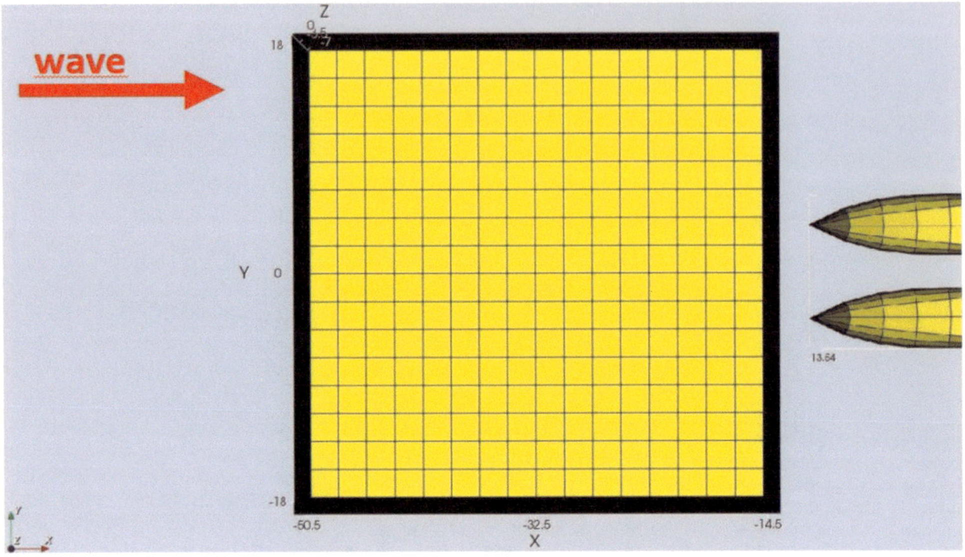

Figure 13.
CTV berthing against 36 m side parallelepipedal floater (plane view).

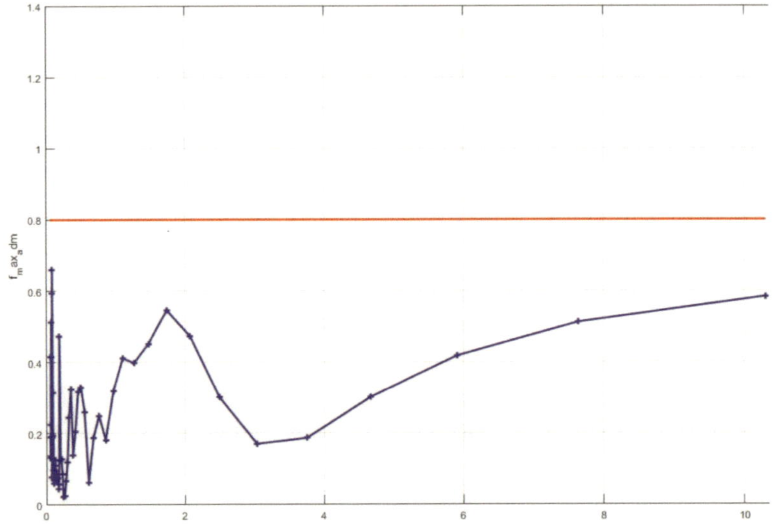

Figure 14.
Curves T/N and f_{grip} versus λ/B for 2 m Hs.

Figure 17 compares for 2 m Hs the calculated ratio T/N with f_{grip} versus λ/B. One more time berthing may take place for 2 m Hs whatever the wavelength.

8. Results

Table 1 sums up the Hs found for berthing to occur whatever the wavelength.

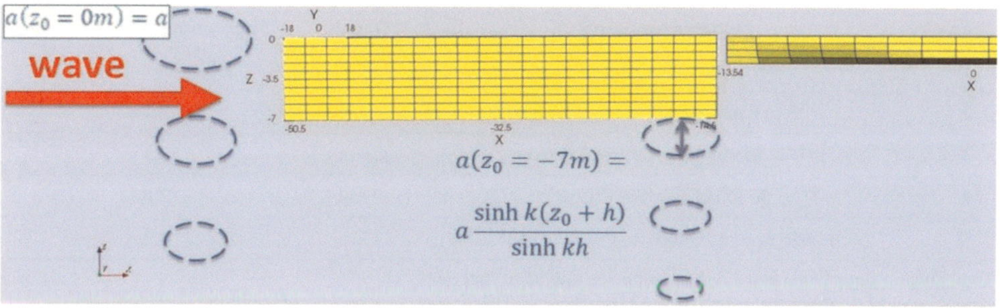

Figure 15.
CTV berthing against FLOATGEN floater (elevation).

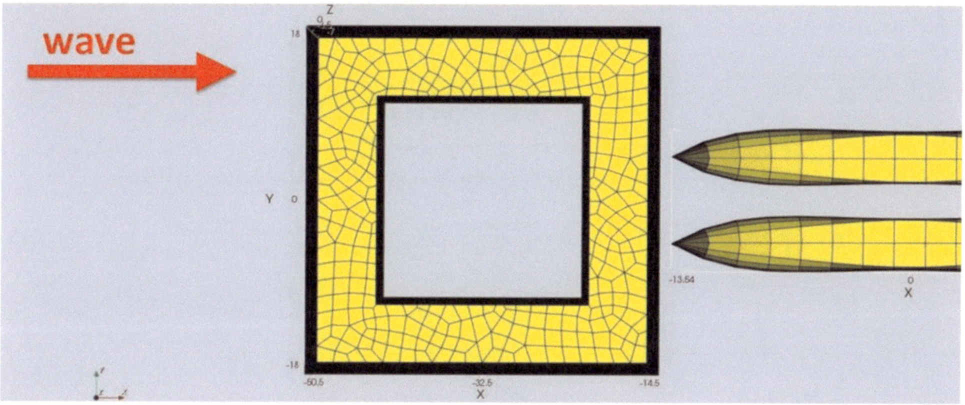

Figure 16.
CTV berthing against FLOATGEN floater (plane view).

9. Interpretations

9.1 General

FLOATGEN 2.3 m Hs berthing limit calculations compare precisely with feedback from offshore site test with full scale prototype [11]:

"Transfer up to 2.3 m significant wave height with no motion compensation".

From **Table 1**, it may be noted that berthing limits are influenced by:

- Hardly by floater geometry if they have the same displacement.
- Significantly by their displacement: the greater the displacement, the greater the berthing limit.

It is possible to propose an approximative analytical of the berthing limit due to the floater masking effect. Indeed, the calculated ratio T/N is (see Eq. (8)):

$$f(T_\pm) = \frac{AT_\pm^2 + 2BT_\pm - A}{(C - \mathcal{G})T_\pm^2 + 2DT_\pm - (C + \mathcal{G})} \tag{16}$$

Case	Floater geometry	Depth	Maximum Hs for berthing
1	5 m diameter monopile	23 m	1.5 m (no masking)
2	13 m diameter cylindrical floater	70 m	2.1 m
3	41 m diameter cylindrical floater[1]	23 m	2.5 m
4	36 m side square floater[1]	23 m	2.6 m
5	36 m side square hollow floater (FLOATGEN)	23 m	2.3 m

[1]*Those floaters have been chosen to have same displacement and draft.*

Table 1.
CAT CTV berthing limits for monopile various floater designs.

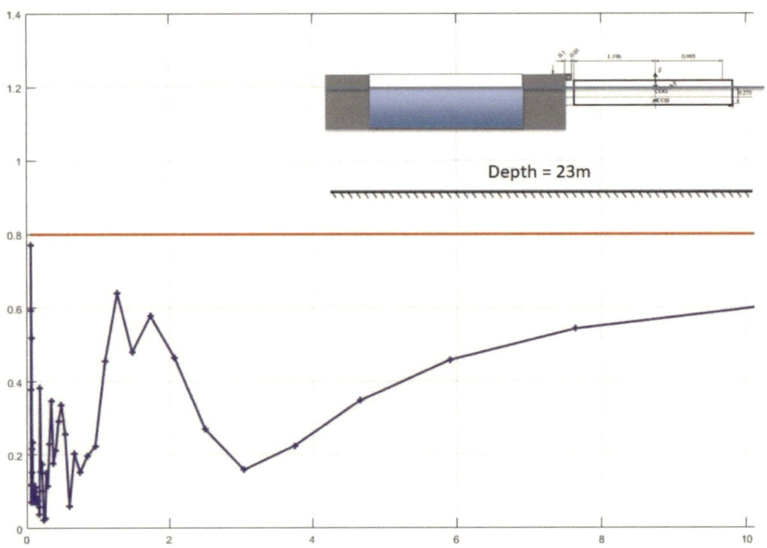

Figure 17.
Curves T/N and f_{grip} versus λ/B for 2 m Hs.

where:

$$
\begin{aligned}
&T_{\pm} \stackrel{\text{def}}{=} \tan\left(\omega t_{\pm}/2\right),\\
&A \stackrel{\text{def}}{=} Z_m \cos\left(-\omega t_T + \varphi_z\right),\\
&B \stackrel{\text{def}}{=} Z_m \sin\left(-\omega t_T + \varphi_z\right),\\
&C \stackrel{\text{def}}{=} X_m,\ \cos\left(-\omega t_N + \varphi_x\right) + Z_G\theta_m \cos\left(-\omega t_N + \varphi_\theta\right),\\
&D \stackrel{\text{def}}{=} X_m \sin\left(-\omega t_N + \varphi_x\right) + Z_G\theta_m \sin\left(-\omega t_N + \varphi_\theta\right).\\
&\mathcal{G} = -\min\left(L, |P_{max}|/[m\omega^2]\right)\\
&P_{max} = \max\left\{(m + m_a)a\sqrt{g/h}\bullet\omega, \text{CTV maximum thruster force}\right\}
\end{aligned}
\tag{17}
$$

At the large wave periods we have the following behavior:

$$
\lim_{\omega \to 0} |f_{max}(T)| = \max\left\{\lim_{\omega \to 0} |f(T_-)|, \lim_{\omega \to 0} |f(T_+)|\right\} \tag{18}
$$

Where (see Eq. (11)):

$$T_{\pm} = \left(A\mathcal{G} \pm \sqrt{\{A^2 + B^2\}\mathcal{G}^2 - (AD - BC)^2}\right)/(AD - BC + B\mathcal{G}) \quad (19)$$

In the case of box barge of same displacement, length and draft, we have:

9.2 Surge amplitude at large wave periods

$$\frac{x_m}{a} = \frac{\sqrt{(\eta\omega^3 - \mathrm{Im}(\vartheta)\omega^2 + \mathrm{Re}(\kappa)\omega)^2 + (\zeta\omega^4 + \mathrm{Re}(\vartheta)\omega^2 + \mathrm{Im}(\kappa)\omega + \mathrm{Re}(\mu))^2}}{\sqrt{(\alpha\omega^6 + \gamma\omega^4 + \varepsilon\omega^2)^2 + (\beta\omega^5 + \delta\omega^3 + \epsilon\omega)^2}}$$

$$\mathrm{Re}(\kappa) \stackrel{\text{def}}{=} \left[K_{55}b_3 + (\Lambda\omega H^3/3)k_3\right](\mathcal{F}_{xm}/a) + \left[(\Lambda\omega H^2/2)k_3\right](\mathcal{M}_{ym}/a)$$

$$\mathrm{Im}(\kappa) \stackrel{\text{def}}{=} + \left[(\Lambda\omega H^2/2)P\right](\mathcal{F}_{zm}/a)$$

$$\mathrm{Re}(\mu) \stackrel{\text{def}}{=} + \left(P^2 + K_{55}k_3\right)(\mathcal{F}_{xm}/a)$$

$$\varepsilon \stackrel{\text{def}}{=} -(m + m_a)P^2 - b_3K_{55}(\Lambda H\omega) - k_3\left\{(m + m_a)K_{55} + \left[(\Lambda\omega H^2)^2/12\right]\right\}$$

$$\epsilon \stackrel{\text{def}}{=} -(\Lambda H\omega)\left(P^2 + k_3K_{55}\right)$$

$$K_{55} \stackrel{\text{def}}{=} -(\Delta g \bullet GC) + \Delta g\left[B^2/(12H)\right] + \left(KZ_A^{+2}\right)$$

$$k_3 \stackrel{\text{def}}{=} (\Delta/H)g$$

$$\lim_{k \to 0} \frac{x_m}{a} = \lim_{\omega \to 0} \frac{\sqrt{(\mathrm{Re}(\kappa)\omega)^2 + (\mathrm{Im}(\kappa)\omega + \mathrm{Re}(\mu))^2}}{\sqrt{(\varepsilon\omega^2)^2 + (\epsilon\omega)^2}}$$

$$\lim_{\omega \to 0} \mathcal{F}_{xm}/a = O(\omega^3)$$

$$\lim_{\omega \to 0} \mathcal{M}_{ym}/a = \lim_{k \to 0} \left[\left\{(m + m_a)\frac{H}{2} - \Delta\frac{B^2}{12H}\right\}\sqrt{\frac{g}{h}}\omega + (m + m_a)\sqrt{\frac{g}{h}}\omega Z_a^+ + O(\omega^3)\right]$$

$$\lim_{\omega \to 0} \mathrm{Im}\mathcal{F}_{zm}/a = -(\Delta g/H)\{\sinh[k(z_0 + h)]/\sinh(kh)\} + O(\omega) \quad (20)$$

Note: since the vertical incident wave loads are only the ones passing below the floater keel then, if its draft is $|z_0|$, $z_m/a = z_m(z = z_0) = \sinh[k(z_0 + h)]/\sinh(kh)$ [12] (refer to equations written in **Figures 6, 9, 12**, and **15**).

Therefore:

$$\lim_{\omega \to 0} \mathrm{Re}(\kappa) = \lim_{\omega \to 0} O(\omega^3) + \left[\frac{\Lambda H^2}{2}k_3\right]\left\{(m + m_a)\left[Z_a^+ + \frac{H}{2}\right] - \Delta\frac{B^2}{12H}\right\}\sqrt{\frac{g}{h}}\omega^2$$

$$\lim_{\omega \to 0} \mathrm{Im}(\kappa) = \lim_{\omega \to 0} O(\omega^2)$$

$$\begin{aligned}
&\lim_{\omega \to 0} \mathrm{Re}(\mu) = \lim_{\omega \to 0} \left[O(\omega^2) + K_{55}k_3\right]O(\omega^3)\\
&\lim_{\omega \to 0} \varepsilon = \lim_{\omega \to 0} - k_3(m + m_a)K_{55} + O(\omega)\\
&\lim_{\omega \to 0} \epsilon = \lim_{\omega \to 0} - (\Lambda H\omega)(P^2 + k_3K_{55})\\
&\lim_{k \to 0} \frac{x_m}{a} = \lim_{\omega \to 0} \frac{\sqrt{[O(\omega^3)]^2 + [O(\omega^3) + (O(\omega^2) + K_{55}k_3)O(\omega^3)]^2}}{\sqrt{[-k_3(m + m_a)K_{55}\omega^2]^2 + [(\Lambda H)(O(\omega^2) + k_3K_{55})\omega^2]^2}}\\
&\lim_{k \to 0} x_m/a = O(\omega)
\end{aligned} \tag{21}$$

9.3 Surge phase angle at large wave periods

$$\begin{aligned}
\cos\varphi_x = &\left\{+[\eta\omega^3 - \mathrm{Im}(\vartheta)\omega^2 + \mathrm{Re}(\kappa)\omega]\left(\alpha\omega^6 + \gamma\omega^4 + \varepsilon\omega^2\right)\right.\\
&\left.+[\zeta\omega^4 + \mathrm{Re}(\vartheta)\omega^2 + \mathrm{Im}(\kappa)\omega + \mathrm{Re}(\mu)](\beta\omega^5 + \delta\omega^3 + \epsilon\omega)\right\}\\
&/\sqrt{(\alpha\omega^6 + \gamma\omega^4 + \varepsilon\omega^2\iota)^2 + (\beta\omega^5 + \delta\omega^3 + \epsilon\omega)^2}\\
&/\sqrt{[\eta\omega^3 - \mathrm{Im}(\vartheta)\omega^2 + \mathrm{Re}(\kappa)\omega]^2 + [\zeta\omega^4 + \mathrm{Re}(\vartheta)\omega^2 + \mathrm{Im}(\kappa)\omega + \mathrm{Re}(\mu)]^2}\\
\sin\varphi_x = &\left\{+[\eta\omega^3 - \mathrm{Im}(\vartheta)\omega^2 + \mathrm{Re}(\kappa)\omega](\beta\omega^5 + \delta\omega^3 + \epsilon\omega)\right.\\
&\left.-[\zeta\omega^4 + \mathrm{Re}(\vartheta)\omega^2 + \mathrm{Im}(\kappa)\omega + \mathrm{Re}(\mu)]\left(\alpha\omega^6 + \gamma\omega^4 + \varepsilon\omega^2 + \iota\right)\right\}\\
&/\sqrt{(\alpha\omega^6 + \gamma\omega^4 + \varepsilon\omega^2\iota)^2 + (\beta\omega^5 + \delta\omega^3 + \epsilon\omega)^2}\\
&/\sqrt{[\eta\omega^3 - \mathrm{Im}(\vartheta)\omega^2 + \mathrm{Re}(\kappa)\omega]^2 + [\zeta\omega^4 + \mathrm{Re}(\vartheta)\omega^2 + \mathrm{Im}(\kappa)\omega + \mathrm{Re}(\mu)]^2}\\
&c_\Lambda \stackrel{\mathrm{def}}{=} \Lambda H/(m + m_a)\\
&\lim_{T\to+\infty} \cos\varphi_x = -c_\Lambda/\sqrt{1 + c_\Lambda{}^2}\\
&\lim_{T\to+\infty} \sin\varphi_x = 1/\sqrt{1 + c_\Lambda{}^2}
\end{aligned} \tag{22}$$

9.4 Heave amplitude at large wave periods

$$\frac{z_m}{a} = \frac{\sqrt{[-\mathrm{Im}(\Gamma)\omega^4 - \mathrm{Im}(\Pi)\omega^2 + \mathrm{Re}(\Sigma)\omega]^2 + [\mathrm{Im}(\Xi)\omega^3 + \mathrm{Re}(\Pi)\omega^2 + \mathrm{Im}(\Sigma)\omega]^2}}{\sqrt{[\alpha\omega^6 + \gamma\omega^4 + \varepsilon\omega^2]^2 + [\beta\omega^5 + \delta\omega^3 + \epsilon\omega]^2}}$$

$$\begin{aligned}
&\mathrm{Re}(\Pi) \stackrel{\mathrm{def}}{=} [I_{GA15}P](\mathcal{F}_{xm}/a) - [(m + m_a)P]\left(\mathcal{M}_{ym}/a\right)\\
&\mathrm{Im}(\Pi) \stackrel{\mathrm{def}}{=} \left\{(m + m_a)K_{55} + B_{11}B_{55} - B_{15}^2\right\}\mathrm{Im}(\mathcal{F}_{zm}/a)\\
&\mathrm{Re}(\Sigma) \stackrel{\mathrm{def}}{=} - [B_{15}P](\mathcal{F}_{xm}/a) + [B_{11}P]\left(\mathcal{M}_{ym}/a\right)\\
&\mathrm{Im}(\Sigma) \stackrel{\mathrm{def}}{=} - [KB_{55} + K_{55}B_{11}]\mathrm{Im}(\mathcal{F}_{zm}/a)
\end{aligned}$$

$$
\begin{aligned}
&I_{GA15} \stackrel{\text{def}}{=} mz_G - (m_{a1}H/2)\\
&B_{11} \stackrel{\text{def}}{=} \lambda H\\
&B_{15} \stackrel{\text{def}}{=} -(\lambda H^2/2)\\
&B_{55} \stackrel{\text{def}}{=} \Lambda\omega H^3/3\\
&\lim_{k \to 0} z_m/a = \sinh[k(z_0+h)]/\sinh(kh)
\end{aligned}
\tag{23}
$$

9.5 Heave phase angle at large wave periods

$$
\begin{aligned}
\cos\varphi_z = &\{-[-\text{Im}(\Gamma)\omega^4 - \text{Im}(\Pi)\omega^2 + \text{Re}(\Sigma)\omega](\alpha\omega^6 + \gamma\omega^4 + \varepsilon\omega^2)\\
&-[\text{Im}(\Xi)\omega^3 + \text{Re}(\Pi)\omega^2 + \text{Im}(\Sigma)\omega](\beta\omega^5 + \delta\omega^3 + \epsilon\omega)\}\\
&/\sqrt{[-\text{Im}(\Gamma)\omega^4 - \text{Im}(\Pi)\omega^2 + \text{Re}(\Sigma)\omega]^2 + [\text{Im}(\Xi)\omega^3 + \text{Re}(\Pi)\omega^2 + \text{Im}(\Sigma)\omega]^2}\\
&/\sqrt{[\alpha\omega^6 + \gamma\omega^4 + \varepsilon\omega^2]^2 + [\beta\omega^5 + \delta\omega^3 + \epsilon\omega]^2}\\
\sin\varphi_z = &\{+[-\text{Im}(\Gamma)\omega^4 - \text{Im}(\Pi)\omega^2 + \text{Re}(\Sigma)\omega](\beta\omega^5 + \delta\omega^3 + \epsilon\omega)\\
&+[\text{Im}(\Xi)\omega^3 + \text{Re}(\Pi)\omega^2 + \text{Im}(\Sigma)\omega](\alpha\omega^6 + \gamma\omega^4 + \varepsilon\omega^2)\}\\
&/\sqrt{[-\text{Im}(\Gamma)\omega^4 - \text{Im}(\Pi)\omega^2 + \text{Re}(\Sigma)\omega]^2 + [\text{Im}(\Xi)\omega^3 + \text{Re}(\Pi)\omega^2 + \text{Im}(\Sigma)\omega]^2}\\
&/\sqrt{[\alpha\omega^6 + \gamma\omega^4 + \varepsilon\omega^2]^2 + [\beta\omega^5 + \delta\omega^3 + \epsilon\omega]^2}\\
\lim_{T \to +\infty} &\varphi_z = 0^\circ
\end{aligned}
\tag{24}
$$

9.6 Pitch amplitude at large wave periods

$$
\begin{aligned}
&\frac{\theta_m}{a} = \frac{\sqrt{\{\text{Re}(\mathbb{D})\omega^3 + \text{Re}(\mathbb{E})\omega^2 + \text{Re}(\mathbb{F})\omega\}^2 + \{\text{Im}(\mathbb{C})\omega^4 + \text{Im}(\mathbb{E})\omega^2 + \text{Im}(\mathbb{F})\omega\}^2}}{\sqrt{\{\alpha\omega^6 + \gamma\omega^4 + \varepsilon\omega^2\}^2 + \{\beta\omega^5 + \delta\omega^3 + \epsilon\omega\}^2}}\\
&\text{Re}(\mathbb{E}) \stackrel{\text{def}}{=} -\{-[(m+m_a)P]\text{Im}(\mathcal{F}_{zm}/a)\} = -(m+m_a)P\Delta g/H\\
&\text{Im}(\mathbb{E}) \stackrel{\text{def}}{=} [I_{GA15} - (m+m_a)Z_a^+]k_3(P/a)\\
&\text{Re}(\mathbb{F}) \stackrel{\text{def}}{=} -\{[k_3B_{15}](\mathcal{F}_{xm}/a) - [k_3B_{11}](\mathcal{M}_{ym}/a)\}\\
&\text{Im}(\mathbb{F}) \stackrel{\text{def}}{=} -\{-[B_{11}P]\text{Im}(\mathcal{F}_{zm}/a)\}\\
&\frac{\lim}{k \to 0}\frac{\theta_m}{a} = \frac{\lim}{\omega \to 0}\frac{(m+m_a)\sqrt{g/h}\omega}{K_{55}}\sqrt{\left\{\left[\frac{H}{2} + Z_a^+\right] - [\Delta/(m+m_a)]\frac{B^2}{12H}\right\}^2 + a^2}\\
&\lim_{k \to 0} \theta_m/a = 0
\end{aligned}
\tag{25}
$$

9.7 Pitch phase angle at large wave periods

$$\cos\varphi_\theta = \{+[\mathrm{Re}(\mathbb{D})\omega^3 + \mathrm{Re}(\mathbb{E})\omega^2 + \mathrm{Re}(\mathbb{F})\omega](\alpha\omega^6 + \gamma\omega^4 + \varepsilon\omega^2)$$
$$+[\mathrm{Im}(\mathbb{C})\omega^4 + \mathrm{Im}(\mathbb{D})\omega^3 + \mathrm{Im}(\mathbb{E})\omega^2 + \mathrm{Im}(\mathbb{F})\omega](\beta\omega^5 + \delta\omega^3 + \epsilon\omega)\}$$
$$/\sqrt{\{\mathrm{Re}(\mathbb{D})\omega^3 + \mathrm{Re}(\mathbb{E})\omega^2 + \mathrm{Re}(\mathbb{F})\omega\}^2 + \{\mathrm{Im}(\mathbb{C})\omega^4 + \mathrm{Im}(\mathbb{D})\omega^3 + \mathrm{Im}(\mathbb{E})\omega^2 + \mathrm{Im}(\mathbb{F})\omega\}^2}$$
$$/\sqrt{\{\alpha\omega^6 + \gamma\omega^4 + \varepsilon\omega^2\}^2 + \{\beta\omega^5 + \delta\omega^3 + \epsilon\omega\}^2}$$

$$\sin\varphi_\theta = \{+[\mathrm{Re}(\mathbb{D})\omega^3 + \mathrm{Re}(\mathbb{E})\omega^2 + \mathrm{Re}(\mathbb{F})\omega](\beta\omega^5 + \delta\omega^3 + \epsilon\omega)$$
$$-[\mathrm{Im}(\mathbb{C})\omega^4 + \mathrm{Im}(\mathbb{D})\omega^3 + \mathrm{Im}(\mathbb{E})\omega^2 + \mathrm{Im}(\mathbb{F})\omega](\alpha\omega^6 + \gamma\omega^4 + \varepsilon\omega^2)\}$$
$$/\sqrt{\{\mathrm{Re}(\mathbb{D})\omega^3 + \mathrm{Re}(\mathbb{E})\omega^2 + \mathrm{Re}(\mathbb{F})\omega\}^2 + \{\mathrm{Im}(\mathbb{C})\omega^4 + \mathrm{Im}(\mathbb{D})\omega^3 + \mathrm{Im}(\mathbb{E})\omega^2 + \mathrm{Im}(\mathbb{F})\omega\}^2}$$
$$/\sqrt{\{\alpha\omega^6 + \gamma\omega^4 + \varepsilon\omega^2\}^2 + \{\beta\omega^5 + \delta\omega^3 + \epsilon\omega\}^2}$$

$$\begin{cases} \lim\limits_{T\to+\infty} \cos\varphi_\theta = \dfrac{1}{\sqrt{1 + \left\{-(Z_a^+/a) - \left[(1/2) - \left\{B^2/\left[12\underline{H}^2(1+C_{m1})\right]\right\}\right](H/a)\right\}^2}} \\ \lim\limits_{T\to+\infty} \sin\varphi_\theta = \dfrac{-(Z_a^+/a) - \left[(1/2) - \left\{B^2/\left[12\underline{H}^2(1+C_{m1})\right]\right\}\right]}{\sqrt{1 + \left\{-(Z_a^+/a) - \left[(1/2) - \left\{B^2/\left[12\underline{H}^2(1+C_{m1})\right]\right\}\right](H/a)\right\}^2}} \end{cases} \quad (26)$$

$$\varphi_{x1}''' \overset{\text{def}}{=} (Z_a^+/a) + \{[(1/2) - \{B^2/[12H^2(1+C_{m1})]\}](H/a)\}$$
$$\lim_{\omega\to 0}\cos\varphi_\theta = 1/\sqrt{1+\varphi_{x1}'''^2}$$
$$\lim_{\omega\to 0}\sin\varphi_\theta = -\varphi_{x1}'''/\sqrt{1+\varphi_{x1}'''^2}$$

9.8 Friction coefficient at large wave periods

Therefore:

$$\lim_{\omega\to 0} A = \lim_{\omega\to 0} a\{\sinh[k(z_0+h)]/\sinh(kh)\} \bullet \cos 360° = a(z_0+h)/h,$$
$$\lim_{\omega\to 0} B = \lim_{\omega\to 0} a\{\sinh[k(z_0+h)]/\sinh(kh)\} \bullet (-\omega t_T) = O(\omega)$$
$$\lim_{\omega\to 0} C = O(\omega)(-c_\Lambda/\sqrt{1+c_\Lambda{}^2}) + Z_G O(\omega)\left(1/\sqrt{1+\varphi_{x1}'''^2}\right) = O(\omega),$$
$$\lim_{\omega\to 0} D = O(\omega)(1/\sqrt{1+c_\Lambda{}^2}) + Z_G O(\omega)\left(-\varphi_{x1}'''/\sqrt{1+\varphi_{x1}'''^2}\right) = O(\omega)$$
$$\lim_{\omega\to 0} \mathcal{G} = (-L) \quad (27)$$

$$\lim_{\omega\to 0} T_\pm = \lim_{\omega\to 0} \frac{A(-L) \mp \sqrt{\left\{A^2 + O(\omega)^2\right\}L^2 - [AO(\omega) - O(\omega)O(\omega)]^2}}{[AO(\omega) - O(\omega)O(\omega) + O(\omega)(-L) + O(\omega)(-L)]}$$
$$\lim_{\omega\to 0} T_\pm = \lim_{\omega\to 0} \left[-AL \mp \{AL + O(\omega^2)\}\right]/O(\omega)$$
$$\lim_{\omega\to 0} T_- = \mathrm{LIM}_{\omega\to 0}[-[a(z_0+h)/h]L - [a(z_0+h)/h]L + O(\omega)]/O(\omega) = \infty$$

Case	Floater geometry	Depth	Maximum Hs for berthing (Hs = 2a)[1]
1	5 m diameter monopile	23 m	1.5 m (no masking)
2	13 m diameter cylindrical floater	70 m	2.0 m
3	41 m diameter cylindrical floater	23 m	2.3 m
4	36 m side square floater	23 m	2.3 m
5	36 m side square hollow floater (FLOATGEN)	23 m	2.3 m

[1]*The CAT CTV fender length is L = 1 m [1].*

Table 2.
CAT CTV analytical berthing limits for monopile various floater designs.

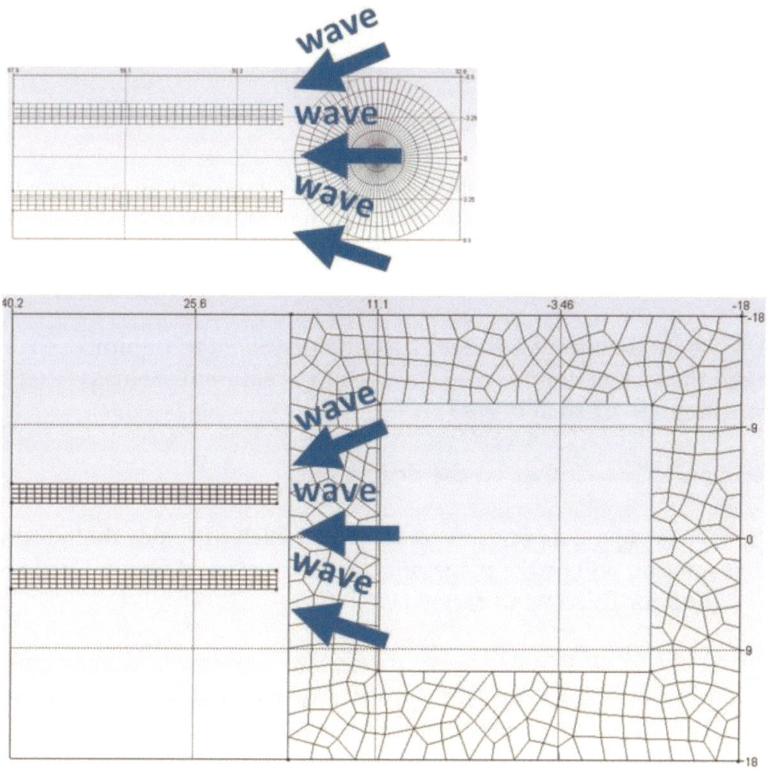

Figure 18.
Floater wave masking performances (cylinder versus FLOATGEN).

$$\lim_{k\to 0}|f(T_-)| = \left|\frac{-A - 2\bullet 0 + \{A\bullet 0^2\}}{-(0-[-L]) - 2\bullet 0 + [(0+[-L])\bullet 0^2]}\right| = \frac{[a(z_0+h)/h]}{L}$$

$$\lim_{\omega\to 0} T_+ = \underset{\omega\to 0}{\mathrm{LIM}}[-AL + AL + O(\omega^2)]/O(\omega) = O(\omega) = 0 \quad (28)$$

$$\lim_{k\to 0}|f(T_+)| = \left|\frac{-A\bullet 0^2 - 2\bullet 0\bullet 0 + A}{-(0-[-L])\bullet 0^2 - 2\bullet \mathrm{D}\bullet 0 + (0+[-L])}\right| = \frac{[a(z_0+h)/h]}{L}$$

Eventually, we get the following formula for berthing to possible [1]:

$$\lim_{k\to 0} |f(T_{\pm})| = [a(z_0 + h)/h]/L < f_{grip} \quad (29)$$

$$\text{if } f_{grip} = 0,8 \ (rubber - very\ wet\ soil) \ (\text{refer to Eq.(9)})$$

Table 2 sums up the analytical Hs found for berthing to occur whatever the wavelength, by using Eq. (29).

As can be seen results from **Tables 1** and **2** meet with less than 12% discrepancy. The reason why is that the analytical calculation does not account for the diffraction forces.

Eventually, some practical considerations must be accounted for (see **Figure 18**):

The more the wave direction varies, the more a large floater width becomes necessary to allow berthing by masking the waves.

10. Conclusions and recommendations

The present study results find that berthing a CTV against an offshore wind turbine is not yet optimized, at least from a marine maintenance point of view: most designs do not provide sheltered waters.

Some proposals focus on improving the CTV:

- ESNA promote the use of Surface Effect Ships (SES), to minimize their heave. However, the lack of commercial success of the solution seems related to the high fuel consumption of such boats [7].

Other proposals focus rather on the floater side:

- IDEOL have designed a pontoon-like floater which provides sheltered waters. At best that solution will prove successful for floating wind farms. Obviously, that is not a solution for fixed wind farms [11].

- FLOATING POWER POINT propose a floating wind turbine combined with a wave energy convertor (WEC), to provide an artificial harbor downstream, thanks to the wave energy extracted by the WEC [13].

Another possible axis of development would be to design an additional wall to existing boat landings, providing a sheltered water.

Eventually, since the present study only applies to a unidirectional wave, the next studies will focus on multidirectional waves, in order represent a more realistic sea state.

Acknowledgements

The author gratefully acknowledges the support from ENSM and its director of research and industrial relations, Mr. Dominique FOLLUT, for throughout the present research work.

Acronyms and abbreviations

Abbreviation and definition

Cat	Catamaran
CTV	crew transfer vessel
DP	dynamic positioning
H_s	significant wave height
HSVA	Hamburgische Schiffbau-Versuchsanstalt GmbH (Hamburg Ship Model Tank Test Facilities)
O&M	operation & maintenance
RAO's	response amplitude operators
3D	three dimensional
2D	two dimensional
WT	wind turbine

Terminology and designation

O, x, y, z	absolute reference frame
M	operating point fixed to the CTV
ρ	water specific gravity
h	water depth
B	ship length
H	ship draft
m (or Δ)	ship mass (or displacement)
G	ship centre of gravity
I_G	ship inertia at G
C	ship centre of buoyancy
I	matrix of ship own inertia
I_a	matrix of ship added inertia
K	matrix of stiffnesses
B	matrix of dampings
F_{excit}	vector of wave loads
λ	wavelength
A-	berthing point
O, X, Y, Z	reference frame attached to the ship
F	centre of floatation
g	gravitational acceleration
x_G, z_G	coordinates of ship gravity centre
x_C, z_C	coordinates of ship buoyancy centre
b_3	ship heave damping factor
k_3	ship vertical hydrostatic stiffness
C_d	shipdrag coefficient
C_{m1}	added mass coefficient in x direction
m_a	added mass in x direction
X	vector of ship motions
τ_x	ship surge
τ_z	ship heave
θ	ship pitch angle
T	regular wave period

k	wave number
a	wave amplitude (half crest to through)
x (or b_c)	flat rate heave damping coefficient (%-age of critical damping)
x_{max} or τ_{xm}	max. surge estimate (for quadratic damping force)
ω	wave pulsation
θ_{max}	adimensional max. pitch estimate (for quadratic damping force)
Λ	$4\rho CC_d(x_{max} + H\theta_{max})/(3\pi)$
c_Λ	$\Lambda H/(m + m_a)$
$\mathcal{G}$	CTV surge over which the CTV captain has the time to adjust the propeller thrust P, in order for the fender never to lose contact with the boat landing
x_m	max. CTV surge
z_m	max. CTV heave
θ_m	max. CTV pitch
φ_x	max. CTV surge
φ_z	max. CTV heave
φ_θ	max. CTV pitch
L	fender length
Z_a^+	CTV propeller elevation
Z_0	floater keel elevation

Author details

Laurent Barthélemy
Ecole Nationale Supérieure Maritime, Nantes, France

*Address all correspondence to: laurent.barthelemy@supmaritime.fr

References

[1] Barthélemy. Berthing criteria for wind turbine crew transfer vessel with low or high friction fender. In: Proceedings of the 5th World Conference on Smart Trends in Systems, Security and Sustainability (World S4 '21); 29–30 July 2021. London, UK: IEEE/ Springer; forthcoming

[2] König M, Ferreira González D, Abdel-Maksoud M, Düster A. Numerical investigation of the landing manoeuvre of a crew transfer vessel to an offshore wind turbine. Ships and Offshore Structures. 2017;**12**(Suppl. 1):S115-S133. DOI: 10.1080/17445302.2016.1265883

[3] Carbon Thrust. Crew Transfer Vessel (CTV) Performance Plot (P-Plot) Development Notice to the Offshore Wind Energy Sector. 2017. Available from: https://www.carbontrust.com/media/674745/carbon-trust-p-plot-development-researchsummary-june-2017.pdf [Accessed: 13 May 2019]

[4] Available from: http://gmsh.info/ [Accessed: 01 August 2019]

[5] Babarit A, Delhommeau G. Theoretical and numerical aspects of the open source BEM solver NEMOH. In: Ecole Centrale Nantes, 11th European Wave and Tidal Energy Conference (EWTEC 2015); Nantes, France. 2015. Available from: https://lheea.ec-nantes.fr/logiciels-et-brevets/nemoh-presentation-192863.kjsp [Accessed: 12 July 2019]

[6] Muller J. Formulaire technique de Mécanique Générale 16è édition—Théorie et dimensionnement. DUNOD. 2015. Available from: http://maron.perso.univ-pau.fr/meca_old/ch3coef.htm [Accessed: 01 May 2019]

[7] Skomedal NG, Espeland TH. Cost-effective surface effect ships for offshore wind. Esna AS, Kristiansand S, Norway, FAST 2017 Conference; Nantes, France. 2017

[8] Available from: https://info-efgl.fr/le-projet/le-flotteur-ppi-eiffage/ [Accessed: 29 October 2021]

[9] Available from: https://www.energiesdelamer.eu/bin/statsnews.php?lst=2&nid=395&img=publications/4085-bouygues-travaux-publics-presente-le-flotteur-floatgen-au-concours-tekla [Accessed: 03 June 2018]

[10] Weller S. Fibre rope selection for offshore renewable energy: Current status and future needs. In: Tension Technology International Ltd EUROMECH 607, Brest; 29th August 2019. Available from: https://wwz.ifremer.fr/rd_technologiques/content/download/136107/file/Session%204%20Weller.pdf [Accessed: 02 January 2020]

[11] Ideol TC. IDEOL'S French and Japanese demonstrators: A necessary stepping stone on the way to commercial-scale projects. In: Conference FOWT 2019; Montpellier, France. 2019

[12] Le Boulluec M. «CRE-1-Houle-MLB». Cours 2015-2016 de Mastère Spécialisé Energies Marines Renouvelables. Ecole Nationale Superieure Des Techniques Avancees; Brest, France. 2015

[13] Available from: https://www.floatingpowerplant.com/ [Accessed: 28 October 2021]

Chapter 7

Prediction of Environmental Sound Pressure Levels from Wind Farms: A Simple but Accurate Model

Alice Elizabeth González and José Cataldo Ottieri

Abstract

The prediction of the environmental sound pressure levels of wind farms does not yet have a global consensus on how to achieve an easy-to-build model with accurate results. This chapter aims to present the sound pressure level prediction model developed at Universidad de la República (Uruguay). Its main characteristics are: taking into account atmospheric stability to determine the acoustic power of the machines; describing the generation of noise along the blades due to turbulent phenomena; and calculating the noise propagation at different distances from the tower taking into account atmospheric absorption, turbulence energy dissipation, and geometric divergence. It does not have any special IT requirements. A new approach to calculating geometric divergence is considered. The results show more than 80% of the cases within the range of ±3 dB, for wind farms throughout our country. The need to use not only A-weighted sound pressure levels but also octave frequency band levels is indicated.

Keywords: wind turbines noise, wind farms noise, wind energy, noise prediction model, aerodynamic noise

1. Introduction

Uruguay is a small country located in South America. According to the 2010 Strategic Energy Development Plan for Uruguay, the diversifying of the energy matrix should be prioritized. Alternative and renewable energy sources rose to be exploited at a major scale. Because of its windy climate, wind energy is now one of the most important sources in our country: currently, Uruguay has more than 1500 MW of installed power lying on this green energy source.

The fast growing development of wind power in Uruguay has encouraged research on many environmental issues, especially those related to wind turbines operation. In many countries—Uruguay included—the method of ISO Standard 9613-2 is the preferred tool for predicting environmental sound pressure levels due to stationary noise sources. However, it is well known that it can incur on great underestimations when sources are large wind turbines, especially under certain atmospheric conditions [1].

 IntechOpen

This paper is focused on the prediction of environmental sound pressure levels due to the operation of large wind turbines, emphasizing in the prediction model developed by the Research Group on Noise Pollution at the Faculty of Engineering (UdelaR). For its development, the theoretical analysis of the phenomena involved on noise emissions was complemented with tests at the University wind tunnel and, of course, with sound pressure levels measurements at some wind farms in Uruguay. The aerodynamic phenomena involved in acoustic emissions were analyzed with the Research Group on Wind Energy at the Faculty, which has been working on wind energy for nearly 30 years.

The application of the prediction model allows obtaining the expected sound pressure levels at different points. Only airborne sound propagation is considered, as the importance of ground propagation does not involve an important amount of acoustic energy for onshore wind farms.

2. Some limitations of the usual prediction methods to represent wind turbines noise generation and propagation

Environmental sound pressure levels related to stationary sources are usually predicted as prescribed by the ISO Standard 9613-2. It is not only a standardized method of calculation, but it has been for a long time the recommended one in the European Union [2]. This is a strong argument at some developing countries. Convincing the decision-makers about the need of developing another prediction method to achieve more reliable results in the case of wind turbines is not an easy task.

This section aims to point out the main hypothesis of the ISO Standard 9613-2 [3] and to discuss their applicability to wind turbine noise.

2.1 The origin of the calculation method

In 1981, CONCAWE (a group of oil companies, aiming toward the research on the conservation of water and air quality in Europe) hired C. J. Manning for developing a prediction model of environmental sound pressure levels [4]. Some novel prediction methods were inspired on it, as the ISO Standard 9613-2 was.

According to CONCAWE, the environmental sound pressure levels at remote places due to a noise source can be obtained by solving the following expression:

$$L_p = L_W + D - K \tag{1}$$

Where L_p is the sound pressure level in the short time for the octave band i, L_W is the acoustic power level for the octave band i, D is the correction due to directivity of the source, and K is the sum of the attenuation terms.

The ISO Standard 9613-2 general expression is just the same:

$$L_{pi} = L_{Wi} + D - A_i \tag{2}$$

The definitions of Eq. (1) are valid for Eq. (2). A_i are the attenuation terms (atmosphere absorption, ground absorption, presence of obstacles or noise barriers, etc.). The subscript i refers to the values for the octave band i.

Both calculation methods assume the divergence law to be quadratics, thus the emitter is supposed to be a point source. Also, both calculation methods promote their application by frequency octave bands. Nevertheless, if there is not enough information to work by bands, ISO Standard 9613-2 will accept calculating in A-weighted sound pressure levels using all formulae and coefficients corresponding to the octave band centered at 500 Hz. If the acoustic emissions have high energy content in low frequencies, this way of calculating will cause a great underestimation of immission sound pressure levels.

CONCAWE's model uses the meteorological categories proposed by Parkin and Scholes instead of the currently preferred Pasquill-Gifford ones. ISO Standard 9613-2 does not consider calculating differences due to different meteorological conditions when the main calculation hypothesis is satisfied: wind speed between 1 and 5 m/s at a height between 3 and 11 m above the ground and averaged over a short period of time or moderate temperature inversion with its base at ground level. These conditions are not always met when the source is a wind turbine.

In this century, it has been verified that the differences between environmental sound pressure levels predicted by ISO Standard 9613-2 and those that do occur due to the operation of wind turbines would be very important: underestimations of 15 dB or more have been reported during the occurrence of certain combination of environmental conditions [5, 6].

Then, ISO Standard 9613-2 calculation method has been submitted to a deeper analysis.

2.2 Understanding the ISO Standard 9613-2 limitations

Aerodynamic noise generation during operation of wind turbines is inherent to them in nature: the major acoustic emissions from large wind turbines are caused by the interaction between the air flow and the blades. The acoustic emissions occur all along each blade, most of them at low frequencies. Then, the height of the noise source is from about 40–130 m above the ground. The incident wind speed largely varies between these two heights, so that the wind turbine becomes a heterogeneous and complex sound emission source.

There are some limitations for the use of ISO Standard 9613-2 to predict environmental sound pressure levels due to large size wind turbines [1]. Some of the general ones are the following:

- The hypothesis about wind speed and atmospheric stability is not always fulfilled.

- Atmospheric conditions (neutral, instability, or under an inversion layer) have a great incidence both on generation and on propagation of noise [6].

- At the typical distances of interest, a wind turbine cannot be supposed to be a point source [7].

- The Standard supposes the distance from source to receiver to be between 100 and 1000 m.

- The average height of source and receiver should be between 0 and 30 m (then, the maximum height of the source will not exceed 60 m if the receiver is at 0 m height).

- Source and receiver should be placed over a plain surface (a surface with a continuous slope, i.e., the method is not valid for complex terrain).
- If the calculations are done based on A-weighted sound pressure levels (as it is allowed by the Standard), a greater underestimation should be done at low frequencies.

Some experimental findings also refer to better results when not considering ground attenuation effects during propagation [8].

But there are also two of the major assumptions that are at the very beginning of the conceptual framework of environmental acoustics that are not fulfilled by the physic/fluid mechanic phenomena involved in the aerodynamic sound generation from wind turbines [9]:

- The hypothesis that acoustic processes are adiabatic, because they occur very fast and involve only very small amounts of energy. Most of vibration phenomena can be well described as adiabatic ones, but this is not the case of the aerodynamic noise related to wind turbines' operation. Noise generation is related to turbulent phenomena, which are not adiabatic but very dissipative ones.
- The hypothesis of ideal fluid, which is opposite to the main phenomena that are related to release of eddies from a boundary layer; these phenomena only can occur if the air is considered as a viscous or real fluid.

These are thought to be the root causes for both CONCAWE and ISO methods not to being appropriate for predicting the environmental sound pressure levels related to wind turbines' operation, as they cannot describe the main involved phenomena on a right way [1, 9–11].

3. Improving the prediction method

In order to improve the current prediction method, we have proposed several modifications. We have focused on the noise generation phenomena, but we have also worked on two other points: the explicit consideration of the atmospheric stability condition and the dissipative nature of the main phenomena during propagation.

3.1 Wind velocity at the hub height: Considering atmospheric stability class

The wind velocity is usually measured at 10 m height above the ground. One of the main causes of underestimating the environmental sound pressure levels is related to calculating the wind speed at the hub height using a neutral atmospheric profile with basis on its value at 10 m. To avoid this problem, the atmospheric stability class (according to Pasquill-Gifford) has to be explicitly taken into account for this calculation. The atmospheric stability does not only influence the wind speed profile but also the turbulence intensity and, therefore, the acoustic energy depletion law.

If a stable or thermal inversion atmospheric condition occurs, not including it in the prediction method will conduct to:

Pasquill stability class		m	
Class	**Description**	**Usual bibliography values**	**Van den Berg experimental values [from 6]**
A	Highly unstable	0.09	0.15
D	Neutral	0.28	0.40
F	Highly stable	0.41	0.65

Table 1.
Values of m *according to Pasquill-Gifford stability class.*

- A great underestimation of the wind speed at the hub height, which would result in the underestimation of the emitted acoustic power level.
- A great overestimation of the sound pressure levels depletion, due to the lower atmospheric turbulence and hence, the lower energy dissipation during propagation.

If the atmospheric thermic profile is not known, the wind speed at the hub height should be obtained by supposing a strong atmospheric stability profile (class F according to Pasquill-Gifford), to be in the most demanding hypothesis for protecting the health of noise receivers.

Then, the wind speed at the hub height should be met by converting the measured wind speed data—that are usually taken at 10 m over the ground—using a proper method.

Using the logarithmic profile approach for wind velocity (Eq. (3)) is better than using the potential profile approach (Eq. (4)), even though there are good experimental values for the potential approach. Indeed, since several authors refer that the usual values of m may lead to underestimation of the acoustic power, using the experimental values met by Van den Berg [6] is strongly recommended when using the potential approach, in order to remain on the safe side (see **Table 1**).

Logarithmic profile approach:

$$u(h_{hub}) = \frac{u_*}{k}\left[ln\left(\frac{h_{hub}}{z_0}\right) - \psi_m\left(\frac{h_{hub}}{L_*}\right)\right] \quad (3)$$

Where:
$u(h_{hub})$ wind velocity at hub height
u_* friction velocity
k von Karman's constant
z_0 roughness length
ψ_m thermal stratification function
L_* Monin-Obukhov length
Potential profile approach:

$$u(h_{hub}) = u_{ref}\left(\frac{h_{hub}}{h_{ref}}\right)^m \quad (4)$$

Where:
u_{hub} wind velocity at hub height h_{hub}
u_{ref} measured wind velocity at a reference height h_{ref}

m coefficient depending on Pasquill-Gifford class of atmospheric stability (see **Table 1**)

The calculation procedure that we recommend to meet the wind velocity at h_{hub} height, taking into account its value at any other height h_{ref}, is as follows [10–12]:

1. If the stability class to which u_{ref} corresponds is known, the velocity at the hub height can be met by applying either Eqs. (3) or (4).

2. If the stability class to which u_{ref} corresponds is not known, a stable atmospheric profile should be assumed for calculating the velocity at hub height with basis on the wind velocity at a reference height (usually 10 m above the ground).

3. Once the wind velocity at the hub height (u_{hub}) has been obtained, a "corrected" wind velocity at 10 m in height should be calculated. This is the 10 m height wind velocity to be used for meeting the acoustic power level of the wind turbine from tables or charts provided by the manufacturer. **Figure 1** sketches the procedure.

Please note:

1. If the wind speed at the hub height is known, the wind velocity at 10 m must always be obtained assuming a neutral atmosphere (even when the stability class is known not to be neutral).

2. For obtaining the sound pressure level resulting from a wind velocity value measured at a height "H" (other from h_{hub}) in any given atmospheric condition "X": u_{hub} should be calculated assuming the class of stability "X"; then, the "corrected" wind speed at 10 m in height should also be calculated by assuming neutral atmosphere (class D). The acoustic power shall be read from the datasheet provided by the manufacturer; it will also be associated with that atmospheric stability class: $L_W^{"X"}$.

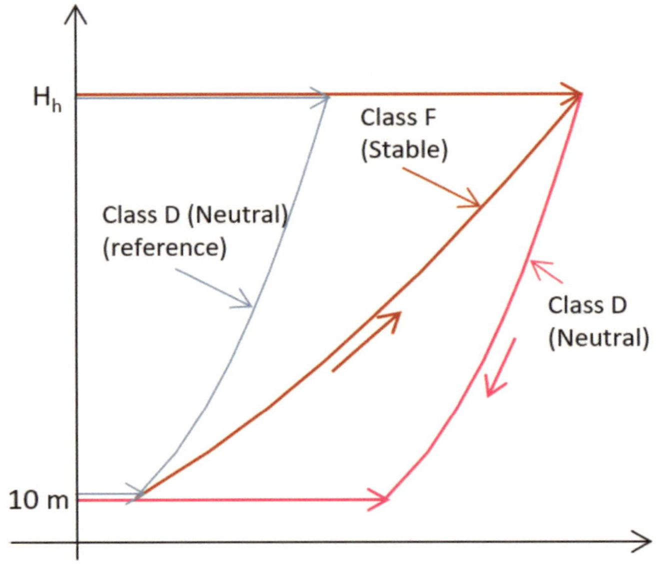

Figure 1.
How to reach the wind speed at 10 m height to obtain $L_{W,A}$ (redrawn from [10]).

f (Hz)	16	31.5	63	125	250	500	1000	2000	4000	8000
Add to L_{WA} (dB)	−44	−26	−21	−14	−7	−6	−6	−9	−12	−22

Table 2.
Reference spectrum of acoustic power of 2 MW wind turbines in octave bands (based on [13]).

Wind turbine manufacturers often provide tables or graphs relating the wind velocity at 10 m in height (u_{10}) to the acoustic power level (in dBA) emitted by the wind turbine in neutral atmosphere conditions. However, providing emission spectra in frequency bands is not so common. If this information is not available, a reference spectrum should be used, e.g., spectrum in **Table 2**.

Table 2 (based on [13]) presents the values to be added arithmetically to the acoustic power level of the wind turbine (L_{WA}) to obtain the acoustic power levels in each octave band, also in dBA ($L_{W,f,A}$).

3.2 Modeling noise generation phenomena

We aim to obtain the sound pressure levels due to the operation of a typical three-blade wind turbine, at a generic receiver point located downwind at a distance *d*.

Aerodynamic noise is generated by the interaction of wind with the blades of the machine. Most of the acoustic emissions occur in low frequencies, so the acoustic print of wind turbines can be found at large distances from the sources, making the problem more complex to manage.

3.2.1 General background

There are three main processes causing the fluctuation of the pressure field and then the acoustic emissions [14]:

1. The turbulence of the incoming wind, which causes pressure fluctuations around the blades; it is variable over time and it is called "incoming flow noise."

2. The viscous forces in the boundary layer over the solid surfaces of the turbine, such as blades, tower, and hub. Viscous forces in this layer are not negligible compared with the inertial forces (related to the medium air flow). The release of eddies with negative gauge pressure at their cores, developed on solid surfaces such as blades, tower, and hub due to viscous stresses, causes a continuous noise called "trailing edge noise."

3. The power exchange between the wind and the machine that produces the release of two families of eddies linked to each blade; one of them has helical motion and the other one is centered on the rotation axis, and its length scale is about the length of the diameter of the rotor.

These phenomena are related to three different geometric scales [14, 15]:

1. Macroscale: it is the scale related to the largest eddies. If U, L, and T are the scales of velocity, length, and time associated to these eddies, the Reynolds number of the biggest eddies is the same as for the main flow.

2. Intermediate scale: it includes lower scales than the macroscale ones; there is still no power dissipation. The range of scales included here is called “inertial range.”

3. Microscale: it is the lowest scale, in which the energy dissipation occurs. Unlike what happens in the macroscale, the smallest eddies are isotropic, as if the flow has “forgotten” where it comes from.

The turbulent cascade hypothesis is then to be considered. According to it, the larger eddies are dissipating into smaller scale eddies with increased kinetic energy. However, there is a length scale at which the power transfer to a smaller eddies scale is not possible. At this point, the turbulent cascade ends and the energy from the last eddies is finally dissipated. The smallest eddies scale is the Kolmogorov scale; the so-called Kolmogorov frequency or dissipation frequency is the generation frequency of these smallest eddies [15]. According to their frequency and energy, the released eddies are able to produce audible phenomena, i.e., they can become noise sources (**Figure 2**).

The passage of the blades ahead the tower imposes a fluctuation of the sound level pressures emitted by the abovementioned phenomena. It results in an amplitude-modulated noise, called the “blade passage noise.” It has a double nature, one related to the flow and one related to the geometry of the source. The modulation is the most related process to annoyance in wind turbine noise. This process is not modeled in detail: the informed sound pressure levels are the highest of those corresponding to the fluctuation.

3.2.2 Basic concepts concerning wind turbines

A first approach to describe the wind turbines operation is to model the rotor as an active disk, which absorbs kinetic energy from the incoming wind, resulting in a reduction in the flow speed downstream of the turbine.

If v_1 is the incoming velocity and v_2 is the outcoming velocity, the velocity induction coefficient “a” is defined according to Eq. (5):

$$v_2 = v_1\,(1\text{–}2a) \tag{5}$$

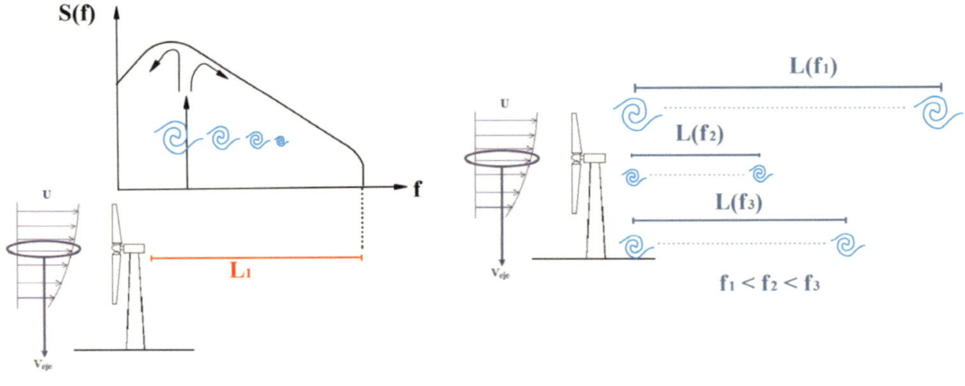

Figure 2.
Atmospheric conditions for propagation. From: [15].

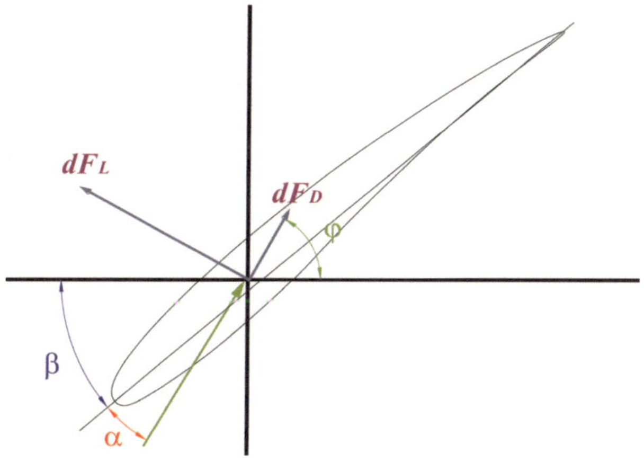

Figure 3.
Drag F_D and lift F_L efforts over the blade (from [15]).

Applying mass and energy balances to the incoming flow, and supposing an adiabatic and incompressible flow, the maximum amount of power absorbed by the disk is:

$$W = \frac{1}{4}\rho A(v_1 + v_2)\left(v_1^2 - v_2^2\right) \tag{6}$$

In Eq. (6), ρ is the air mass density, and A is the swept area or rotor area.

The power absorbed is maximized when $a = 1/3$. Then, Eq. (7) is the expression of the so-called "Betz power":

$$W_{Max} = \frac{16}{27}\bullet\left(\frac{1}{2}\rho A v_1^3\right) \tag{7}$$

The wind turbine operates at its maximum power for each wind velocity, while the wind velocity is under the rate value. As consequence of the power exchange process between the wind and the machine, the flow downwind the rotor rotates around the turbine axis.

The force over the blade, consequence of the interaction between the flow and the blade, is split in two components: the drag effort (D) and the lift one (L) (**Figure 3**). The magnitude of these components strongly depends on the angle of attack (α), which is the angle between the chord of the blade and the incoming flow relative to the blade. When the drag component increases, the noise generation increases too. This usually occurs when the angle of attack α increases.

3.2.3 Theory of turbulence

There are several analytical expressions to describe the universal shape of the turbulence spectrum. One of that, the Von Karman's spectrum, expresses the spectrum as function of a nondimensional ratio built with turbulence integral length scale L_u and the main flow speed v (Eq. (8)):

$$X = \frac{L_u f}{v} \tag{8}$$

According to Von Karman spectrum, we propose to estimate the energy content in a third-octave band centered at a frequency f according to Eq. (9), in which σ is the standard deviation of the flow speed:

$$S = 4\,X\,\frac{\left(2^{1/6} - 2^{-1/6}\right)\sigma^2}{\left(1 + 70,8\,X^2\right)^{5/6}} \tag{9}$$

3.2.4 Emitted acoustic power level

To meet the acoustic power level, we accept that each blade is composed of a group of discrete thin elements or slices. Each one would be sufficiently thin to be thought as a noise point source. Then, the total acoustic power emitted by one blade element should be obtained as the superposition of the acoustic power emitted by the incoming flow ($L_{W,IF}$) and the trailing edge ($L_{W,TE}$) as following (Eq. (10)):

$$L_W = 10\ log\left(10^{\frac{L_{W,IF}}{10}} + 10^{\frac{L_{W,TE}}{10}}\right) \tag{10}$$

The incoming flow noise is the result of the fluctuation of the lift effort on the blade, which makes the drag effort to fluctuate. We propose estimate of the pressure field fluctuation using McLaurin series, where first-order terms are related to the turbulent fluctuation.

The length scale of interest, for the incoming turbulence, is similar to the length of the blade chord, which corresponds to the length scale of the eddies that produce the greatest amplitude fluctuation on the pressure field. Van den Berg proposes to use a length scale equal to 60% of the blade chord for this length [8]. Smaller eddies would produce lower fluctuations on the pressure field. Then, the shape of the spectrum of the incoming edge noise, associated to eddies with length scale larger than the blade chord, will be the same as Von Karman's spectrum.

The trailing edge noise is due to the turbulent boundary layer separation over the blade. The length scale of interest in this case is about the boundary layer thickness.

We built a routine for obtaining the emitted acoustic power level by third-band octave. Its aim is to obtain the predicted sound pressure levels at a point placed at 100 m downwind of the wind turbine tower. We discretize the blade in infinitesimal length blade elements. The coordinates of each slice in every moment could be thought as [x(t), y(t), z (t)]. Then, the propagation into the first 100 m from the tower axis is done assuming that each one of the blade elements is a nonstationary noise point source (**Figure 4**).

For the propagation from each blade element to the receiver location, our routine only considers the geometrical divergence and the atmospheric sound absorption as indicated at ISO Standard 9613-1 [16]. The output of this routine is the input for the propagation module [10, 11].

3.3 Modeling noise propagation

For computing sound propagation, the input data are the results of the computing at 100 m far from the tower of the wind turbine. Not only geometric divergence but also atmospheric absorption and turbulent dissipation are considered; both phenomena depend on the frequency. The final sound pressure levels at a given reception point are obtained by superposing the sound pressure levels due to different wind

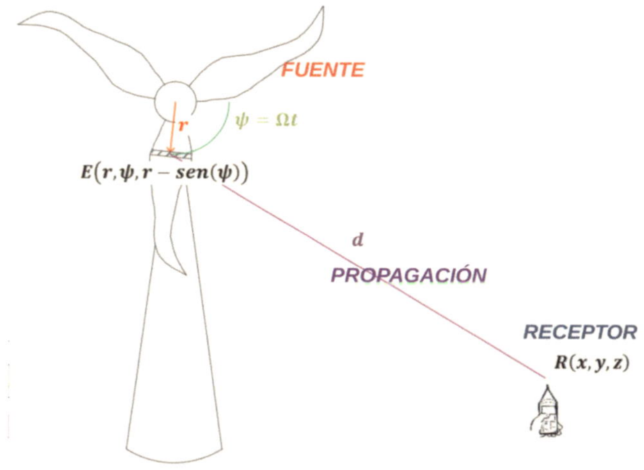

Figure 4.
Sketch of calculation of sound pressure levels in the first 100 m (from [10]).

turbines operation. All computations are done in octave bands, and the final results are expressed as L_{Aeq} values [11].

3.3.1 Atmospheric conditions and audibility of acoustic emissions

The analysis of the evolution of eddies generated due to wind turbine operation requires the use of the cascade process as it is usual in turbulent flow studies. According to it, the larger eddies are melting into smaller ones, increasing its kinetic energy. At some point, small eddies cannot continue to transfer power to smaller ones; so, they dissipate their remaining energy, thus ending the cascade process. The scale of these last eddies is the order of the Kolmogorov's scale.

Under atmospheric instability condition, the turbulence is very high and the eddies scale interval is broad; the cascade process is very efficient to dissipate the produced turbulence. For distances greater than the one at which that dissipation occurs, it shall be assumed that the flow conditions are the same as upstream the wind turbine. The ratio between the current wind velocity up and downstream the machine tends to 1 for greater distances, and the difference between them is practically negligible at a distance of about 6 or 7 rotor diameters downstream of the wind turbine (i.e., about 600 m).

In strong atmospheric instability conditions, the prior distance is the shortest one to fully carry out the whole energy cascade process. For any other atmospheric conditions, the dissipation process occurs in greater distances.

Under strong atmospheric stability conditions (i.e., class "F" according to Pasquill-Gilfford stability classes), the effect of turbulence should be negligible. The only mechanism that affects the energy depletion process in any frequency band—in addition to the geometric divergence or attenuation by distance—is the atmospheric absorption.

3.3.2 The atmospheric absorption

The effect of atmospheric absorption can be considered as the depletion of the acoustic energy of a wave over a given distance, due to energy loss caused by the viscosity of the propagation medium (currently, the atmosphere). To estimate the effect of the atmospheric absorption, the computation method of the ISO Standard 9613-1 was used [16].

The attenuation due to atmospheric absorption in dB of a pure tone with frequency f, from its initial level at a distance $d = 0$ to its level at d, can be obtained according to Eq. (11):

$$Abs = \Gamma_i(f) \cdot \frac{d}{1000} \tag{11}$$

Where Γ_i is the atmospheric absorption in the i-th frequency band in dB/km, and d is the distance from the base in m. The absorption coefficient Γ_i is a function of the relaxation frequencies from oxygen and nitrogen [16].

The generation and propagation phenomena of eddies can be described from a wave approach. Then, close to the source, the sound pressure levels should be estimated considering energy depletion by geometric divergence (*Div*) and by atmospheric absorption as shown in Eq. (12).

$$L_p = L_W - Div - \Gamma_i(f) \cdot \frac{d}{1000} \tag{12}$$

The threshold of perception at each frequency band should be another criterion for determining the distance upon which the sound is still audible. Hearing threshold levels were retrieved from ISO Standard 226 [17].

3.3.3 Geometric divergence

For the depletion of sound pressure levels due to distance, the adjustment is focused on the exponent (n) of the divergence law, which is neither squared nor linear as many measured sound pressure levels close to operating wind farms have shown.

As one of the main hypotheses of linear acoustics was broken (nonviscous effect), we intend not to be mandatory for n to be constant across every one of the considered third-octave bands: the exponents may be related to the distance scale at which eddies are expected to dissipate all their turbulence energy. Then:

$$Div = 10\, log \left(\frac{d}{d_0}\right)^{n(\,f_i)} \tag{13}$$

Here, $n = n(f_i, d, u)$ depends on the central frequency f_i of each octave band.

For the calculations of geometric divergence, the turbulent cascade approach is taken into account (**Figure 2**). The released eddies can propagate along great distances while the turbulent cascade occurs [15]. These distances are related to a certain energy level and a length scale. They are also closely related to atmospheric stability.

We calculated the length scale where the turbulent cascade is expected to end by considering it to depend on the incoming wind velocity and the frequency of the released eddies. This is based on prior consideration of the atmospheric stability during the calculation of noise emissions [10, 11].

Different sets of n values were achieved by fitting measured data. At first, we consider only the dependence of n related to the frequency. Then, we explored the dependences on the distance d, the wind speed u, and the atmospheric stability. The

f (Hz)	16	31.5	63	125	250	500	1000	2000	4000	8000
Closer than 750 m, less than 6.5 m/s	0.37	0.01	0.05	0.66	1.60	1.67	1.94	1.61	0.85	0.43
Closer than 750 m, 6.5 m/s or more	0.52	0.02	0.01	0.22	0.93	0.79	1.04	1.07	1.06	0.65
Further than 750 m	0.45	0.01	0.06	0.72	1.58	1.71	2.03	1.60	0.67	0.34

Table 3.
Divergence coefficient "n" values for octave bands, according to the wind speed u *and the distance* d.

best set of n values was selected by the application of the statistic Friedman test and the comparison of residues (differences) between measured to predicted sound pressure levels expressed as A-weighted broadband levels.

Field data were taken at a height of 1.2 m with class 1 sound pressure meters, close to three different wind farms over plain terrain; the distances of measurements covered from 100 m to about 2000 m.

The best set of values was found to be the one obtained for $n = n(\text{f, d, u})$. The explicit consideration of atmospheric stability in the propagation term did not result in an improvement in the simulated sound pressure levels.

The values of n(f, d, u) are given in octave bands according to whether the calculation distances are closer or further than 750 m and that the wind speed at the hub height is less or greater or equal to than 6.5 m/s (**Table 3**).

4. Validation of the model

4.1 Field measurements

Several sound pressure level measurement campaigns were conducted at three different wind farms with large wind turbines (rate power of 1.8 MW), covering several operating conditions.

Sound pressure level records were taken at 1.2 m height, simultaneously with records of wind speed and direction at 10 m height. In addition, the records of wind speed and direction were obtained from the wind farm anemometer, located at 66 m high and close to the turbines. To carry out the measurements, we used two sound level meters class 1 (Brüel and Kjaer 2250 and Casella 633C), an anemometer (Extech EN-300), a GPS, and two computers.

The measurement points covered four different geographical locations:

- A hilly zone, far from external sources such as houses or roads, to avoid introducing disturbances to the data obtained.
- A flat zone close to the sea, where 10 large wind turbines are installed.
- Another flat zone where two 1.8 MW wind turbines are installed.
- A location in the countryside close to a private company, which has only one wind turbine. This location was particularly interesting for this study, since the records are not affected by other turbines or any external sources.

A set of 59 measurements was used during the calibration and validation processes. The main findings showed that the model gave a good approach for the environmental sound pressure levels related to the operation of wind turbines for wind speeds over 5 m/s at the hub height.

In order to validate the model, another set of field data was used. It was another set of data of 49 cases from 10 wind farms in different locations in Uruguay:

- The four abovementioned places.
- A flat zone in the northern of the country, where 35 wind turbines are installed.
- Two hilly wind farms placed on the Southern part of the country, each one with around 25 wind turbines
- Three rather flat zones in the center/south-western part of the country, having from 20 to 35 wind turbines each one.

Some adjustments were needed for improving the prediction of noise propagation from wind farms built on uneven terrain.

4.2 Accuracy of predictions

The results obtained in the verification of the performance of the model are presented in **Tables 4–6**. Almost 80% of the cases are reproduced within ±3 dB range.

5. Further discussion

5.1 Computed spectra

As important as the percentage of accurate predictions is to state that not only the levels in scale A are predicted in a reasonably adjusted way, but particularly that

	Number of cases	**Cases in ±3 dB**	
		Number	**%**
d closer than 750 m	30	24	80
d further than 750 m	18	14	78
Total	48	38	79

Table 4.
Quality of results according to distance to the wind turbine.

	Number of cases	**Cases in ±3 dB**	
		Number	**%**
u lower than 6.5 m/s	12	10	83
u higher than 6.5 m/s	36	28	78
Total	48	38	79

Table 5.
Quality of results according to wind velocity at the hub height.

	Number of cases	Cases in ±3 dB	
		Number	%
d closer than 750 m, u lower than 6.5 m	8	7	88
d closer than 750 m, u higher than 6.5 m	22	17	77
d further than 750 m, u lower than 6.5 m	4	3	75
d further than 750 m, u higher than 6.5 m	14	11	79
Total	48	38	79

Table 6.
Quality of results according to distance to the wind turbine and wind velocity at the hub height.

the spectra obtained with the proposed model are also rightly adjusted to the measured ones [10].

Figure 5 shows some results for short and long distances. The blue bar is the measured sound pressure level; the pink bar is the computed sound pressure level using ISO 9613-2 with attenuation only due to geometric divergence and atmospheric absorption; and the dark red bar is the result of our prediction proposal. As it can be seen, our model achieves a good performance as a prediction tool.

5.2 Comparison with a nonnegative matrix factorization (NMF) estimation

It is not usual to find references that use a variable depletion law according to the frequency.

We compared our attenuation curves with those presented in a paper published in 2021 [18]. The authors estimate the sound pressure level related to wind turbines with a nonnegative matrix factorization (NMF), a machine learning technique.

They present some attenuation filters in third-octave bands from 31.3 to 2000 Hz, for attenuation only and for attenuation considering three kinds of residual noise designed by the authors with basis on real noise samples. The filters were published in graphic format for three distances: 500, 1000, and 1500 m. We read the graphs and compared the attenuations proposed in [18] with the attenuation achieved for our prediction model in the same frequencies range.

The comparison was done using the Wilcoxon's test for differences between pairs. H_0 was the equivalence of the compared curves; accepting H_0 at 95% confidence means that our attenuation curves are equivalent to those from [18]. Test results are summarized in **Table 7**. We conclude that each one of our attenuation curves reasonably fit at least one case of the filters suggested by [18], the filter with residual noise 1 being the most similar to our proposal.

6. Summarizing our proposal for predicting sound pressure levels related to the operation of large wind turbines

Our proposed calculation process has two steps: at first, modeling the noise generation and its propagation in the short scale (less than 100 m); and propagating the output of the first step from short to large distances far away from the source. At the beginning of the process, the atmospheric stability class is to be taken into account by a correction to the wind velocity; this is very important, to avoid underestimations.

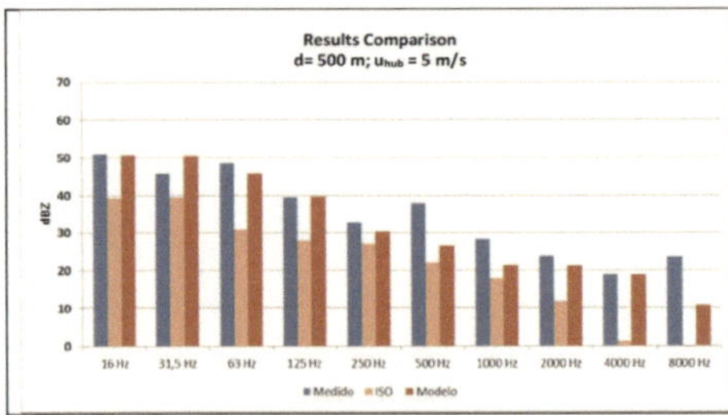

Case 1: d = 500 m; u = 5 m/s

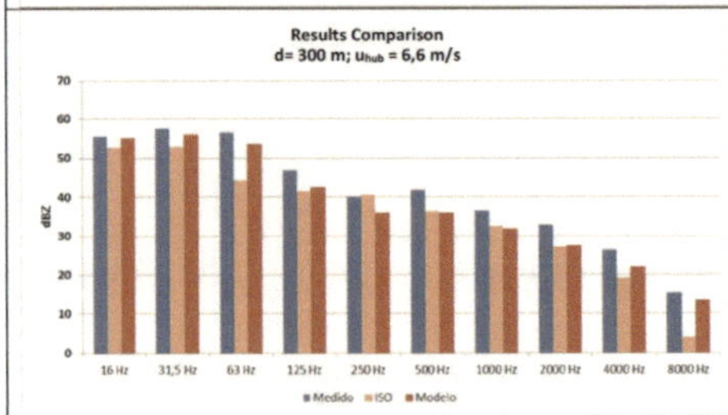

Case 2: d = 300 m; u = 6,6 m/s

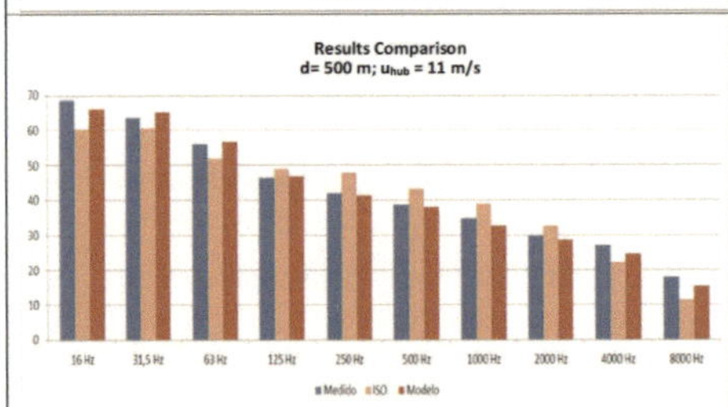

Case 3: d = 500 m; u = 11 m/s

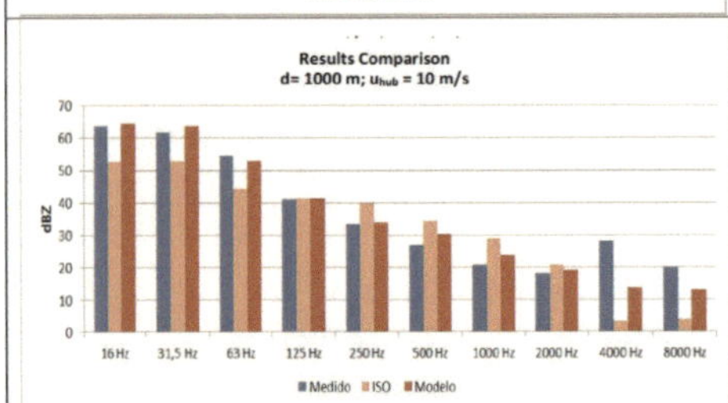

Case 4: d = 1000 m; u = 10 m/s

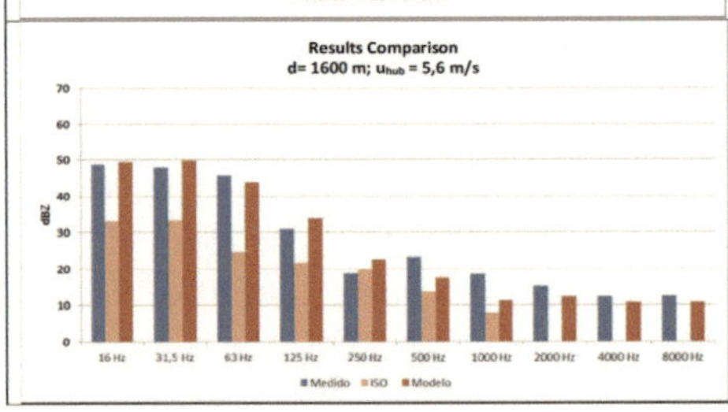

Case 5: d = 1600 m; u = 5,6 m/s

Figure 5.
Comparison of results; left bar: Measured sound pressure levels; center bar: ISO 9613-2 predicted sound pressure levels; left bar: Our prediction proposal. All sound pressure levels are in dBZ. (Adapted from [10]).

	Only attenuation	**With residual noise 1**	**With residual noise 2**	**With residual noise 3**
500 m, u less than 6.5 m/s	Accept H_0	Reject H_0	Reject H_0	Reject H_0
500 m, u 6.5 m/s or higher	Reject H_0	Accept H_0	Reject H_0	Reject H_0
500 m (all together)	Reject H_0	Accept H_0	Reject H_0	Reject H_0

	Only attenuation	With residual noise 1	With residual noise 2	With residual noise 3
1000 m	Reject H_0	Accept H_0	Accept H_0	Accept H_0
1500 m	Reject H_0	Accept H_0	Accept H_0	Accept H_0

Table 7.
Comparison between our attenuation curves and those from [18].

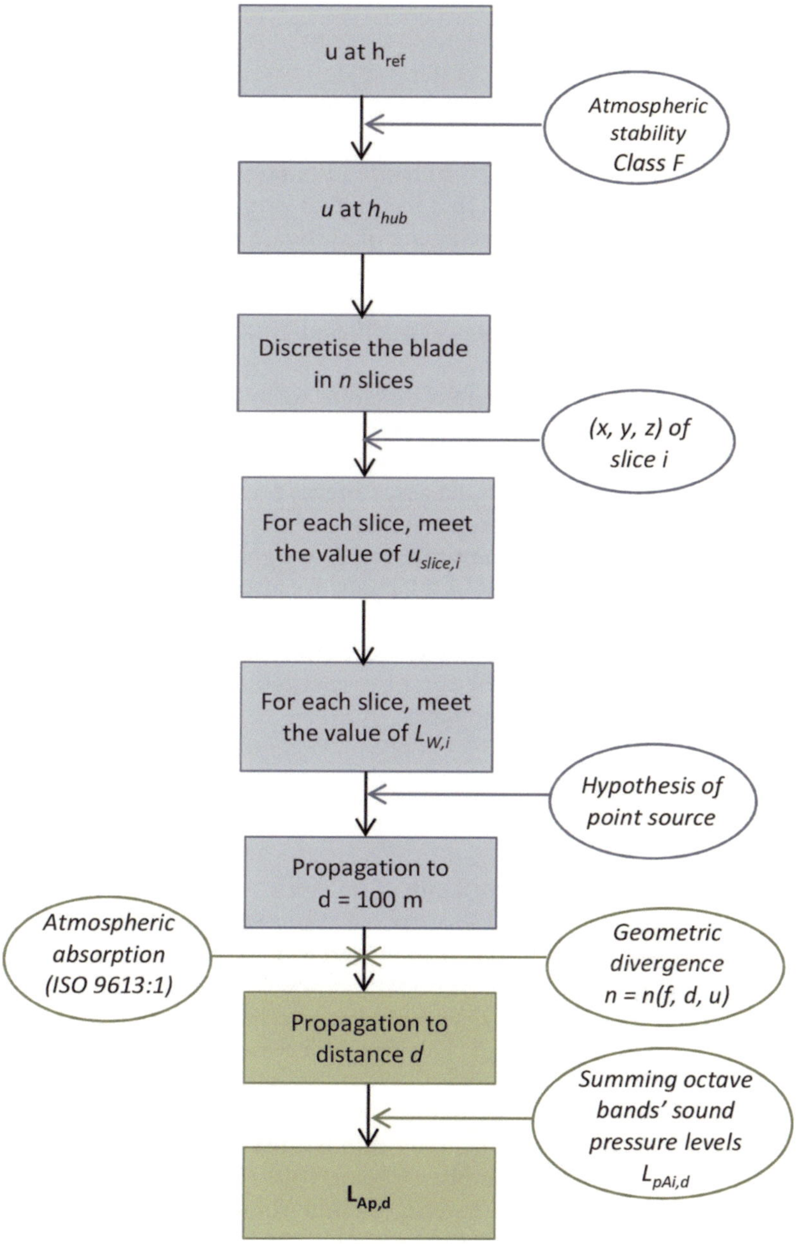

Figure 6.
Calculation process of this prediction proposal (adapted from [10]).

The first step of the calculation process divides the blade in infinitesimal length slices that behave as point sources, and for them the sound pressure level associated to the three mentioned generation processes for each source is calculated and summed logarithmically.

The propagation model takes into account the geometric divergence and the atmospheric absorption, considering n as $n(f_i, d\ u)$. Our proposed "n" values have been presented in Section 3. The fact of not working under the usual hypothesis of environmental acoustics allows considering the coefficient "n" as variable.

Figure 6 shows a sketch of the procedure.

7. Final remarks

We presented an alternative proposal for predicting sound pressure levels from wind turbines. It is a simple method that has shown very good results, it is easy to build, and it does not need special hardware or software requirements.

The whole model was calibrated, validated, and verified, with field measurements made in different wind farms located in Uruguay. Field data for calibration and validation were taken at distances between 300 and 2000 m from the tower of the wind turbine.

Although this method can be used for different turbines, our results were obtained for 2 MW power rate turbines with between 80 and 90 m of hub height.

More, our prediction model seems to be a good one to be used in noise impact studies related to environmental impact assessments for getting the environmental license of wind farms before their construction.

We presented the accuracy of the predicted A-weighted sound pressure levels, which are good or very good in most of the cases; we also showed the obtained spectra fit accurately to the measured ones.

This fitting is more noticeable at low frequencies, the most problematic ones for noise phenomena in wind turbine, because the energetic content in those frequencies is potentially related to people's annoyance.

Acknowledgements

This chapter is the result of many years of research. It has been supported by funds from cooperation with the National Program of Wind Energy (National Energy Directory of the Ministry of Industry, Energy and Mining of Uruguay, DNE – MIEM), by the Sectorial Energy Fund of the National Agency of Research and Innovation (Project ANII_FSE_1_2013_1_10942) and the Research Commission of the Universidad de la República through their Fund for Research and Development Groups (CSIC I + D Groups 2014).

Many people have been involved in our team along the years. Here they are, in alphabetical order: Fabiana Bianchi Falco, Pablo Bonilla Medina (in memoriam), Nicolás Cunha Apatie, Matteo Deambrosi Papini, Pablo Gianoli Kovar, Matías Hernández Castellani, Marcos Raúl Lisboa, Joaquín Montero Croucciée, Luciana Olazábal Barrios, Juan Ignacio Pais, Martín Paz Urban, Nicolás Rezzano Tizze, Felipe Silva Rodríguez, and Guillermo Sugasti Sánchez. We thank all of them for taking part in this research.

Conflict of interest

The authors declare they have no conflict of interest.

Author details

Alice Elizabeth González* and José Cataldo Ottieri
IMFIA, Faculty of Engineering, Universidad de la República UdelaR (University of the Republic), Montevideo, Uruguay

*Address all correspondence to: elizabet@fing.edu.uy

References

[1] González AE, Rezzano Tizze N, Bianchi Falco F. Algunas limitaciones de la Norma ISO 9613—Parte 2 para el estudio de propagación de ruido de aerogeneradores de gran porte (Some limitations of ISO 9613-Part 2 for studying noise propagation from big wind turbines). In: Reunión Regional de Acústica, 11–12 October 2011; Montevideo, Uruguay. Montevideo: Asociación Uruguaya de Acústica–Departamento de Ingeniería Ambiental; 2011. p. 54. ISBN: 978-9974-0-0785-7

[2] Official European Union Journal. Directive 2002/49/EC of the European Parliament and of the Council relating to the assessment and management of environmental noise. 25 June 2002

[3] International Standard Organization. International Standard 9613. Attenuation of sound during propagation outdoors—Part 2: General method of calculation. 1996

[4] Manning CJ. The Propagation of Noise from Petroleum and Petrochemical Complexes to Neighbouring Communities. Report 4/81, CONCAWE. 1981

[5] IMFIA. Impacto acústico de aerogeneradores. Subproyecto A. Informe Final (Acoustic Impact of Wind Turbines. Subproject A. Final Report). Specific Agreement between the Universidad de la República–Faculty of Engineering and the Ministry of Industry, Energy and Mining–National Energy Directorate. December 2012

[6] Van den Berg GP. The sound of high winds: the effect of atmospheric stability on wind turbine sound and microphone noise [doctoral thesis]. Netherlands: University of Groningen; 2006

[7] Martín B, María Á, Tarrero AI, Bravo D, Copete M, González J, Machimbarrena M, García L. Impacto acústico de los parques eólicos y su evolución (Acoustic impact of wind parks and their evolving). In: Acústica 2008. Coimbra, Portugal. 2008

[8] Kaliski K, Duncan E. Propagation Modelling Parameters for Wind Power Projects. (Based on a paper presented at Noise-Con 2007. Reno, NV: Institute of Noise Control Engineering; 2008

[9] Nuñez Pereira I. Elementos de Acústica, Environmental Engineering Magister Program. Montevideo: Faculty of Engineering, Universidad de la República; 2013

[10] González AE, Gianoli Kovar P, Deambrosi Papini M, Hernández Castellani M, Paz Urban M. Síntesis de Resultados del Proyecto ANII FSE_2013_1_10942 "Impacto Acústico de Aerogeneradores de Gran Porte": Modelo predictivo de niveles ambientales de presión sonora. [Summary of Outcomes from the Research Project ANII FSE_2013_1_10942 "Acoustic Impact of Large Wind Turbines": Prediction Model for Environmental Sound Pressure Levels.]. September 2016. p. 23

[11] González AE, Gianoli-Kovar P, Deambrosi-Papini M, Hernández-Castellani M, Paz-Urban M. Environmental noise due to large wind turbines: What we have learnt. In: International Congress on Acoustics ICA 2016. Buenos Aires, Argentina. 2016

[12] González AE, Rezzano Tizze N, Lisboa MR, Gianoli Kovar P. Environmental noise from wind farms: a prediction proposal for flat areas. In:

International Congress of Sound and Vibration ICSV22, Florence, Italy. 2015

[13] Jørgen, Kragh Birger Plovsing, DELTA, Acoustics & Vibration; Svein Å. Storeheier, SINTEF, Telecom and Informatics; Hans G. Jonasson, SP Swedish National Testing and Research Institute, Acoustics. Nordic environmental noise prediction methods, Nord 2000. Summary report. General Nordic sound propagation model and applications in source-related prediction methods. Client: Nordic Noise Group; 2002

[14] Van den Berg GP. Effects of the wind profile at night on wind turbine sound. Journal of Sound and Vibration. 2004;**277**:955–970. DOI: 10.1016/j.jsv.2003.09.050

[15] Cataldo J. Introducción a la Turbulencia" (Introduction to turbulence) [postgraduate course notes]. Uruguay: Universidad de la República; 2016

[16] International Standard Organization. International Standard 9613. Attenuation of sound during propagation outdoors—Part 1: Calculation of the absorption of sound by the atmosphere. 1993

[17] International Standard Organization. ISO 226. Acoustics – Normal Equal-Loudness-Level Contours. 2nd ed. 2003

[18] Gloaguen J-R, Ecotière D, Gauvreau B, Finez A, Petit A, Le Bourdat C. Automatic estimation of the sound emergence of wind turbine noise with nonnegative matrix factorization. Journal of the Acoustical Society of America. October 2021;**150**(4):3127-3138

Chapter 8

Wind Power Forecasting Models

Enas Raafat Maamoun Shouman

Abstract

The rising costs and undesirable environmental effects of traditional, nonrenewable energy sources have led to increased research regarding the viability of renewable energy sources. Wind has been the fastest-growing source of electricity generation in the world since the 1990s. One of the primary limiting constraints of wind energy is its reliability and there is no cost-effective mechanism for storing wind energy generated by a wind turbine, thus it must be quickly integrated into the electrical grid. The financial implications of wind forecasting are also of great consequence. A 1% error in forecasted wind speeds can result in a loss of $12,000,000 during the facility's life time. As more wind power is incorporated into electricity markets, the capacity to correctly and precisely estimate wind speeds has become increasingly vital. Hence, the importance of this chapter by addressing the different divisions related to wind speed prediction into two overall groups. The first group is based upon analysis of historical time series of wind energy forecasting explanatory variables which are generated from a meteorological model of wind dynamics and the second uses forecasted values from a numerical weather prediction (NWP) model as an input to utilize a statistical approach to anticipate energy predication.

Keywords: wind energy, forecasting models, short-term time prediction, long-term forecasting, physical forecasting, statistical wind forecasting

1. Introduction

More than ever, the world has to work together to find renewable energy solutions to combat the Climate Crisis. Since 2015, all UN Member States have committed to ensuring that everyone has access to cheap, dependable, sustainable, and contemporary clean energy by 2030. Clean energy is derived from renewable natural resources such as the sun, wind, tides and waves, and geothermal [1].

Renewable energy sources, such as solar and wind energy, are more volatile than traditional energy sources since they are weather-dependent. As many countries throughout the world expand their renewable energy supply [2, 3], it is critical to ensure that these clean energy sources offer a consistent supply while replacing fossil fuel-based energy sources. The renewable energy applications range from large-scale and off-grid electricity generating (for rural and remote areas) [1] to heating/cooling systems and transportation.

Wind energy is one of the most widely used renewable energy sources, accounting for 4.8 percent of global electricity production in 2018 [4, 5] and 15 percent of Europe's electricity consumption in 2019 [6]. The mechanical power of the wind is used to power turbines that generate electricity, which generates wind energy. Because wind has a fluctuating intensity over time and might stop blowing at any time, electricity generated by this source is frequently coupled with other power sources to improve reliability and stability.

Wind energy is one of the RES with the lowest electricity production costs and the largest available resource. As a result, a growing number of countries are realizing that wind power offers a great future power generation opportunity.

By dealing with the intermittence characteristic of wind, forecasting methods can improve wind position. Although wind energy cannot currently be dispatched, the financial impacts of wind can be greatly decreased if wind energy can be scheduled using precise wind predictions. As a result, improving wind power output and developing a wind speed forecasting tool has a huge economic and technical impact on the system, **Figure 1** detailed classification of deterministic wind speed and power forecasting.

A number of institutes and organizations with extensive experience in the subject have dedicated numerous studies to the advancement of wind forecasting techniques. Models like WPMS, WPPT, Prediktor, ARMINES, Previento, and others have been developed and deployed in wind farms all around the world. Physical, statistical, and hybrid methodologies were used to develop these models, **Table 1** presents a list of wind power software prediction models developed internationally.

In general, wind forecasting is mostly concerned with the immediate-short-term of minutes to hours to commonly up to 1 day and the long-term of up to 2 days. WPMS, as an example of immediate-short-term models, currently predicts wind generation

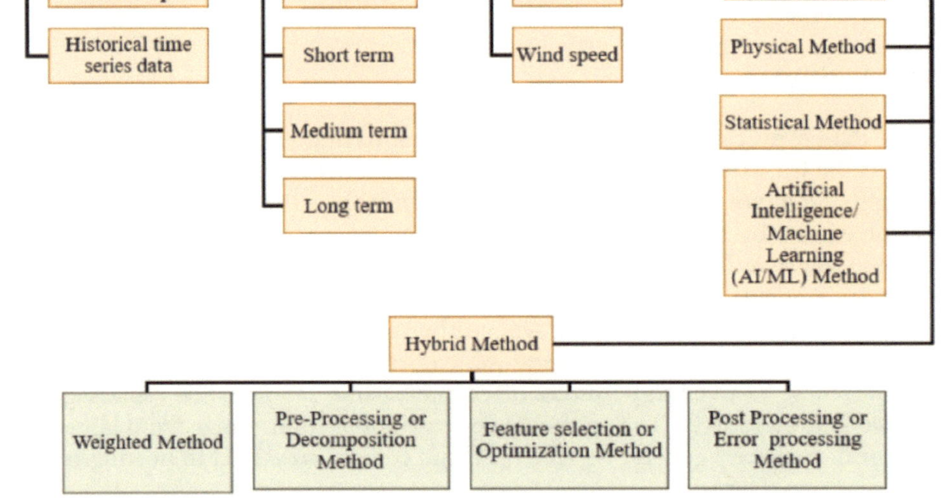

Figure 1.
Detailed classification of deterministic wind speed and power forecasting. Image taken from: ref. [7]. Creative Commons Attribution 4.0 International (CC BY 4.0).

Model name	Developer(s)	Method	Some geographical locations of applications
Prediktor	L. Landberg at Risø, Denmark	Physical	Spain, Denmark, Republic of Ireland, Northern Ireland, France, Germany, USA, Scotland & Japan
WPPT	Eltra/Elsam collaboration with Informatics and Mathematical Modeling at Denmarks Tekniske Universities (DTU), Denmark	Statistical	Denmark, Australia, Canada, Republic of Ireland, Holland, Sweden, Greece & Northern Ireland
Zephyr	Risø & IMM ay DTU, Denmark	Hybrid	Denmark& Australia
Previento	Oldenburg University	Hybrid	Germany, Northern Ireland
e-WindTM	True Wind Inc., USA	Hybrid	USA
Sipreólico	University Carlos III, Madrid, Spain	Statistical	Spain
WPMS	Institute of solar energy technology (ISET), Germany	Statistical	Germany
WEPROG	J. Jorgensen & C. Möhrlen at University College Cork	Hybrid	Ireland, Denmark and Germany
GH Forecaster	Garrad Hassan	Statistical	Greece, Great Britain & USA
AWPPS	École des Mines, Paris	Statistical	Crete, Madeira, Azores & Ireland
LocalPred&RegioPred	M. Perez at center national energy renewable (CENER)	Hybrid	Spain and Ireland
Alea Wind	Aleasoft at the Polytechnic University of Catalonia Spain (UPC)	Statistical	Spain SOWIE Eurowind GmbH, Germany Physical Germany, Austria & Switzerland
EPREV	Institute of Systems and Computer Engineering of Porto (INESC), Institute of Mechanical Engineering and Industrial Management (INEGI) and Center for the Study of Wind Energy and Atmospheric Flows (CEsA) in Portugal	Statistical	Portugal
Scirocco	Aeolis Forecasting Services, Netherlands	Hybrid	Netherlands, Germany & Spain

Table 1.
Presents wind power software models prediction internationally.

for over 95 percent of Germany's territory. Reference [5] discusses immediate short-term wind forecasting models. In addition, various models for short-term wind forecasting have been created, such as Predictor, Zephyr, AWPPS, and Ewind, which are all based on high precision numerical weather prediction (NWP) [6, 8]. Previento, which employs a hybrid technique, can anticipate wind for up to 48 hours. References [9, 10] include more studies on long-term forecasting models.

Variations in energy production (induced by variations in wind speed) will become more noticeable on the electrical system as the penetration of wind power generation grows (in terms of the overall energy mix). To avoid balancing concerns, Transmission System Operators (TSOs) operating to balance supply and demand on regional or national grid systems will need to foresee and manage this unpredictability. The moment at which this is necessary varies by system, although it has been noted that it becomes critical when wind energy penetration reaches roughly 5% of installed capacity.

As wind energy's penetration into individual networks grows, it will be important to make wind farms look more like conventional plants, necessitating the ability to estimate how much energy will be produced over short to medium periods (1 hour to 7 days). Operators, managers, and TSOs commonly anticipate the output from their wind farms in European nations where there is already a substantial level of penetration, such as Spain, Germany, and Denmark. These estimates are used to plan the operations of other factories and for trading.

As the amount of installed capacity develops, forecasting wind energy generation will become more important. The wind industry must expect to do everything possible to enable TSOs to use wind energy to its full potential, which necessitates reliable aggregated output estimates from wind farms.

At the same time as improving the predictability of wind energy plant production through better forecasting tools, it is important to be aware of the true behavior of conventional plants. All of the different energy forms must be considered on an equal level in order to produce the best mix of plants and technologies. As a result, a comprehensive statistical analysis of renewable and conventional plants is critical. This task should be viewed as a critical component of a wind energy development plan, and it should be approached from a comprehensive power system standpoint.

Electricity producers, which include corporations that run wind farms, sell predetermined amounts of energy (measured in kWh) to regional or national energy companies (in the case of wind energy). Because the grid is intended to provide a constant supply of electricity, governments may punish energy producers with large fines if there are power outages.

Energy trade businesses play a critical role in assessing the risk of energy transaction shortfalls by assisting in the forecasting of expected energy production (especially in the case of wind, as a non-steady energy source). Energy dealers, on behalf of energy producers, forecast energy production (in our case, wind energy) using two scenarios:

- If there is a shortfall below the forecast, electricity is purchased on the spot market to keep the system running (with prices above the average energy price).

- Energy producers are not rewarded for surplus energy produced in excess of predicted output.

In this regard, accurate energy output forecasting is critical to the financial performance of wind farms (i.e. wind energy producers).

2. The prediction of a wind farm's energy production

The following steps must be completed in order to anticipate the wind farm's electricity production:

- Predict the variation in long-term wind speed over the site at the machine hub height based on long-term wind speeds at the mast locations;

- Predict the wake losses that occur when one turbine operates behind another.

3. Information required for an analysis

In addition to the wind data, the inputs to this process are typically as follows:

- Wind farm layout and hub height;
- Wind turbine characteristics, such as the power curve (which depicts a turbine's power production as a function of wind speed) and the thrust curve (which plots the force exerted by the wind at the top of the tower as a function of wind speed);
- Air density and turbulence intensity at the place with time (the "gustiness" of the wind).
- The topography of the place and its environs; and
- Overlay of surface ground cover on the site and in the adjacent region.

4. Variability and predictability of wind energy

Low variability and great predictability are required for a reliable energy source. While the modest variance is acceptable, poor predictability is not, and can result in significant revenue loss. Wind energy fluctuation is caused by a heavy reliance on weather, which varies during the day and annually. As a result, precise weather forecasting is required to produce a useful wind power forecast, **Figure 2** shows Dual-step wind power prediction approach based on a hybrid wavelet transform (WT)-ant colony optimization algorithm (ACO)-feedforward artificial neural network (FFANN).

It is commonly known that the accuracy of weather forecasts improves as the forecast horizon shortens. Combining forecasts from multiple numerical techniques can also be advantageous. As a result, wind farms rely on a variety of weather forecasts given by different models at different times of day or week. Although the weather cannot be controlled, the wind sector may take advantage of advances in artificial intelligence to increase the predictability of the energy supply.

In the chapter, there are several ways for forecasting wind power are categorized as Physical models, statistical models, and hybrid.

5. Current forecasting and prediction methods

Models for wind energy forecasting can be classified into two categories. The first is based on historical wind time series analysis, while the second is based on anticipated values from a numerical weather prediction (NWP) model. However, physical methods, classic statistical or 'black box' methods, and more recently, so-called learning approaches, artificial intelligence, or "gray box" methods are used to characterize wind power forecasts. All of these can be included into hybrid approaches.

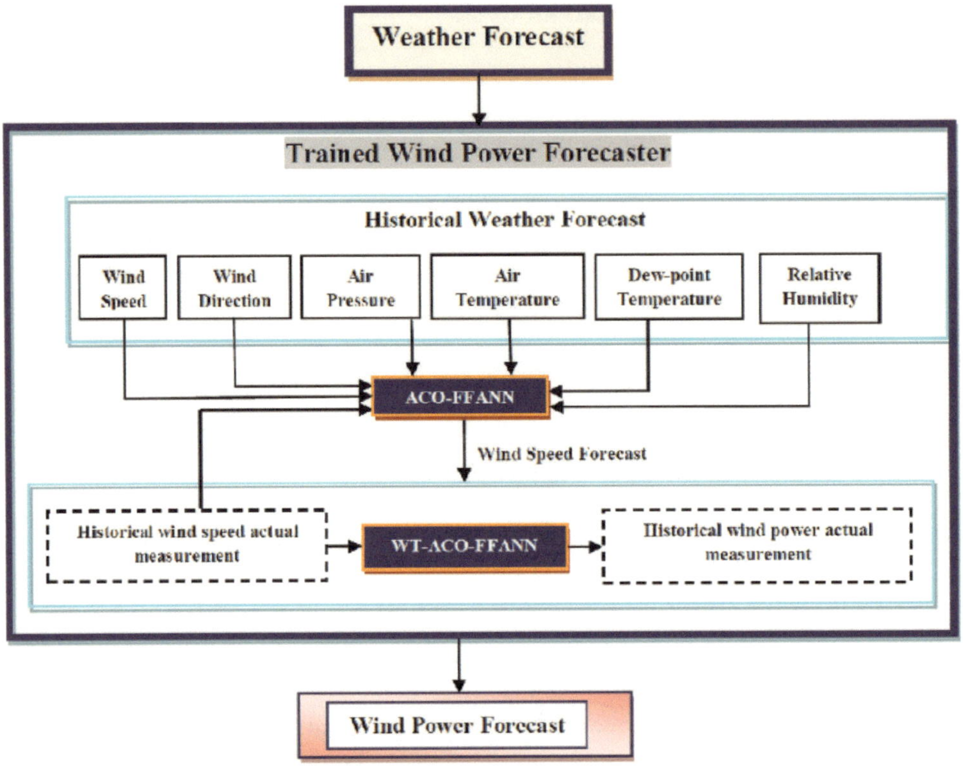

Figure 2.
Dual-step wind power prediction approach based on hybrid wavelet transform (WT)-ant colony optimization algorithm (ACO)-feedforward artificial neural network (FFANN).

The first category of models utilizes a statistical approach to anticipate mean hourly wind speed or directly forecast electric power production. To anticipate wind power N-steps ahead, the models in the second category use explanatory variables (often hourly mean wind speed and direction) generated from a meteorological model of wind dynamics. In the majority of cases, the models in the first group produce good results in the estimation of mean monthly or even higher temporal scale (quarterly, annual) wind speed.

However, the influence of atmospheric dynamics becomes more important in the short term (mean daily or hourly wind speed predictions), making the adoption of the second group's models necessary [11].

In wind power forecasting, there are three steps: first, determining wind speed from a model; second, calculating the wind power output forecast or prediction; and finally, regional forecasting or upscaling or downscaling, which can be implemented over various time horizons. Statistical models are typically used in very short-term forecasting. Ensemble forecasting is utilized to overcome these statistical and learning method conditions [12].

Nielsen et al. [12] demonstrated that if several NWP forecasts are used the forecast error decreases. Louka et al. [13] showed that the Kalman filter can remove systematic forecast errors in NWP wind speed forecasts. Wind forecasting can be separated based on the prediction horizon into three categories:

- **Very short-term forecasts** (from seconds to minutes) are used for real-time turbine control and electrical grid management, as well as for market-clearing. Very short-term forecasts (from 30 minutes to hours) are used for dispatch planning and intelligent load shedding decisions; –

- **Short-term forecasts** (from 30 minutes to hours) are used for dispatch planning and intelligent load shedding decisions [14].

For short-term forecasting, several tools have been created, including WPPT, Predictor, Zephyr, Ewind, WPFS Ver1.0, and AWPPS. A number of case studies in Spain, Germany, Denmark, Ireland, Greece, and France have used these models [15, 16].

The Wind Power Prediction Tool is a well-known model with a wide range of applications for this time frame (WPPT). It can be used to generate short-term (say, up to 120 hours, or 36 hours) wind power output projections. Because the system can provide prediction values as a total including not only a single wind farm, but also a region, it is extremely flexible. The system also gives accurate estimates of the tools' uncertainty, which is critical for efficient scheduling or trading. WPPT uses advanced nonlinear techniques.

Because it may produce prediction values as a total spanning not just a particular wind farm, but also a region, the technique is extremely versatile. The system also provides accurate estimations of the tools' uncertainty, which is critical for optimum trading or scheduling. Advanced nonlinear statistical models underpin WPPT. A semi-parametric power curve model for wind farms that take both wind speed and direction into account, as well as dynamical forecasting models that describe the dynamics of wind power and any diurnal variations, are among the models included in the package. Self-calibrating and self-adaptive models have been developed.

As a result, they update parameters automatically in response to changes in the number of turbines and their features, the environment, the NWP models, and non-explicit model attributes like roughness and filthy blades. WPPT can automatically calibrate to the observed circumstances using artificial intelligence [15]. The system requires online wind power measurements in its simplest configuration. However, the following data is taken into account depending on the configuration: Wind power measurements are now available online. Energy readings from all (or almost all) turbines in a region aggregated (for regional forecasting). Wind speed and direction forecasts by meteorologists for wind farms and regions.

Other measurements or predictions, such as local wind speed, stability, and the number of active turbines, are available. Prediktor, a tool developed by the meteorology research program, is another useful tool (MET). Unlike WPPT, however, Prediktor's main goal is to represent as much as possible using physical models. Every 6 hours, the system provides the predicted production of wind farms for up to 48 hours. All it requires is online access to NWP model output.

The basic processes are as follows: a NWP model predicts overall weather patterns. Only the entire wind can be predicted by such a model, and only correct forecasts can be made at a given site. Then, if needed, these projections are tailored. The WAsP model tailors the wind turbines to each other by modeling local characteristics such as roughness, horography (ridges and hills), and obstructions, as well as the influence of the wind turbines on each other.

Since no model can simulate nature perfectly, two MOS (model output statistics) filters are used in Prediktor to correct shortcomings. The wind power observed is used to adjust the parameters of these filters. The final output of the model is the expected production of the wind farm every 3 hours over the next 48 hours. Furthermore, Prediktor forecasts or will forecast in the near future for up to 50 wind farms in Ireland, Denmark, Germany, France, and Spain in 2025 [15].

The AWPPS is the only instrument available that estimates confidence intervals for wind power predictions at a predetermined level of certainty (i.e. 85 percent, 90 percent, and 95 percent). The intervals are generated using an important international dedicated to the problem of wind prediction. The Prediction Risk Module allows to forecast uncertainty for the next 24 hours based on projected weather stability. Furthermore, the online use of this module allows for the development of appropriate techniques for optimizing the value of power forecasts [17, 18].

A general overview of wind forecasting models is presented in **Table 2**. This section is divided into three parts based on the time-scales, and for each of them and its applications.

- Immediate short-term forecasting Models

Medium-term forecasts (from 6 hours up to a day) are used to make decisions for switching the turbine on or off for safety or conditions on the market.

WPMS has been adapted for performance in the ICT settings of various grid operators and carriers of major wind parks, as one prominent example of immediate-short-term wind forecasting [20].

WPMS deployed artificial neural networks (ANN) in wind farms that were trained using a large amount of historical data. A preprocessor translated input data, output data measured in wind farms, and forecasted meteorological parameters into XML-format before being sent to the program core, which consists of prediction and transformation modules.

- Long-term forecasting

Long-term wind forecasting methods have been studied in a few researches. And there aren't many prediction tools available for this timeframe. Simple models can no

Time-scale	Range	Applications
Immediate-short-term	8 hours-ahead	• Real-time grid operations
		• Regulation actions
Short-term	Day-ahead	• Economic load dispatch planning
		• Load reasonable decisions
		• Operational security in electricity market
Long-term	Multiple-days-ahead	• Maintenance planning
		• Operation management
		• Optimal operating cost

Table 2.
Time-scale classification for wind forecasting [19].

longer match the criteria because to the extended ahead-forecasting time, hence NWP or hybrid NWP models are being investigated. Modern wind power forecasting methods, which are typically based on NWP, provide forecasts over a time range of up to several days. To put it another way, the NWP is the source of all information about the future of wind forecasting.

The national weather service or private weather data provider supplies a collection of NWP data that can be used to predict wind speed and power. In the future, it is becoming more common to use NWP for long-term forecasting [21]. Previento is comparable to Prediktor, but it utilizes more severe physical downscaling and specific upscaling techniques. It provides a reliable forecast of projected wind power for any locations and regions in Germany, Europe, and the rest of the world up to 10 days ahead of time, with a temporal resolution of up to 15 minutes. The wind power forecast is based on the best possible mix of meteorological models, as well as the local conditions of the wind farm's surrounds and the NWP [22].

The Previento system involves a physical approach with data from a large-scale weather prediction model, such as the German Weather Service's Lokalmodell. It simulates roughness, horography, and wake effects in the boundary layer. The daily variation of the thermal stratification of the atmosphere, which is employed to adjust the logarithmic profile, is critical for calculating wind speed at hub height. The expected power output for single sites is derived using the turbine's particular power characteristic. The total amount of power generated by wind in a certain region is computed using data from chosen wind farms.

For long-term planning, long-term forecasts (from a day to a week or even a year) are utilized (to schedule the maintenance or unit commitment, optimize the cost of operation). Maintenance of offshore wind farms can be extremely costly, thus proper planning of maintenance activities is essential. Wind power predictions have a temporal resolution of 10 minutes to a few hours (depending on the forecast length). Wind power forecasting improvements are concentrating on using additional data as input to the models involved, as well as offering uncertainty estimates alongside the standard predictions.

Wind forecasting schemes as **Figure 3** can also be classified based on their methodology into many categories:

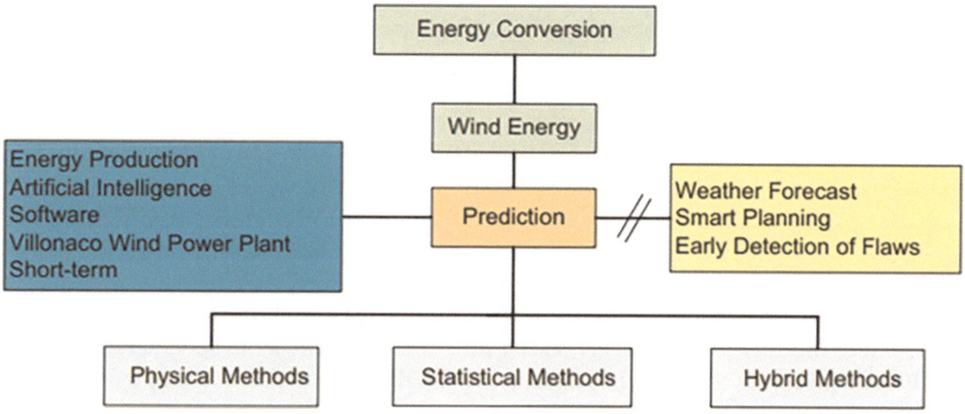

Figure 3.
The conceptual mind on wind energy prediction.

5.1 Physical approach to wind power forecasting

Approaching the situation physically (deterministic approach), the physical approach, also known as the deterministic technique, is based on weather forecast data such as temperature, pressure, surface roughness, and obstructions in the lower atmosphere, or numerical weather prediction (NWP).

Established several physical models based on weather data to predict wind speed and estimated wind power [23]. Physical models often rely on global databases of meteorological information or atmospheric mesoscale models, but to provide accurate results, they require massive computer systems [24].

To estimate wind power production, the physical method uses a thorough description of the lower atmosphere. Cellura et al. [23] provide an overview of some of the neural, geostatistical, and hybrid models that have been applied in space-temporal wind forecasting. Dynamic models (also known as prognostic) and kinematic models (also known as diagnostic) are the two main forms of numerical codes for wind field modeling across rugged terrain [25, 26]. The momentum and energy equations are not explicitly solved in these models; instead, parametric relations and/or wind data are used to examine them implicitly [27].

To account for the local circumstances of the physical topography, computational fluid dynamics (CFD) is utilized as an alternative to the power law [28]. Model output statistics (MOS) are frequently employed to reduce systematic forecasting mistakes and to compensate for unknowns in the expected power output [29].

Forecasts are provided at specified nodes on a grid that covers a certain area. Due to the fact that wind farms are not located on these nodes, these estimates must be extrapolated to the required location and turbine hub height. Physical-based forecasting methods are comprised of multiple sub-models that work together to translate wind forecasts at various grid points and model levels to power forecasts at the actual site.

Converting wind speed to power at the level of the wind farm and at hub height depending on the using theoretical power curves supplied by the wind turbine manufacturer. However, since multiple studies have demonstrated a preference for empirically obtained power curves over theoretical ones; theoretical power curves are becoming less and less important. When using a physical methodology, the function that calculates wind generation from NWPs at various locations around the wind farm is modeled once and for all. The calculated transfer function is then applied to the current weather predictions. Physical simulations frequently integrate Model Output Statistics (MOS) for post-processing power forecasts to account for systematic forecasting errors that may be due to the NWP model or modeling approach, **Figure 4** shows steps forecasting wind farm with NPW.

5.2 Statistical approach to wind power forecasting

Statistical approach statistical method is based on the vast amount of historical data without considering meteorological conditions. It usually involved artificial intelligence (neural networks, neuron-fuzzy networks) and time series analysis approaches [30, 31]. Statistical models, the set of models includes a semi-parametric power curve model for wind farms taking into account both wind speed and direction, and dynamical forecasting models describing the dynamics of the wind power and any weather variation, etc.

Statistical forecasting approaches are based on one or more models that establish the relationship between historical power values, historical and future values of

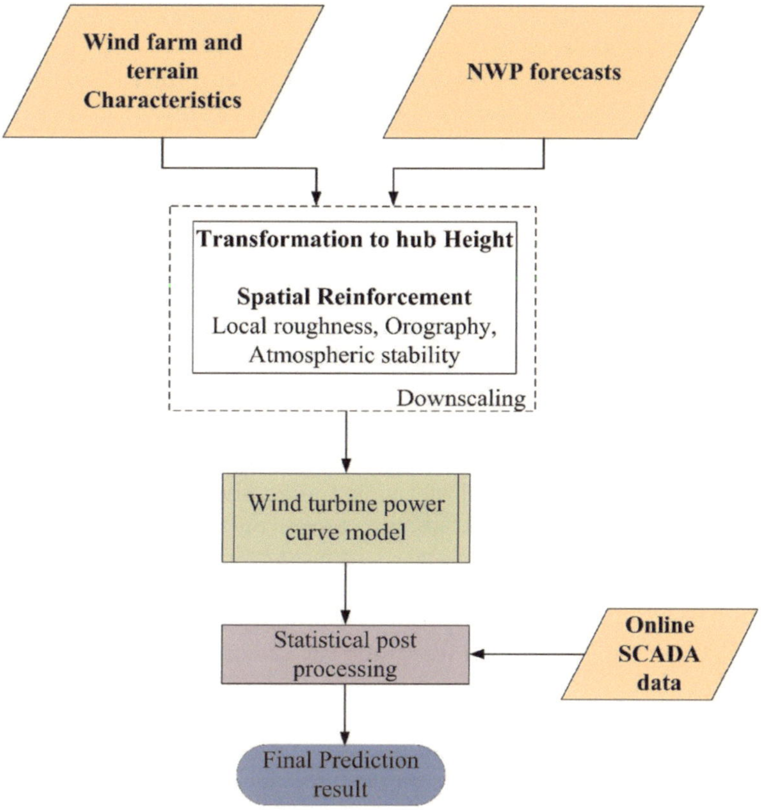

Figure 4.
Steps forecasting wind farm with NPW.

meteorological variables, and wind power measurements. The physical events are not deconstructed and accounted for, despite the fact that problem expertise is required for selecting the appropriate meteorological variables and developing appropriate models.

Model parameters are calculated using a collection of previously known data, and they are updated on a frequent basis during online operation to account for any new information that becomes available (i.e. meteorological forecasts and power measurements).

Linear and nonlinear statistical models, as well as structural and black-box models, are all examples of statistical models. Structural models rely on the analyst's knowledge of the phenomenon of interest, whereas black-box models are built from data in a fairly mechanical manner and require little subject-matter knowledge.

Structural models for wind power forecasting would include diurnal wind speed changes modeling or an explicit function of meteorological variable predictions. Neural-Networks (NNs) and Support Vector Machines are examples of black-box models (SVMs). Some models, on the other hand, are 'in-between' the extremes of being entirely structural or completely black-box. Expert systems, for example, learn from experience (from a dataset) and can be programmed with prior information. The subject of gray-box modeling is then discussed.

Statistical models are often made up of two parts: an autoregressive portion for capturing the wind's persistent behavior, and a "meteorological" part for nonlinear transformation of meteorological variable projections. The autoregressive component provides for considerable gains in forecast accuracy across horizons up to 6–10 hours ahead, when the use of meteorological forecast information alone may not be adequate to exceed persistence.

Statistical approaches to wind power prediction are currently focusing on the use of multiple meteorological forecasts as input and forecast combination, as well as the best use of spatially distributed measurement data for prediction error correction or issuing warnings on potentially large uncertainty.

Calculate a statistical relationship between the essential input data and the generation of wind energy. They entail utilizing a statistical model to directly turn the input factors into wind generation. With these models, a one-step direct calculation of wind power from input parameters is achievable. Most data mining-based models (e.g., ANN, SVM, fuzzy model, model trees), as well as time series analysis methods, can be used as output models (e.g. ARIMA, fractional ARIMA).

A massive quantity of data is processed in the statistical technique, and meteorological processes are not clearly represented. The relationship between historical power output and weather is established, and this information is then used to anticipate future power output. Statistical methods, unlike physical methods, simply require one step to convert input variables to power output. As a result, the procedures used are referred to as "black box." In most cases, a statistical relationship is established between the weather forecast or projection and the wind farm's prospective power output. Other statistical approaches employed include the Box-Jenkins methodology, the use of the Kalman filter, and the use of autoregressive (AR), moving average (MA), autoregressive moving average model (ARMA), and autoregressive integrated moving average model (ARIMA).

Torres et al. [30] discovered that compared to persistence, it was possible to get a 20% error reduction when forecasting average hourly wind speed for a 10 h forecast horizon at a number of locations using nine variables.

Classical time series analysis is not the only approach to model a statistical relationship between data points. Artificial neural networks (ANN) and fuzzy systems are the most common soft computing (or machine learning) techniques utilized, however other models such as gray predictors and support vector machines (SVM) have also been used. Artificial intelligence (AI) approaches are another term for learning approaches. They're known as learning techniques since they take historical time series to learn about the relationship between projected wind and predicted power production. They've been dubbed "gray box" approaches in recent years.

5.2.1 Parametric methods

The presentation of parametric statistical methods directly inspired from the physical equation. Parametric modeling according to the wind speed only, the investigated the simplest parametric models, namely linear regression and logistic regression, with the wind speed as the unique explanatory variable. If the predicted power at time t is denoted by Yˆt, these models are given by

$$\hat{Y} t = a_0 + a_1 W_t, \text{ and} \quad (1)$$

$$\hat{Y} t = C_1 + \exp\left(a_0 + a_1 W_t\right), \quad (2)$$

where the parameters a0, a1, C are estimated using the associated methodology.

5.2.2 Logistic regression

Logistic regression has also been considered to mimic more closely Eq. (1). More precisely, the model is then defined by:

$$Y\hat{}\ t = C/\left(1 + \exp\left(a_0 + a_1 W_t + a_2 {W_t}^2 + a_3 {W_t}^3\right)\right), \tag{3}$$

where ai, i = 0, ..., 3 and C are estimated parameters.

This model is using not only wind speed as a predictor, but also wind direction, (coded by its cosine and sine: Dcos and Dsin), temperature T, and the variances of the wind speed WS and direction, DS, Re and DS, Im.

5.2.3 Lasso model

The Lasso method, which simultaneously performs variable selection and regularization through the least squares criterion penalized by the ` 1 norm of the regression coefficients has been investigated as well (see for instance [21]). The model is defined by.

$$Y\hat{}\ t = a0 + a_1 W^t + a_2 {D_t}^{\cos} + a_3 {D_t}^{\sin} + a_4 Tt + a_5 WS_t + a_6 {D_t}^{S,Re} + a_7 {D_t}^{S,Im}, \tag{4}$$

with a0, ..., a7 minimizing.

$$1/n \sum_{n}^{i=1} \left(Y_i - a_0 - a_1 W_i - a_2 {D_i}^{\cos} - a_3 {D_i}^{\sin} - a_4 T_i - a_5 {W_i}^S - a_6 {D_i}^{S,Re} - a_7 {D_i}^{S,Im}\right)^2 + \lambda \sum_{j=1}^{7} |aj|. \tag{5}$$

5.3 Hybrid approach energy power forecasting

Hybrid method, which combines physical methods and statistical methods particularly uses weather forecasts and time series analysis.

ANEMOS is a hybrid wind forecast tool that takes into account a variety of time horizons. The development of combining high-resolution meteorological predictions and appropriate prediction models for the offshore is emphasized [19, 31].

Hybrid models aim to combine the advantages of each model in order to produce the best predicting results possible. Because the information provided in individual forecasting techniques is restricted, a hybrid approach can take use of the available data, integrate individual model data, and maximize the benefits of many forecasting methods, improving prediction accuracy [32].

Many techniques, such as mixing physical and statistical procedures or short-term and medium-term models, are included in hybrid methods. A number of hybrid models were utilized to anticipate wind power. Here are some examples of potential combinations:

- A combination of physical and artificial intelligence approaches.
- Using a combination of artificial intelligence models Zhao et al. [33] looked into a hybrid wind forecasting system that included both NWP and ANN models. To anticipate meteorological characteristics, the NWP model combines the

Global Forecasting System (GFS) with the Weather Research and Forecasting (WRF) system. **Figure 5** shows an example of an ANN structure with 4 inputs and 2 hidden layer.

Shi et al. [34] proposed two hybrid models for wind speed and power forecasting: ARIMA-ANN and ARIMA-SVM. Based on two case studies on wind speed and wind power generation, this research analyses the application of the suggested hybrid models in a systematic and thorough manner. The findings imply that hybrid approaches are feasible alternatives for predicting both wind speed and wind power generation time series, but that they do not always provide better forecasting performance for all forecasting time horizons investigated.

Guo et al. [35] proposed a novel hybrid wind speed forecasting method based on a back propagation neural network and the notion of seasonal exponential adjustment to exclude seasonal effects from real wind speed datasets. A proposed technique outperformed the single back propagation neural network in the tests.

For short-term wind power forecasting in Portugal, Catalo et al. [19] presented a hybrid approach based on the combination of ANN and wavelet transform. To deconstruct the wind power series into a set of better-behaved constituent series, the wavelet transform is applied. The test findings show that the proposed hybrid technique for forecasting wind output has a lot of potential.

Finally, hybrid models (e.g. [19, 36]) are based on the combination of the physical and statistical models, the combination of models with several time horizons, and the combination of alternative statistical models

5.4 Spatial correlation models

The spatial correlation models take into account the spatial link between wind speeds at different sites. The wind speed time-series of the projected point and its neighbors is used to predict the wind speed in spatial correlation models [36]. When predicting wind speed at one location based on observations taken at another, a spatial correlation model is used. Data obtained over a seven-year period [37] was used to test its behavior and provide adequate verification.

Based on cross-correlation at surrounding sites, Alexiadis et al. [38] demonstrated a technique for forecasting wind speed and power output up to several hours ahead.

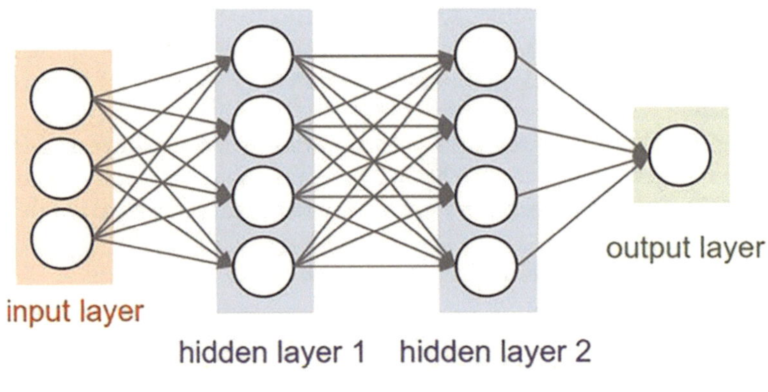

Figure 5.
ANN structure with 4 inputs and 2 hidden layer.

This research established an ANN technique based on spatial correlation models that outperform the persistence forecasting model in terms of forecasting accuracy [39].

Barbounis and Theocharis [40] proposed the use of local feedback dynamic fuzzy neural network (LF-DFNN) to forecast wind speed using spatial correlation. Remote meteorological stations are installed at two reference sites in accordance with the location of the base site so that the three sites are aligned along the prevailing wind direction. Using spatial information from remote meteorological stations, the LF-DFNN is used in this paper to predict multi-step forward wind speed in the base site. The LF-DFNN outperforms other network models tested in this application, according to simulation data.

5.5 Artificial intelligence methods

Various novel AI algorithms for wind speed and power prediction have recently been developed as a result of the advancement of artificial intelligence (AI). Artificial neural networks (ANN), adaptive neuro-fuzzy inference system (ANFIS), fuzzy logic approaches, support vector machine (SVM), neuro-fuzzy network, and evolutionary optimization algorithms are among the newly developed methodologies.

Through the training process, ANN models can represent a complex nonlinear relationship and extract the dependency between variables [40]. Back propagation neural networks, recurrent neural networks, radial basis function (RBF) neural networks, ridgelet neural networks, and adaptive linear element neural networks are examples of ANN-based methods. The application of an ANN-based method to the problem of wind power forecasting is appropriate.

ANN might handle nonlinear and complex scenarios in terms of categorization or forecasting. ANN models can depict a complex nonlinear relationship and extract the link between variables through the training phase [40]. Examples of ANN-based techniques include back propagation neural networks, recurrent neural networks, radial basis function (RBF) neural networks, ridgelet neural networks, and adaptive linear element neural networks. It appears that applying an ANN-based technique to the problem of wind power forecasting is a good idea.

Using time series analysis, Sfetsos [41] proposed an ANN technique for forecasting mean hourly wind speed data. The proposed methodology also has a benefit for utilities that have a high level of wind penetration and utilize hourly intervals for power system operational procedures like economic dispatch and unit commitment.

Chang [42] discussed back propagation neural network-based wind power forecasting algorithms. The created model for short-term wind forecasting demonstrated excellent accuracy when utilized to supply energy to a 2400 kW (WECS) on the Taichung coast. Back propagation neural networks and recurrent neural networks were used in More and Deo's [43] wind forecasting methodology. Traditional statistical time series analysis has been found to be less accurate than neural network forecasting [44].

Chang [45] described a method for forecasting wind power generation time series using an RBF neural network. The numerical results show that the suggested forecasting method is accurate and dependable, with good matches between realistic values and predicting values.

Guo et al. [46] studied a feed-forward neural network (FNN) wind forecasting approach based on modified empirical mode decomposition (EMD). Through multi-step forecasting of mean monthly and daily wind speeds in Zhangye, China, the proposed technique outperforms basic FNN and unmodified EMD-based FNN [47].

Li and Shi [48] used three types of conventional ANNs to anticipate wind speed: adaptive linear element, back propagation, and radial basis function.

The outcomes of comparing three types of ANN reveal that no single ANN model outperforms another universally in terms of all evaluation measures, even for the same wind dataset. Furthermore, the type of ANN to use for the best results is determined by the data sources.

Yang et al. [49] proposed an ANFIS approach for interpolating missing and incorrect wind data. Twelve measured wind data sets from a wind farm in North China are interpolated and examined for performance testing. The ANFIS method's effectiveness was demonstrated by the test results. A SVM-based technique for wind power forecasting was described by Zeng and Qiao [32]. Real wind speed and wind power data obtained from the National Renewable Energy Laboratory are used in simulation research.

The suggested SVM method outperforms the persistence model and the RBF neural network-based model, according to the results. For one-step ahead, wind speed forecasting, Zhou et al. [50] described a systematic investigation on fine-tuning least-squares support vector machines (LSSVM) model parameters. Three SVM kernels are implemented: linear, Gaussian, and polynomial kernels. LSSVM approaches are proven to outperform the persistence model in the vast majority of scenarios. For short-term wind power forecasting, Xia et al. [51] introduced a neuro-fuzzy network technique.

For the wind power forecasting of a practical wind farm in China, the forecasting approach is used. The results of the tests revealed that the trained neuro-fuzzy networks are capable of predicting and forecasting wind power.

Jursa and Rohrig [52] proposed a new short-term prediction technique based on the automated specification of neural networks and the nearest **neighbor** search using evolutionary optimization algorithms. The test results demonstrated that by employing the proposed automated specification method, the wind power forecast error can be decreased.

6. Conclusion and future advances for wind power prediction

Wind Forecasting in the Future The forecast accuracy of wind power prediction systems is becoming increasingly significant due to the high penetration of wind power in the energy grid. Many academics have been working on wind power forecasting in recent years. Forecast accuracy has steadily increased and intensive research and development efforts are projected to be underway soon. In order to improve wind power projections even more, various literature [33, 53] suggest that future studies should focus on the following areas:

- Research new artificial intelligence approaches and enhance training algorithms in order to increase forecast accuracy. Future studies will also focus on new strategies for dealing with complicated terrain.

- More research into hybrid methods is needed to combine different approaches, such as combining physical and statistical approaches, to achieve good results in both long-term and short-term prediction.

- In actual WECS, the existing forecast approach should be used. Continue your investigation into the practical application of the methodologies rather than just the theory.

- Develop a more precise assessment methodology and a standard for measuring technique performance.
- Improved input data for wind power forecasting will come from improved NWP models and more regular weather forecast updates.
- Expand research into the use of online wind data, particularly for short-term wind forecasting.
- More study on adaptive parameter estimation is needed. The models can respond to changes in the farms and their surroundings automatically.
- Conduct additional research into the NWP models designed for use in an offshore environment. Improve meteorological data available to evaluate NWP results for offshore areas.

Author details

Enas Raafat Maamoun Shouman
Engineering Division, Information Systems Department, National Research Center, Giza, Egypt

*Address all correspondence to: enascora@gmail.com

References

[1] Goldemberg J. World Energy Assessment: Energy and the Challenge of Sustainability. New York: UNDP; 2001

[2] Kyriakopoulos G. Low Carbon Energy Technologies in Sustainable Energy Systems. 1st ed. San Diego: Elsevier Science & Technology; 2021

[3] Renewable Electricity Capacity and Generation Statistics. International Renewable Energy Agency (IRENA); 2018

[4] Holttinen H, Lemström B, Meibom P, Bindner H, Orths A, Van Hulle F, et al. Design and Operation of Power Systems with Large Amounts of Wind Power, State of the Art Report. Espoo, Finland: VTT Working Papers 82; 2017

[5] Landberg L. Short-term prediction of local wind conditions. Journal of Wind Engineering and Industrial Aerodynamics. 2009;**89**:235-245

[6] Kariniotakis G, Marti I, et al. What Performance Can Be Expected by Short-term Wind Power Prediction Models Depending on Site Characteristics. London, UK: CD-Rom proceedings of the European Wind Energy Conference EWEC 2004; 22-25 Nov. 2019

[7] Yousuf MU, Al-Bahadly I, Avci E. Current Perspective on the Accuracy of Deterministic Wind Speed and Power Forecasting [Internet]. Vol. 7, IEEE Access. Institute of Electrical and Electronics Engineers (IEEE); 2019. pp. 159547-159564. DOI: 10.1109/ACCESS.2019.2951153

[8] Barbounis TG, Theocharis JB, Alexiadis MC, Dokopoulos PS. Long-term wind speed and power forecasting using local recurrent neural network models. IEEE Trans. Energy Convers. March 2006;**21**(1):273-284

[9] Sailor DJ, Smith M, Hart M. Climate change implications for wind power resources in the Northwest United States. Renewable Energy. 2018;**33**(11): 2393-2406

[10] Landberg L. Short-term prediction of local wind conditions. Journal of Wind Engineering and Industrial Aerodynamics. 2001;**89**(3e4):235e45

[11] Taylor JW, Buizza R. Using weather ensemble predictions in electricity demand forecasting. International Journal of Forecasting. 2003;**19**(1):57-70

[12] Nielsen HA, Nielsen TS, Madsen H, MJSI P, Martí I. Optimal combination of wind power forecasts. Wind Energy. 2007;**10**(5):471

[13] Louka P, Galanis G, Siebert N, Kariniotakis G, Katsafados P, Kallos G, et al. Improvements in wind speed forecasts for wind power prediction purposes using Kalman filtering. Journal of Wind Engineering and Industrial Aerodynamics. 2008;**96**(12):2348

[14] Candy B, English SJ, Keogh SJ. A Comparison of the impact of QuikScat and WindSat wind vector products on met office analyses and forecasts. IEEE Trans. Geosci. Remote Sens. June 2019; **47**(6):1632-1640

[15] Cadenas E, Rivera W. Wind speed forecasting in the South Coast of Oaxaca, Mexico. Renewable Energy. 2007;**32**(12): 2116-2128

[16] Cadenas E, Rivera W. Short term wind speed forecasting in La Venta, Oaxaca, Mexico, using artificial neural

networks. Renewable Energy. 2009; **34**(1):274-278

[17] Zhao H, Guo S. Wind speed forecasting in China: A review. Science Journal of Energy Engineering. 2015; **3**(4-1):14-21

[18] ISET. Wind Energy Report Germany 1999/2000 – Annual Evaluation of WMEP. Kassel: ISET; 2000

[19] Giebel G, Badger J, Perez IM, Louka P, Kallos G. Short-term forecasting using advanced physical modelling-the results of the anemos project. Results from mesoscale, microscale and cfd modelling. Proc. of the European Wind Energy Conference 2006; 2006

[20] Kavasseri RG, Seetharaman K. Day-ahead wind speed forecasting using f-ARIMA models. Renewable Energy. 2009;**34**(5):1388-1393

[21] Landberg LA. Mathematical look at a physical power prediction model. Wind Energy. 1998;**1**:238

[22] Landberg L, Myllerup L, Rathmann O, Lundtang Petersen E, Hoffmann Jørgensen B, Badger J, et al. Wind resource estimation - an overview. Wind Energy. 2003;**6**(3):261

[23] Cellura M, Cirrincione G, Marvuglia A, Miraoui A. Wind speed spatial estimation for energy planning in Sicily: a neural kriging application. Renewable Energy. 2008;**33**(6):125166

[24] Lalas DP. Wind energy estimation and siting in complex terrain. International Journal Solar Energy. 1985; **3**:4371

[25] Dinar N. Mass consistent models for wind distribution in complex terrain - fast algorithms for three dimensional problems. Boundary Layer Meteorology. 1984;**30**:17799

[26] Magnusson M, Wern L. Wind energy predictions using CFD and HIRLAM forecast. Copenhagen, Denmark: Proceedings of the European wind energy conference EWEC2001; 2001

[27] Glahn HR, Lowry DA. The use of Model Output Statistics (MOS) in objective weather forecasting. Journal of Applied Meteorology. 1972;**11**(8):120311

[28] G. Giebel The State-of-the-Art in Short-Term Prediction of Wind Power—A Literature Overview. Available from: http://www.anemos-project.eu/download//ANEMOS_D1.1_StateOfTheArt_v1.1.pdf

[29] Garcia AR, De-La-Torre-Vega E. A Statistical wind power forecasting system – A Mexican wind-farm case study. Marseille, France: European Wind Energy Conference & Exhibition – EWEC Parc Chanot; 2009

[30] Torres JL, García A, De Blas M, De Francisco A. Forecast of hourly average windspeed with ARMA models in Navarre (Spain). Solar Energy. 2005;**79**(1):65-77

[31] He Q. Neural Network and Its Application in IR, Graduate School of Library and Information Science. Champaign, Ill, USA: University of Illinois at Urbana; 1999

[32] Zeng JW, Qiao W. Support Vector Machine-Based Short-Term Wind Power Forecasting. Phoenix: Proceedings of the IEEE/PES Power Systems Conference and Exposition; 2011, 20-23 March 2011. pp. 1-8

[33] Zhao DM, Zhu YC, Zhang X. Research on Wind Power Forecasting in

Wind Farms. Wuhan: Proceedings of the 2011 IEEE Power Engineering and Automation Conference; 2011, 8-9 September 2011. pp. 175-178. DOI: 10.1109/PEAM.2011.6134829

[34] Shi J, Guo JM, Zheng ST. Evaluation of Hybrid Forecasting Approaches for Wind Speed and Power Generation Time Series. Renewable and Sustainable Energy Reviews. 2012;**16**:3471-3480. DOI: 10.1016/j.rser.2012.02.044

[35] Guo ZH, Wu J, Lu HY, Wang JZ. A Case Study on a Hybrid Wind Speed Forecasting Method Using BP Neural Network. Knowledge-Based Systems. 2011;**24**:1048-1056. DOI: 10.1016/j.knosys.2011.04.019

[36] Hui Z, Bin L, Zhuo-qun Z. Short-term wind speed forecasting simulation research based on arima-lssvm combination method. Vol. 1. ICMREE 2011; 2011. pp. 583-586

[37] Negnevitsky M, Johnson P, Santoso S. Short term wind power forecasting using hybrid intelligent systems. IEEE Power Engineering Society General Meeting; 2007. pp. 1-4

[38] Alexiadis MC, Dokopoulos PS, Sahsamanoglou HS. Wind Speed and Power Forecasting Based on Spatial Correlation Models. IEEE Transactions on Energy Conversion. 1999;**14**:836-842. DOI: 10.1109/60.790962

[39] Alexiadis MC, Dokopoulos PS, Sahsamanoglou HS, Manousaridis IM. Short-Term Forecasting of Wind Speed and Related Electrical Power. Solar Energy. 1998;**63**:61-68. DOI: 10.1016/S0038-092X(98)00032-2

[40] Barbounis TG, Theocharis JB. A Locally Recurrent Fuzzy Neural Network with Application to the Wind Speed Prediction Using Spatial Correlation. Neurocomputing. 2007;**70**: 1525-1542. DOI: 10.1016/j.neucom.2006.01.032

[41] Sfetsos A. A Novel Approach for the Forecasting of Mean Hourly Wind Speed Time Series. Renewable Energy. 2002;**27**: 163-174. DOI: 10.1016/S0960-1481(01)00193-8

[42] Chang WY. Application of Back Propagation Neural Network for Wind Power Generation Forecasting. International Journal of Digital Content Technology and its Application. 2013;7: 502-509

[43] More A, Deo MC. Forecasting Wind with Neural Networks. Marine Structures. 2003;**16**:35-49. DOI: 10.1016/S0951-8339(02)00053-9

[44] Wu YK, Lee CY, Tsai SH, Yu SN. Actual Experience on the Short-Term Wind Power Forecasting at Penghu-From an Island Perspective. Hangzhou: Proceedings of the 2010 International Conference on Power System Technology; 2010, 24-28 October 2010. pp. 1-8. DOI: 10.1109/POWERCON.2010.5666092

[45] Chang WY. Wind Energy Conversion System Power Forecasting Using Radial Basis Function Neural Network. Applied Mechanics and Materials. 2013;**284-287**:1067-1071. DOI: 10.4028/www.scientific.net/AMM.284-287.1067

[46] Guo ZH, Zhao WG, Lu HY, Wang JZ. Multi-Step Forecasting for Wind Speed Using a Modified EMD-Based Artificial Neural Network Model. Renewable Energy. 2012;**37**:241-249. DOI: 10.1016/j.renene.2011.06.023

[47] Zeng JW, Qiao W. Support vector machine-based short-term wind power forecasting. In: Proceedings of the IEEE/

PES Power Systems Conference and Exposition; Phoenix; 20–23 March 2011. 2011. pp. 1-8

[48] Li G, Shi J. On Comparing Three Artificial Nneural Networks for Wind Speed Forecasting. Applied Energy. 2010;**87**:2313-2320. DOI: 10.1016/j.apenergy.2009.12.013

[49] Yang ZL, Liu YQ, Li CR. Interpolation of Missing Wind Data Based on ANFIS. Renewable Energy. 2011;**36**:993-998. DOI: 10.1016/j.renene.2010.08.033

[50] Zhou JY, Shi J, Li G. Fine Tuning Support Vector Machines for Short-Term Wind Speed Forecasting. Energy Conversion and Management. 2011;**52**: 1990-1998. DOI: 10.1016/j.enconman.2010.11.007

[51] Xia JR, Zhao P, Dai YP. Neuro-Fuzzy Networks for Short-Term Wind Power Forecasting. Hangzhou: Proceedings of the International Conference on Power System Technology; 24-28 October 2010. DOI: 10.1115/1.859612

[52] Jursa R, Rohrig K. Short-Term Wind Power Forecasting Using Evolutionary Algorithms for the automated specification of artificial intelligence models. International Journal of Forecasting. 2008;**24**(4):694-709

[53] Foley AM, Leahy PG, Marvuglia A, McKeogh EJ. Current Methods and Advances in Forecasting of Wind Power Generation. Renewable Energy. 2012;**37**: 1-8. DOI: 10.1016/j.renene.2011.05.033

Chapter 9

Wind Power Fluctuation Compensation by Variable-Speed Pumped Storage Plant in a Grid Integrated System

Girmaw Teshager Bitew and Minxiao Han

Abstract

Wind power technology is the fastest-growing technology for electrical energy production due to its potential characteristics. However, due to its randomness, it has an unnecessary impact on the operation and stability of the grid system. For example, the problem of power grid frequency fluctuation is more obvious. Fluctuations in frequency, in turn, can even affect the breakdown of power systems. To minimize these problems, a droop vector control strategy applied to variable-speed pumped storage (VSPS) systems can be implemented. This method should be used as a wind power fluctuation compensation solution in the wind farm grid-connected system. This chapter introduces the droop-fed direct power control strategy in a VSPS-wind farm-grid-connected system. The modeling of the system is based on the phasor model technique. The spectrum analysis method is used in the VSPS system to determine the dynamic performance of the power grid under wind power fluctuation compensation and emergencies and is verified on the MATLAB/Simulink platform. The results show that the spectrum analysis method is effective for determining the fluctuation and stability requirements of wind power in large power grids. The VSPS control strategy in the wind farm grid-connected integrated system achieves good power flow regulation and stable grid frequency, and the deviation is within an acceptable range.

Keywords: wind power fluctuation, power compensation, variable-speed pumped storage, vector control, frequency spectrum analysis

1. Introduction

Due to increasing environmental concerns, wind power generation has undergone rapid growth [1]. In 2013, the world wind power production capacity was 318GW, and it is forecasted to reach 712 GW, 1480GW, 2089GW, and 2672GW by 2020, 2030, 2040, and 2050, respectively, in moderate scenario [2].

In China, the figure is 123GW, 216GW, 414GW, 680GW, and 1000GW in 2015, 2020, 2030, 2040, and 2050, respectively [2, 3].

Wind energy production brings immense advantages such as sustainable, incredible domestic potential, eco-friendly, revitalizing to rural economies, low operational cost, space efficient, etc. Nevertheless, its main drawback is its stochastic over time in nature; hence, significant power fluctuations are observed in the wind farm. For a power system even with moderate wind power penetration, the fluctuations should be mitigated; otherwise, this may lead to substantial deviations in the grid frequency [1], voltage sag and flicker at the grid busses [4], steady-state voltage deviation, even equipment damage, and system collapse at large. A study in busses [4] asserts that the variable-speed pumped storage (VSPS) system is able to improve the dynamic stability and steady-state operations of the power system. Having dependable features and performances, the VSPS can play a crucial role in stability control and frequency regulator and AC voltage control during contingencies.

Previous works have focused on converter topology development [5–11], which has led to improvements in VSPS system performance and penetration of renewable energy development level [12]. Furthermore, control strategies have been intensively studied in doubly fed induction machine (DFIM)-based VSPS applications. These works mainly relate to the vector control category, including Field-Oriented Control (FOC) [9, 13, 14] and Direct Torque Control (DTC) [5, 8, 11, 13, 14]. As a result, the system performance improvement has achieved remarkable results.

In Ref. [8], a direct power control (DPC)-based frequency regulator is implemented in a full-scale converter-fed synchronous machine. DPC is also used in the application of DFIM-based VSPS systems in [11], with a focus on converter topology studies. A grid-integrated VSPS system was proposed in [5] to study the impact of faults on grid frequency stability and VSPS active power flow. In [9], a VSPS-based H-bridge cascaded multilevel converter with a stator voltage FOC strategy was proposed to suppress the effects of wind farm power fluctuations. On the other hand, the work in [1] proposed a power filter algorithm to solve the control method to adjust the frequency deviation of the power grid caused by wind power fluctuations. However, the above work does not consider the phasor model simulation technique because it is very important in the stability and control analysis of large power systems where simulation time and computational storage are critical. Since sinusoidal voltages and currents are replaced by phasors in polar form in the phasor model technique; and where electromagnetic transients are not important, dynamic simulation time is greatly reduced. Implementing a phasor model in any linear system is also an advantage, since the study of small-signal stability is based on the eigenvalue analysis of linearized power systems, which complements the dynamic simulation of nonlinear systems [15]. However, implementing phasor modeling techniques in DFIM applications, grid frequency control has always been a challenge. In the phasor model technique, the grid frequency is assumed to be a constant fundamental frequency. It is impossible to develop a phasor locked loop for synthesizing frequencies. Therefore, grid frequency control is performed through an active power control strategy, which is an open-loop control scheme for frequency, as shown in **Figure 1**. Therefore, the response of grid frequency and AC voltage lacks good dynamic performance, especially during emergencies [16].

On the other hand, grid frequency is very sensitive to load/source changes. If the load/power connected to the AC grid system is suddenly turned on/off, the grid frequency will decrease/increase, causing the power plant to generate more power or stop it. Furthermore, due to the applied frequency drop, the converter will respond to the decrease in frequency by increasing the inverter power or decreasing the rectifier power. Therefore, the frequency drop is estimated per unit increment of rectified

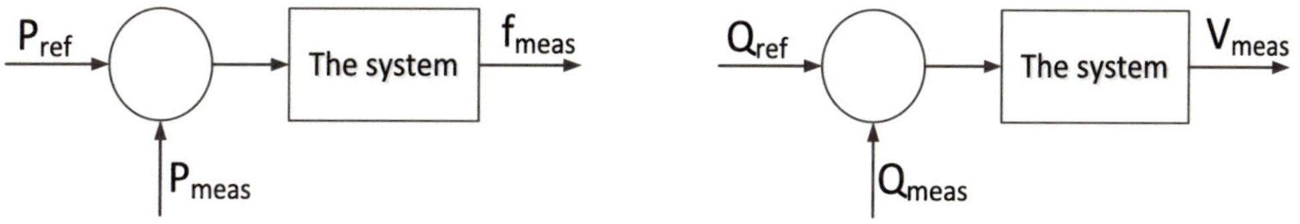

Figure 1.
Open-loop frequency and AC-bus voltage control through the power control strategy.

power fed to the closed control system. Since a change in the converter power results in a change in the DC voltage level, the frequency drop will appear as a drop in the DC voltage level in the AC/DC interface system, which in turn results in the need for power flow compensation [17]. Therefore, the way to solve this problem is that adding a synchronous machine to the grid system helps to obtain the measured and estimated frequency, because the rotor speed of the synchronous machine is absolutely the same as the scale factor of the grid frequency.

However, this chapter aims to propose a frequency droop feeding DPC for VSPS systems to accommodate grid frequency deviations due to wind power fluctuation effects. Spectral analysis is newly used to tune the stability balance and verify the dynamic performance of the proposed control system. The converter used is a three-level back-to-back neutral-point clamped voltage source converter (NPC-VSC).

2. Basic description of frequency control

The replacement system frequency is a measure of the effective power balance in the system. It is a feedback signal that allows the operator to monitor the balance of energy produced and consumed. It is a common characteristic of all location voltages, but varies over time, while system conditions change within an AC power system. Any change in power generation or load will result in a frequency deviation that can lead to system instability and minimal operation of generators and consumers. In extreme cases, system equipment may be damaged.

The concept of power system frequency stability is defined as the requirement to ensure that the system frequency remains within a given safe range during normal operation and emergency situations [1]. This is ensured by using two mechanisms, frequency control and active power balance control [18]. Frequency control is a crucial control problem in system design and operation. This is even more important today due to the complexity of modern power systems driven by ever-increasing scale, environmental constraints, emerging renewable energy sources, and uncertainty.

Based on the dynamic model of generator load based on the correlation between mismatched power (ΔP_m - ΔP_L) and grid frequency deviation, a simplified frequency response model can be defined for an interconnected multigenerator power system. (Δf) and can be expressed by the oscillatory differential equation [19] as

$$\Delta P_m(t) - \Delta P_L(t) = 2H\frac{d\Delta f(t)}{dt} + D\Delta f(t) \qquad (1)$$

where ΔP_m, ΔP_L, D, and H are respectively change of the mechanical power, change of the load, the load damping coefficient, and the inertia constant.

In terms of Laplace transformation function, the expression of the change of frequency in (1) is denoted by

$$\Delta f(s) = \frac{\Delta P_m(s) - \Delta P_L(s)}{2Hs + D} \tag{2}$$

It can be seen from Eq. (2) that the frequency deviation is inversely proportional to the load damping coefficient and inertia constant.

To keep the power system frequency at the nominal value, different mechanisms can be used depending on the magnitude of the frequency deviation faced. Small frequency deviations can be adjusted by primary control and load frequency control systems. When the situation is complex and the deviation is large, emergency control schemes such as low-frequency load shedding should be used to restore the system frequency.

Frequency stability studies can range in time from seconds to minutes as both fast and slow dynamics affect it. For high-power fluctuating systems, the long-term stability issue [19] is considered so in this chapter.

3. The wind power

Wind speed is random in nature and varies erratically over time. As a result, significant wind fluctuations can be observed.

For a given power factor C_p, the mechanical power (P_m) and torque (T_m) produced by the wind turbine rotor can be defined as

$$P_m = 0.5\rho A C_p(\lambda,\beta){v_w}^3 \tag{3}$$

$$T_m = \frac{P_m}{\omega_r} \tag{4}$$

where ρ, v_w, A, and ω_r are respectively the air density, wind speed, swept area, and angular speed of the generator rotor. Maximum mechanical power extraction is possibly made at maximum value C_p and is a function of the pitch angle (β) and tip-speed ratio (λ), which are related as

$$\lambda = \frac{\omega_{rot} R_T}{v_w} \tag{5}$$

The angular speed of the wind turbine ω_{rot} is related to the generator rotor angular speed ω_r and as a function of gear ratio Ng and given by

$$\omega_r = N_g \omega_{rot} \tag{6}$$

The power coefficient expression Cp has been shown to have a unique maximum, $C_{p\text{-}max}$, for λ_{opt} = 8.1 and β_{opt} = 0 [20].

The maximum coefficient of Cp, i.e., the Betz limit is 16/27. But maximum value of C_p is determined in this chapter. Based on the modeling turbine characteristics of [21], a generic Eq. (7) is used to model $C_p(\lambda,\beta)$.

$$C_p(\lambda,\beta) = C_1(C_2/\lambda_i - C_3\beta - C_4)e^{-C_5/\lambda_i} + C_6\lambda$$
$$\Rightarrow \frac{1}{\lambda_i} = \frac{1}{\lambda + 0.08\beta} - \frac{0.035}{\beta^3 + 1} \tag{7}$$

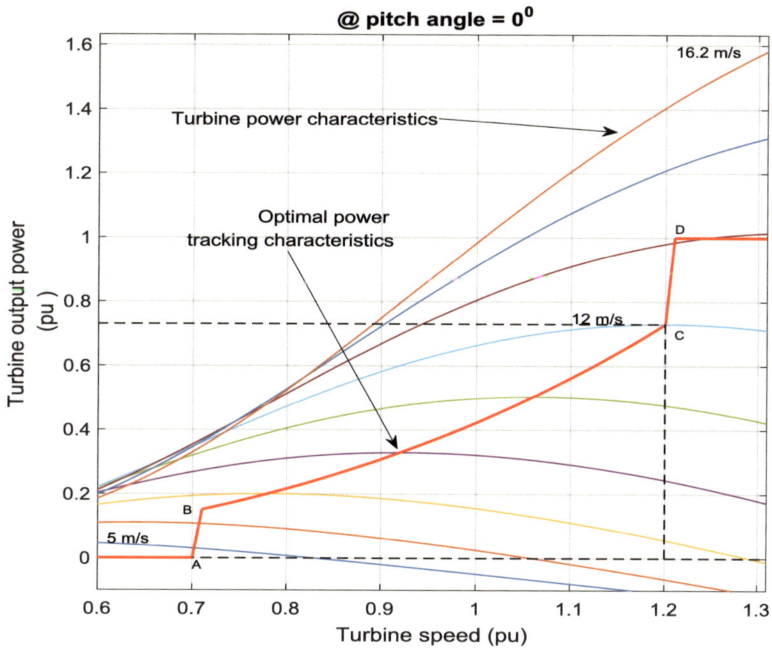

Figure 2.
Wind turbine power versus speed characteristics.

Particular optimal value of λ is defined (λ_{opt} = 8.1). Thus, for β = 0^0 and the coefficients of C_1 = 0.5176, C_2 = 116, C_3 = 0.4, C_4 = 5, C_5 = 21, and C_6 = 0.0068, the maximum value of C_p, which is C_{pmax} = 0.48, is achieved.

Normalizing Eq. (3) in the pu system, we get

$$P_{m-pu} = C_{p-pu}(\lambda, \beta) v_{w-pu}^3 \tag{8}$$

Thus, implementing the above equations, the turbine power is determined and the result is shown in **Figure 2** as power versus speed of the wind turbine.

The electromagnetic torque (T_{em}) is expressed as

$$T_{em} = p\left(\Psi_{ds} i_{qs} - \Psi_{qs} i_{ds}\right) \tag{9}$$

The output active (P_g) and reactive power (Q_g) of a wind turbine supplied to the grid are computed by.

$$\begin{aligned} P_g &= 1.5\left(-v_{ds} i_{ds} + v_{qs} i_{qs} - v_{qr} i_{qr} - v_{dr} i_{dr}\right) \\ Q_g &= 1.5\left(v_{qs} i_{ds} + v_{ds} i_{qs} - v_{dr} i_{qr} - v_{qr} i_{dr}\right) \end{aligned} \tag{10}$$

4. Modeling of the system

4.1 System description

The structure of the proposed system is shown in **Figure 3**. The grid contains conventional synchronous generator-based thermoelectricity, represented by SG1,

SG2, and SG3, and loads. VSPS stands for variable-speed pumped storage powered by a doubly fed induction machine. As shown in **Figure 3**, the VSPS induction motor is configured with the rotor connected to the rotor-side voltage source converter (VSC) and the stator connected to the grid. The grid-side VSC NPCg is connected to the grid through a coupling inductor. Details are explained in [22]. IG stands for induction generator-based wind farm.

4.2 The VSPS machine modeling

As shown in **Figure 3**, the doubly fed induction machine (DFIM)-fed VSPS unit is based on the VSC NPC converter topology. The dynamics model of the three-phase induction machine is expressed in Eqs. (11)–(15) including electrical and mechanical systems. Further details are explained in [22].

$$\frac{d\Psi_{ds}}{dt} = V_{ds} - R_s i_{ds} - \omega_s \Psi_{qs}$$
$$\frac{d\Psi_{qs}}{dt} = V_{qs} - R_s i_{qs} + \omega_s \Psi_{ds} \tag{11}$$

$$\frac{d\Psi_{dr}}{dt} = V_{dr} - R_r i_{dr} + (\omega_s - \omega_r)\Psi_{qr}$$
$$\frac{d\Psi_{qr}}{dt} = V_{qr} - R_r i_{qr} - (\omega_s - \omega_r)\Psi_{dr} \tag{12}$$

$$T_{em} = 1.5p(\Psi_{ds} i_{qs} - \Psi_{qs} i_{ds}) \tag{13}$$

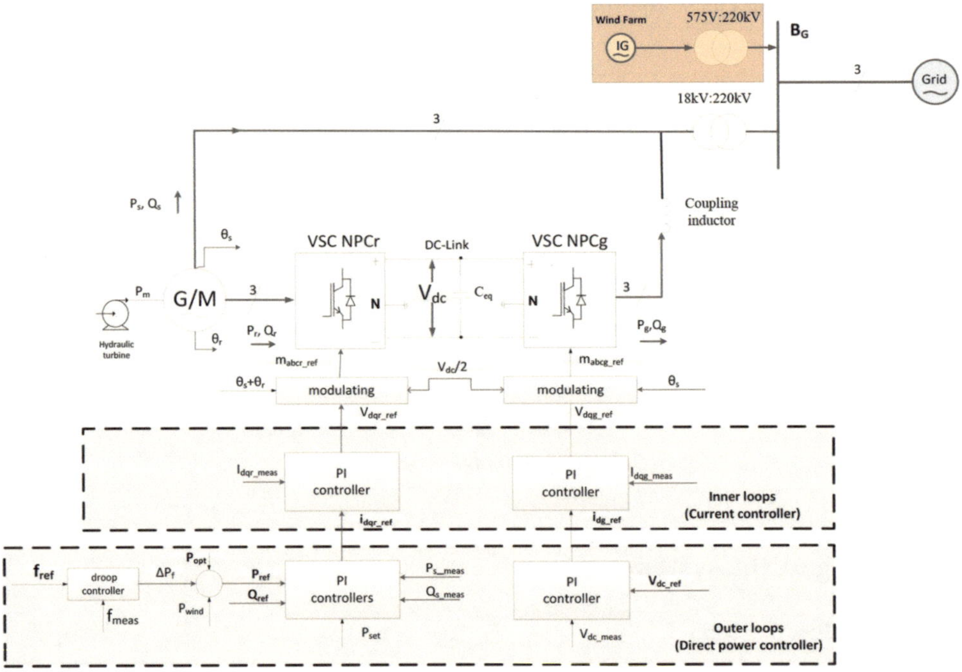

Figure 3.
Overview of a proposed VSPS system based on doubly fed induction machine and VSC converter topology.

$$\begin{aligned} \Psi_{ds} &= L_s i_{ds} + L_m i_{dr} \\ \Psi_{qs} &= L_s i_{qs} + L_m i_{qr} \\ \Psi_{dr} &= L_r i_{dr} + L_m i_{ds} \\ \Psi_{qr} &= L_r i_{qr} + L_m i_{qs} \\ L_s &= L_{ls} + L_m \\ L_r &= L_{lr} + L_m \end{aligned} \tag{14}$$

$$\begin{aligned} \frac{d\omega_m}{dt} &= \frac{1}{2H}(T_{em} - F\omega_m - T_m) \\ \frac{d\theta_m}{dt} &= \omega_m \end{aligned} \tag{15}$$

where, Ψ_{dqs} Ψ_{dqr}, $V_{qds} i_{dqs}$, $L_{ls} L_{lr}$, $V_{qdr} i_{dqr}$, L_m, R_s R_r, $\omega\omega_r$, $T_m T_{em}$, H,F, s, and P are respectively dq-axis stator and rotor fluxes, dq-axis stator voltage and current, stator and rotor leakage reactance, dq-axis rotor voltage and current, magnetizing reactance, stator and rotor resistance, synchronous and rotor angular speeds, turbine shaft and electromagnetic torque, combined inertia constant, combined viscous friction coefficient, slip ratio of the induction machine, and number of pole pairs.

4.3 The VSC-NPC converter modeling for deploying VSPS

The VSC is a self-commutating converter that is fast and controllable for AC/DC interface applications. One type of VSC converter is a three-level NPC power converter. The tertiary block contains three arms, each with its own four switch assemblies with antiparallel diodes and two NPCs.

The VSC self-commutated converter is a fast and controllable converter for AC/DC interface applications. One type of VSC converter is a three-level NPC power converter. The three-level block contains three arms with their own four switch assemblies with antiparallel diodes and two NPCs. Nonetheless, since this chapter is devoted to verifying the performance of the proposed system according to the phasor model technique, since the proposed system is large and should exhibit wind energy compensation and frequency stability not possible to see in the detailed model technique, assumptions are adopted. It is assumed that power losses in the converter are ignored. The switching dynamics can also be ignored because the PWM frequency in the three-level VSC NPC is much larger than the system frequency [23]. Therefore, the model of VSC NPC was developed based on (16).

$$u_{dqr} = \frac{m_{dqr} V_{dc}}{2} \tag{16}$$

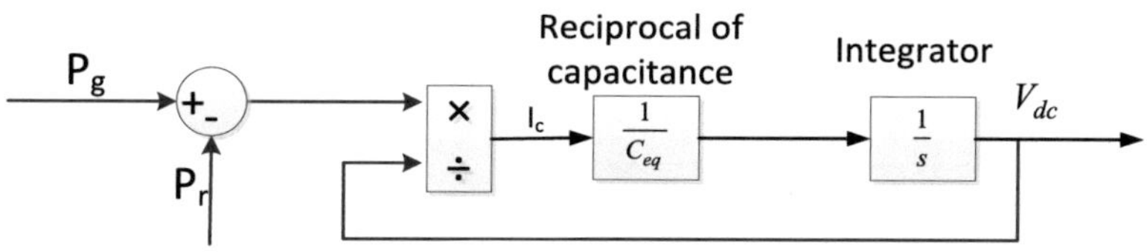

Figure 4.
DC-link phasor model.

where u_{dqr}, u_{dqg} and m_{dqr}, m_{dqg} are converter-injected voltages and modulated signals respectively in the rotor and grid-side converters.

The converter dynamics can be given by.

$$\frac{dV_{dc}}{dt} = \frac{1}{C_{eq}} i_c = \frac{1}{C_{eq} V_{dc}} (P_g - P_r) \tag{17}$$

Or in Laplace transform function

$$V_{dc}(s) = \frac{1}{sC_{eq}} I_c(s) = \frac{1}{sC_{eq} V_{dc}(s)} \left[P_g(s) - P_r(s)\right] \tag{18}$$

where C_{eq} and V_{dc} are the equivalent capacitance and DC voltage, respectively, and i_c is the DC link capacitor charging current. P_g and P_r are active power flow in the grid and rotor side of the converters, respectively. Hence, Eq. (18) can be illustrated by **Figure 4** [22].

4.4 Control strategies for the VSC NPC system

4.4.1 Active power control

Using the DPC strategy, the response speed is fast, and the control system directly and effectively compensates for wind power fluctuations. Even if this is a local control scheme placed in the VSPS unit, as long as the VSPS operates within specified limits, an important output of an adjustable and almost constant power flow can be seen from the network of the power system. The power command P_{opt} of the VSPS is determined by the capacity of the VSPS and its efficiency.

In this regard, an external setpoint P_{set} should be assumed for the VSPS, which is the input to the power control system. The set point depends on the VSPS capacity and grid conditions. The value of this setpoint is specifically used to optimize the system energy balance and is updated at a slower rate. However, the main focus is on controlling wind power fluctuations and regulating VSPS power to mitigate the impact on the grid.

The active power P_s and reactive power Q_s equations provide the basis for the control strategy modeling, and given as.

$$\begin{aligned} P_s &= 1.5\left(v_{qs} i_{qs} + v_{ds} i_{ds}\right) \\ Q_s &= 1.5\left(v_{qs} i_{ds} - v_{ds} i_{qs}\right) \end{aligned} \tag{19}$$

Based on the detail works in [22], we have.

$$\begin{aligned} P_s &= 1.5 V_s (L_m/L_s) i_{qr} \\ Q_s &= 1.5 V_s \{(\Psi_s/L_s) - (L_m/L_s) i_{dr}\} \end{aligned} \tag{20}$$

which implies the independent control of the real power and reactive power in the application of DFIM in VSPS.

Starting from the voltage equations of (11)–(14), assuming that the voltage drop across the stator resistance is very small compared with the grid voltage, and that the magnitude of the stator flux is fairly constant, we have the mathematical equation of (21).

$$V_{qr} = R_r i_{qr} + L_r^* \frac{di_{qr}}{dt} + (\omega_s - \omega_r)(L_r i_{dr} + L_m i_{ds})$$
$$V_{dr} = R_r i_{dr} + L_r^* \frac{di_{dr}}{dt} - (\omega_s - \omega_r)(L_r i_{qr} + L_m i_{qs})$$

where

$$L_r^* = L_r - (L_m^2/L_s) \tag{21}$$

From (21), the rotor dynamics model is determined and defined as

$$\frac{I_{dqr}(s)}{I_{dqr-ref}(s)} = G_i(s) = \frac{1}{\frac{L_r^*}{R_r}s + 1} \tag{22}$$

Thus, the current control (inner loops) structure of **Figure 5** is supposed to be developed. The PI control of $K_i(s)$ is defined by

$$K_i(s) = \frac{P_i s + I_i}{s} \tag{23}$$

where P_i and I_i are proportional and integral gains. P_i and I_i are determined and tuned based on the control stability theory.

From Eq. (23), we obtain

$$I_{qr}(s) = G_i(s) I_{qr-ref}(s) \tag{24}$$

Multiplying both sides of (24) by $1.5V_s L_m/L_s$, we get

$$1.5V_s L_m/L_s I_{qr}(s) = G_i(s) 1.5V_s L_m/L_s I_{qr-ref}(s) \tag{25}$$

Therefore, based on Eq. (20), from (25), we can deduce (26).

$$P_s(s) = G_i(s) P_{s-ref}(s) \tag{26}$$

That is the dynamic model for active power control of rotor-side converter VSC NPCr is the same as the current control dynamic model. The PI control of $K_p(s)$ of **Figure 6** is defined in (30)

$$K_p(s) = \frac{P_p s + I_p}{s} \tag{27}$$

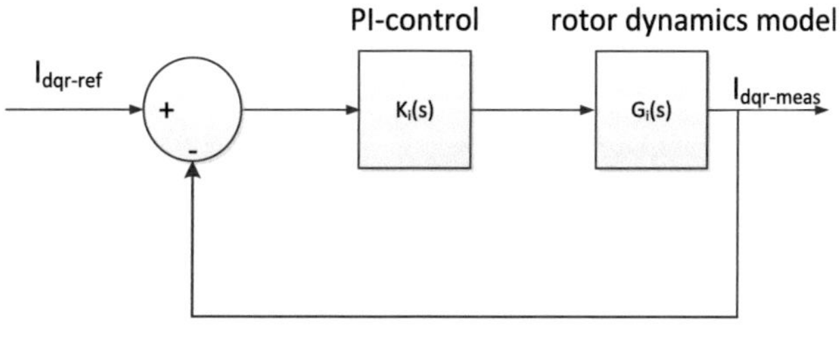

Figure 5.
The current control structure of the VSC NPC1.

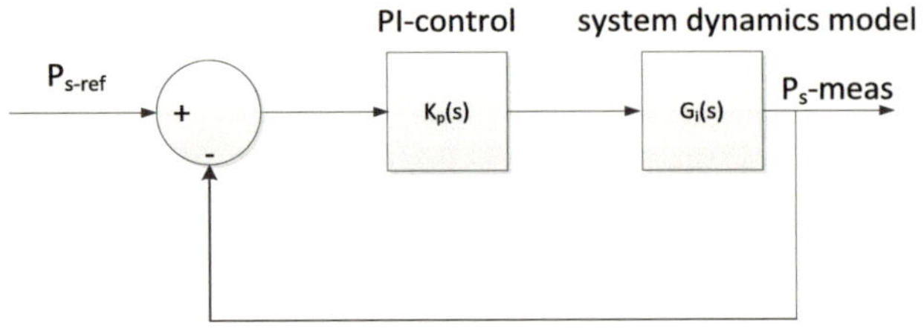

Figure 6.
The active power control structure of the VSC NPC1.

Considering the power command ΔP_{wind} to the VSPS power control unit, it is computed by

$$\Delta P_{wind} = P_{wind} - P_{set} \tag{28}$$

where P_{wind} is the wind farm power measured, and P_{set} an assumed forecasted power output. This P_{wind} is added to the power command P_{opt} to feed the fluctuated wind power to the control loop for regulating.

4.4.2 Frequency droop control for the improvement of active power control performance

Controlling the VSPS unit in primary frequency control is very important in power systems. By implementing a frequency droop control scheme, the response of the power system to emergencies is improved in the case of fluctuations in the dominant wind farm power generation. Power transfer within the DC link can be modulated by frequency droop control. It is specially designed for converters to provide frequency support to the grid. The control structure shown in **Figure** 7 provides a governor-like droop behavior through active power flow. Droop-type control is built on the concept of power synchronization control, where grid synchronization is achieved regardless of dedicated synchronization units [24]. This droop control system provides a DC link with frequency droop characteristics. Load sharing is possible for other generator sets in the grid system.

The introduction of VSPS can support the improvement of system performance to cope with all emergencies of the power system. Therefore, the additional power

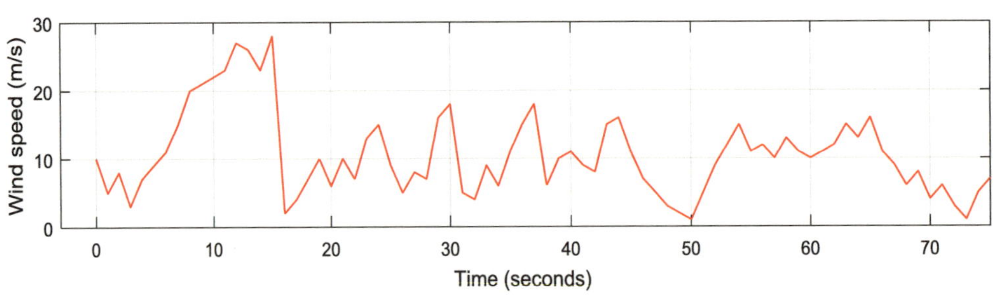

Figure 7.
Typical wind speed characteristics.

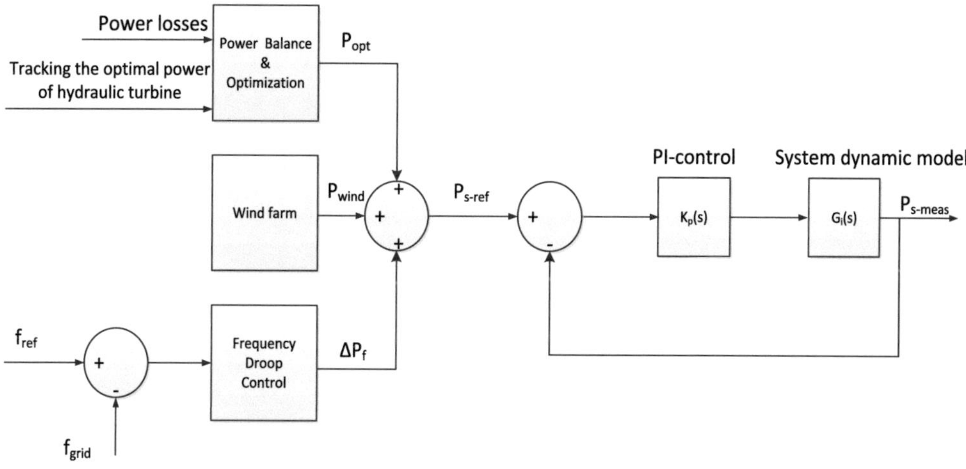

Figure 8.
A frequency droop-fed DPC control strategy for VSPS system.

command ΔP_f due to the frequency variation of the VSPS active power control system (29) is determined.

$$\Delta P_f = -K_{droop}\left(f_{ref} - f_{grid}\right) \tag{29}$$

where K_{droop} is the droop constant, f_{ref} is the 50 Hz reference frequency used in this chapter, and f_{grid} is the grid frequency. Through the active power closed-loop control system, frequency droop control can be used as a support for grid frequency control. Therefore, in order to obtain more possible controllability in the operation of the power system, the frequency droop control of the VSPS system must be introduced significantly. The above strategies are combined for the active power control scheme in the VSPS unit in the proposed system, as shown in **Figure 8**.

The models and details of the reactive power control in both rotor-side and grid-side converters and DC voltage control in the grid-side converter are presented in [22].

4.5 The wind power fluctuation and its spectrum characteristics

The distribution of wind speed and direction varies by region and season overtime. The spectral characteristics of wind power fluctuations recorded over time for a typical wind farm can be obtained using the Fast Fourier Transform (FFT) algorithm. **Figure** 7 shows wind speed fluctuations; the corresponding wind power generation and its spectral characteristics are shown in **Figure 9a** and **b**, respectively.

Spectral analysis is the process of estimating the power spectral density (PSD), which characterizes the frequency content or random processes of a signal, from its time domain representation. Spectrum automatically decomposes a signal or random process into different frequencies and identifies periodicity. FFT-based nonparametric methods make no assumptions about the input data, can be used for any type of signal, produce more accurate spectral estimates, and allow us to convert flow time-domain signals to frequency domain and vice versa. The frequency domain representation of a signal reveals important signal characteristics that are difficult to analyze in the time domain.

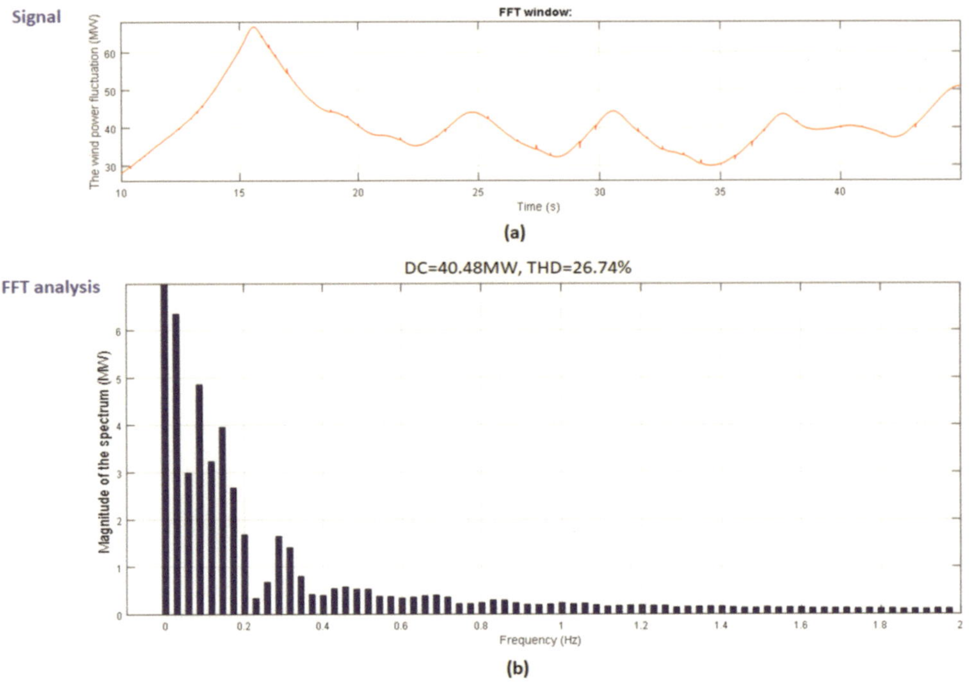

Figure 9.
The wind-turbine output power; (a) in time domain, (b) the frequency spectrum characteristics.

One of the definitions of power spectral density (PSD) function [25] is given by

$$\phi(\omega) = \lim_{N\to\infty} E\left\{\frac{1}{N}\left|\sum_{t=1}^{N} y(t)e^{-i\omega t}\right|^2\right\} \tag{30}$$

Certainly, the PSD $\phi(\omega)$ measures the power at frequency ω in the signal's auto-covariance sequence $r(k)$, i.e., $r(k) = r^*(-k)$. Since $\phi(\omega)$ is a power density, it should be real valued and nonnegative, i.e., $\phi(\omega) \geq 0$ for all ω. Thus, according to [25], the other equivalent definition of the PSD is given by

$$\phi(\omega) = r(0) + 2\sum_{k=1}^{\infty} r(k)\cos(\omega k) \tag{31}$$

where $r(0)$ is the averaging value of the function over the sampled time, and its inverse transformation is

$$r(k) = \frac{1}{2\pi}\int_{-\pi}^{\pi} \phi(\omega)e^{i\omega k}d\omega \tag{32}$$

Therefore, the spectral characteristic of the wind power fluctuation is well described and analyzed by the above PSD equations, and the simulating result of **Figure 9a** is shown in **Figure 9b**.

As shown in **Figure 9b**, the wind power fluctuation is insignificant for the spectral frequency greater than about 1.2 Hz. Hence, the VSPS is engaged to compensate the significant wind power fluctuation components and the spectrum of the VSPS control response is determined accordingly.

According to statistics, the typical frequency of wind power fluctuations is recorded from a few minutes to a few days [9]. For slow fluctuations that are not an obvious problem, such as daily, the grid has adequate means of control. However, it is difficult to adjust minute-to-minute fluctuations only by traditional means such as thermal power. Therefore, to solve such problems caused by the randomness of wind power, VSPS is a possible option for stabilizing and smoothing grid frequency changes.

4.6 The VSPS system spectrum analysis and its implication for wind power fluctuation compensation

The power control in the DFIM-based wind farm VSPS system is essentially the control of the VSC-fed VSPS system. As shown in **Figure 10**, the frequency-droop-fed DPC-controlled VSPS system integrated in the grid tracks the power fluctuation command well. Small distortions are observed when the command signal changes from one state to another.

We know that the dynamic characteristics of the generator set and pump set of VSPS are closely related to the control method and system structure, but it is difficult to obtain a specific value by using the analytical method. Therefore, this chapter uses the spectral analysis method to verify the results obtained by the time domain analysis, as shown in **Figure 11**. The spectral characteristics are determined using the driving mathematically equivalent total active power control system equation of (33). Therefore, the simulation results of the spectral characteristics of the controlled active power response of the VSPS are shown in **Figure 12**. It is observed that the bandwidth of the response of the VSPS control system is large enough to cover the bandwidth power fluctuation of the spectral results of the wind as shown in **Figure 10b**. This means that the frequency-fed DPC control strategy of the VSPS system can use the derivation function of (33) to

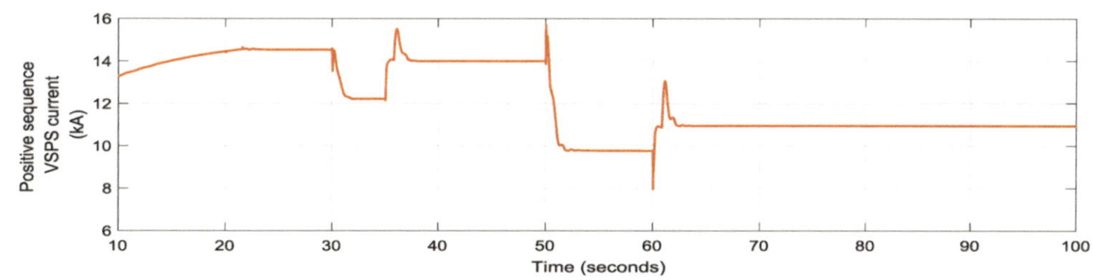

Figure 10.
The tracking of active power instruction in the droop fed DPC-driven VSPS system.

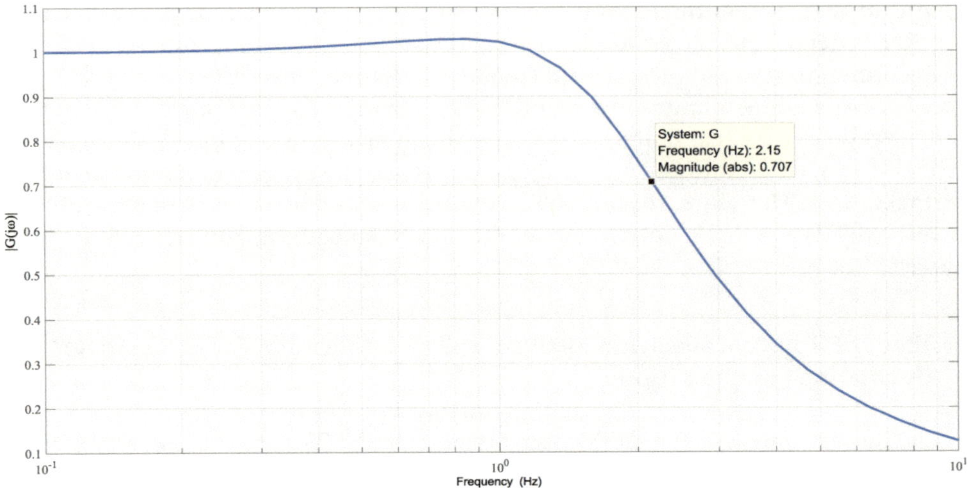

Figure 11.
The frequency spectrum characteristics of the VSPS control system simulated in time response of ***Figure 10***.

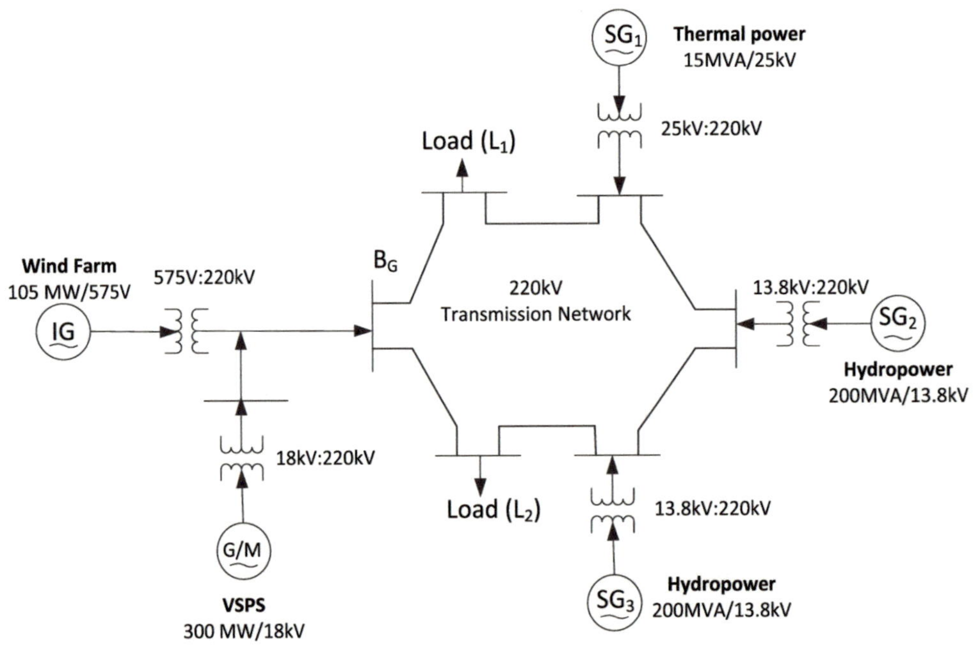

Figure 12.
Single-line diagram of the VSPS-wind farm-grid integrated system simulating setup.

quickly and flexibly adjust the power fluctuations caused by the wind farm energy in the grid integrated system.

$$G(s) \approx \frac{\frac{P_p}{I_p}s + 1}{\frac{\left(\frac{1+P_i}{I_i}\right)k_{droop}}{I_p}s^2 + \frac{(P_p+1)}{I_p}s + 1} \tag{33}$$

5. Simulation of models

In this chapter, a case study is presented involving a 300 MW DFIM-based VSPS and a grid system integrated with a wind farm consisting of seven identical 15 MW wind turbine induction generators. To study the effect of wind fluctuations on grid frequency, a model of a medium-sized power system consisting of two 200MVA hydropower plants and one 15MVA diesel generator set and conventional synchronous generators was established. Phasor modeling techniques in MATLAB/Simulink are applied. For conventional synchronous machines, the primary voltage regulation method in automatic voltage regulators based on the standard IEEE type I and the primary frequency regulation method in turbine speed governors are used. The simulation setup is shown in **Figure 12**.

6. Results and discussion

Figure 13 shows the active power generation characteristics of the VSPS unit. Its power generation trends vary according to the characteristics of wind power fluctuations. **Figure 14** shows a comparison of VSPS active power compensation with wind power fluctuations. The comparison takes into account the power fluctuations of the wind farm from its mean value (40.74 MW) and the difference in VSPS production

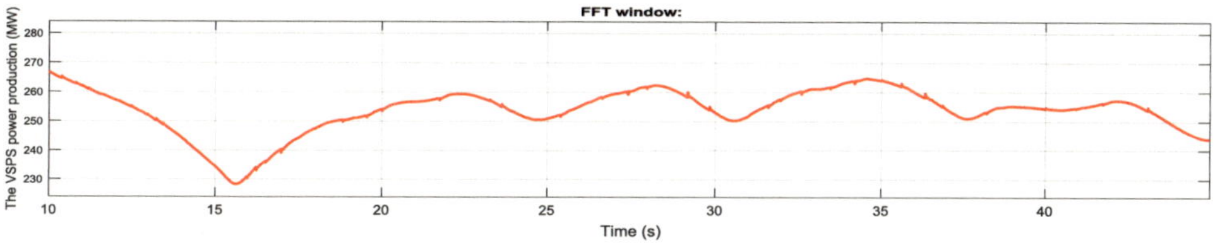

Figure 13.
The active power characteristics of the VSPS.

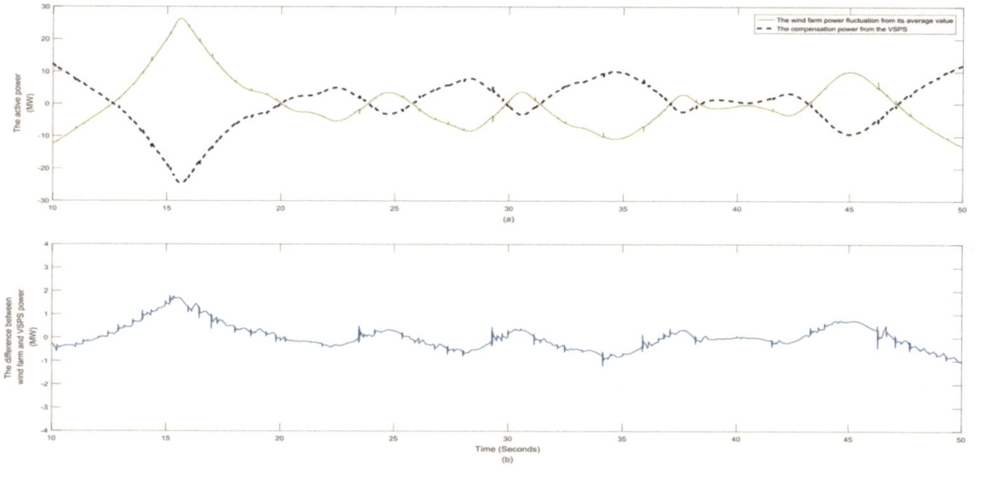

Figure 14.
The fluctuation compensation characteristics. a) the wind farm power fluctuation from its average value and the compensation power generated and regulated by VSPS system. b) the active power difference between the wind fluctuation and the VSPS compensation.

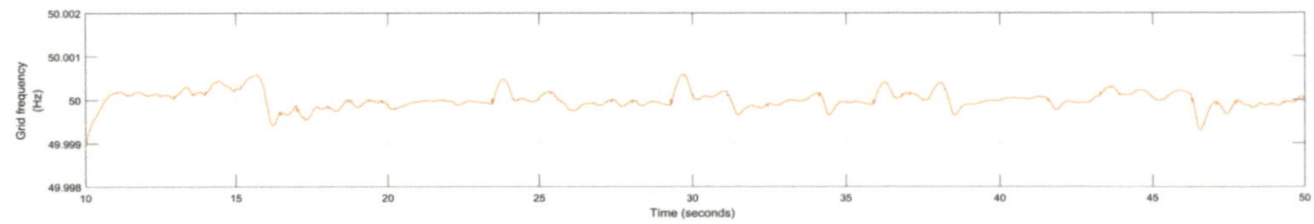

Figure 15.
The grid frequency response associated with the wind power fluctuation.

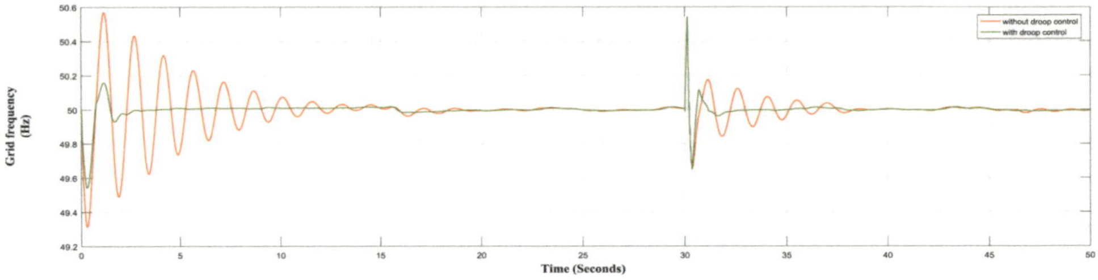

Figure 16.
Evolution of the network in grid frequency response comparing the VSPS active power control with and without droop control during the start-up transient and three-phase faults imposed.

from its mean value (247.1 MW). The error between the wind farm power fluctuation and the power produced by the VSPS is small. Therefore, VSPS can effectively and quickly compensate for large fluctuations in wind energy. This verifies the ability of VSPS to meet wind power fluctuation compensation requirements. Therefore, as shown in **Figure 15**, a very small deviation in grid frequency is observed, less than 0.002%.

Figure 16 shows the evolution of the grid frequency response by comparing VSPS active power control with and without droop control during emergency. Compared with the vector control strategy without droop feed control, the deviation of frequency from the nominal value is greatly reduced in the droop feed vector control strategy. The grid frequency response deviation was observed to be ± 0.04 Hz, while active power was regulated to track wind power fluctuations, start-up transients, and three-phase faults. Importantly, the settling time of the droop control scheme is very short. Therefore, the grid frequency dynamics and steady-state response in the droop-fed vector control strategy are quite stable, and the deviation is within an acceptable range.

7. Conclusion

This chapter discusses VSPS systems in wind farm grid-connected systems, which play a vital role in the context of increasing penetration of renewable energy. It also has an important function in real-time applications for balancing generation and demand as it provides fast-response generation. VSPS active power generation can quickly adjust wind power fluctuations, so that the power flow on the grid bus is stable. This chapter proposes a VSPS system based on a droop-fed DPC strategy as a solution. Spectral analysis is used to verify the performance of the proposed control system as well.

The results show that VSPS has important stability control characteristics to respond to different contingencies, such as generator losses, large load changes, faults, and start-up transient disturbances. The proposed control strategy can also well regulate grid frequency and AC bus voltage. Spectral analysis method implemented in this chapter is also effective for determining the requirements of wind power fluctuations and stability in large power systems.

Author details

Girmaw Teshager Bitew* and Minxiao Han
North China Electric Power University, Beijing, China

*Address all correspondence to: gbitew1983@yahoo.com

References

[1] Li W, Joós G, Abbey C. Wind power impact on system frequency deviation and an ESS based power filtering algorithm solution. In: 2006 IEEE PES Power Systems Conference and Exposition, PSCE 2006—Proceedings. Vol. 2006. pp. 2077-2084

[2] Global Wind Energy Council. Global Wind Energy Outlook. 2014

[3] OECD/International Energy Agency and Energy Research Institute. China Wind Energy Development Roadmap 2050. 2011

[4] Chowdhury MA, Hosseinzadeh N, Shen WX. Smoothing wind power fluctuations by fuzzy logic pitch angle controller. Renewable Energy. 2012; **38**(1):224-233

[5] Song X, Girmaw B, Han M, Meng Z, Xu J, Wang X. Stability and control of a grid integrated DFIM based variable speed pumped storage system. In: 2017 International Electrical and Energy Conference (CIEEC 2017). 2017. pp. 175-181

[6] Bocquel A, Janning J. Analysis of a 300 MW variable speed drive for pump-storage plant applications. In: 2005 European Conference on Power Electronics and Applications. 2005. pp. 1-10

[7] Furuya S, Taguchi T, Kusunoki K, Yanagisawa T, Kageyama T, Kanai T. Successful achievement in a variable speed pumped storage. In: Conference Record of the Power Conversion Conference. 1993. pp. 603-608

[8] Xin C, Minxiao H, Chao Z. Power control analysis for variable speed pumped storage with full-size converter. In: IECON2015-Yokohama. 2015. pp. 1327-1332

[9] Abdalla OH, Han M, Liu C. Multi-level converter based variable speed pump storage for wind power compensation. In: 2014 International Conference on Information Science, Electronics and Electrical Engineering. 2014. pp. 1497-1501

[10] Steimer PK, Senturk O, Aubert S, Linder S. Converter-fed synchronous machine for pumped hydro storage plants. In: 2014 IEEE Energy Conversion Congress and Exposition, ECCE 2014. 2014. pp. 4561-4567

[11] Pronin MV, Shonin OB, Vorontsov AG, Gogolev GA. Features of a drive system for pump-storage plant applications based on the use of double-fed induction machine with a multistage-multilevel frequency converter. In: 15th International Power Electronics and Motion Control Conference. 2012. pp. 1-8

[12] Rehman S, Al-Hadhrami LM, Alam MM. Pumped hydro energy storage system: A technological review. Renewable and Sustainable Energy Reviews. 2015;**44**: 586-598

[13] Djeriri Y, Meroufel A, Massoum A, Boudjema Z. A comparative study between field oriented control strategy and direct power control strategy for DFIG. Journal of Electrical Engineering. 2014;**14**:169-178

[14] Merzoug M, Naceri F. Comparison of field-oriented control and direct torque control for permanent magnet synchronous motor (pmsm). International Journal of Electrical,

Computer, Energetic, Electronic and Communication Engineering. 2008;**2**(9): 299-304

[15] Gihga R, Wu Q, Nielsen AH. Phasor model of full scale converter wind turbine for small-signal stability analysis. The Journal of Engineering. 2017; **2017**(13):978-983

[16] Teshager BG, Han M, Meng Z, Song X, Xu J, Simiyu P. Phasor model simulation of a grid integrated variable speed pumped storage system. The Journal of Engineering. 2017;**2017**(13): 1002-1009

[17] Haileselassie TM, Uhlen K. Primary frequency control of remote grids connected by multi-terminal HVDC. In: IEEE PES General Meeting, PES 2010. 2010

[18] Rusidovic Z. Emergency Frequency Control in Future Networks. Manchester, United Kingdom: University of Manchister; 2015

[19] Bevrani H. Robust Power System Frequency Control. 2nd ed. Springer Cham Heidelberg New York Dordrecht London: Springer International Publishing Switzerland; 2014

[20] Bourdoulis MK, Member S, Alexandridis AT. Direct power control of DFIG wind systems based on nonlinear modeling and analysis. IEEE Journal of Emerging and Selected Topics in Power Electronics. 2014;**2**(4):764-775

[21] Heier S. Grid Integration of Wind Energy: Onshore and Offshore Conversion Systems. 3rd ed. Hoboken, New Jersey, US: John Wiley & Sons Ltd; 2014

[22] Girmaw Teshager B, Han M, Meng Z, Song X, Xu J, Simiyu P. Phasor model simulation of a grid integrated variable speed pumped storage system. Journal of Engineering. 2017;**2017**(13): 1002-1009

[23] Demiray T, Andersson G, Busarello L. Evaluation study for the simulation of power system transients using dynamic phasor models. In: 2008 IEEE/PES Transmission and Distribution Conference and Exposition. 2008. pp. 1-6

[24] Nanou SI, Papathanassiou SA, Member S. Frequency control of Island VSC-HVDC links operating in parallel with AC interconnectors and onsite generation. In: IEEE Transactions on Power Delivery. 2017. pp. 2014-2020

[25] Stoica P, Moses R. Spectral Analysis of Signals. Upper Saddle River, New Jersey, US: Prentice Hall, Inc.; 2005

Chapter 10

Simulation Analysis of DFIG Integrated Wind Turbine Control System

Ramesh Kumar Behara and Kavita Behara

Abstract

Recently, scientists and academics are discovering progressive improvements in the arena of wind power technology economically and reliably, allowing them to produce electricity focusing on renewable energy resources. Wind turbines (WT) using the Doubly Fed Induction Generators (DFIGs) have attracted particular attention because of their advantages such as variable speed constant frequency (VSCF) operation, independent control capabilities for maximum power point tracking (MPPT), active and reactive power controls, and voltage control strategy at the point of common coupling (PCC). When such resources have to be integrated into the existing power system, the operation becomes more challenging, particularly in terms of stability, security, and reliability. A DFIG system with its control strategies is simulated on MATLAB software. This entails the rapid control prototype testing of grid-connected, variable speed DFIG wind turbines to investigate the WT's steady-state and dynamic behavior under normal and disturbed wind conditions. To augment the transient stability of DFIG, the simulation results for the active and reactive power of conventional controllers are compared with the adaptive tracking, self-tuned feed-forward PI controller model for optimum performance. Conclusive outcomes manifest the superior robustness of the feed-forward PI controller in terms of rising time, settling time, and overshoot value.

Keywords: renewable energy sources, DFIG's, MATLAB/SIMULINK, WT, MPPT, distribution network, wind energy

1. Introduction

The increasing demand for electrical power, and the rapid depletion and environmental concerns of fossil fuels have prompted the increased need for alternate forms of clean and sustainable energy sources. Globally there is an unassertive move away from unabated coal, and the rise of renewable energies from under 30% of generation in 2020 to above 40% in 2030 [1]. According to statistics (2021) put out by International Renewable Energy Agency (IRENA), an entire of 2,802,004 MW of electricity generation was produced worldwide in 2020, added by 53,824 MW of electricity generation from Africa [2]. The International Energy Agency (IEA) projected that the worldwide energy mandate would be twofold by 2030 [3]. The global energy generation, ingesting, and financial development (1991–2017) swing displays a direct upsurge associated with the growing development of the global economy [4]. These cited worldwide trends of financial development mutually with industrial development and countryside electrification energy requirements are reflected in South Africa. South Africa's energy source is a mixture of 59% from carbon coal, 16% from petroleum, 3% from gasoline, 2% nuclear, and 20% from renewable resources and leftover [4, 5]. The worldwide cumulative swing of wind energy facility systems has speedily extended the wind energy facilities [6]. The wind is the gross effect of the pressure rise force, gravity, Coriolis, centrifugal and friction forces performing on the troposphere. The wind's aerodynamic features, lift and drag, angle of attack, and the effect of a high lift to drag ratio as the highest draft feature of the turbine rotor blade for effective wind power harvesting as explained in [7].

In [8], it was discussed how the twisted and tapered propeller-type rotor blades, could improve the angle of attack, rotor speed, and efficiency while reducing drag and bending stress. Betz equation evaluation of the wind energy conversion signifies that 59% optimum efficiency in which a conventional wind turbine can extract power from the wind [9]. Wind turbines generator systems are generally classified in lift and drag type, upwind and downwind type, and horizontal and vertical axis turbines. Most power utility networks operate either as N + 1 or N + 2 configuration criteria. N-0 configuration criterion is widely practiced in radial low voltage distribution networks [10]. A grid-connected wind energy system needs to meet certain standards before being integrated into the grid. Wind turbine generators control system (WTGCS) connects wind turbine generators to the grid, with a generation scheduling in place, that regulates the generator speed consequently adjusting the generator frequency, the voltage at the grid, active and reactive power flow using rotor side converter (RSC) and the grid side converter (GSC) and at the same time prefer to disconnect the wind turbines from the grid during faults, resulting in power losses, out of synchronization, and cascaded tripping of generation facilities [11].

The study in [12], clarifies in what way the controlled rotor current on the rotor side converter and a dynamic disruption elimination control by the resources of an extended state observer (ESO) controller the real and reactive stator powers produced by a wind power transformation method. In [13], an MPPT is combined with the DFIG stator flux oriented vector control to disassociate the control of real and reactive power produced by the DFIG centered wind turbine, with the generated power plattering as the dynamic energy reference for the DFIG. In [14], unit 3 directs the symmetrical and asymmetrical voltage rise and fall of the power grid-integrated DFIG, by including further current controller loops by disintegrating the vectorial references into progressive and undesirable signal indications. In [15], an open circuit stator negative sequence rotor current control system is applied, permitting the induced stator voltage to develop as unbalanced as the grid system voltage, henceforth allowing an even linking of DFIG to the power grid system. In [16], a wind speed assessment process centered on particle swarm optimization, and support vector regression, was evaluated to allow the MPPT control. The South African Renewable Energy Grid Code (SAREGC) published narration (2.9) in November 2016 stipulates the necessities for manufacturing standards, networking reliability, and unbalanced admittance to the power grid using RPPs.

This chapter aims to investigate the impact of wind energy penetration into the distribution grid for different percentages for the different scenarios of wind energy integration into the existing grid.

2. Aerodynamics, electrical, and drive train modeling

2.1 Aerodynamics modeling

The wind is the net result of the pressure gradient force, gravity, Coriolis, centrifugal and friction forces acting on the atmosphere. Since wind speed usually varies from one location to another and also fluctuates over time in a stochastic way, J.G. Slootweg [17] proposed a mathematical model that takes some landscape parameters to generate a wind speed $V_m(t)$ in (meter/sec) sequence for any location, as per equation one:

$$V_w(t) = V_{w\ a}(t) + V_{w\ r}(t) + V_{w\ g}(t) + V_{w\ t}(t) \tag{1}$$

Where, $V_{wa}(t)$ is a constant component, $V_{wr}(t)$ is a common ramp component, $V_{wg}(t)$ is a gust component, and $V_{wt}(t)$ is a turbulence component in (meter/sec).

The operation of a wind turbine can be characterized by its mechanical power output P_m through a cross-sectional area A normal to the wind as a function of wind speed V_w [18].

$$P_m = 0.5\rho A V_w^3 C_P(\alpha, \beta) \tag{2}$$

Where ρ is the air mass density, $A = \pi r^2$ is the turbine swept area, r is the turbine radius, and V_w is the wind speed. C_p is a nonlinear function of λ and β is referred to as the performance coefficient and is smaller than 0.59, given by [18, 19].

$$C_p = 0.5\left[\frac{rC_f}{\lambda} - 0.022\beta - 2\right]e^{-0.255\frac{rC_f}{\lambda}} \tag{3}$$

Where β is the turbine pitch angle, λ is the tip-speed ratio, and C_f is the blade design constant λ is defined by:

$$\lambda = \frac{r\omega_{tur}}{V_w} \tag{4}$$

Where ω_{tur} is the rotational angular speed of turbine blades in mechanical rad/sec.

$$P_{m\max} = \left(\frac{0.5\rho A r^3 C_{P\max}}{\lambda_{opt}^3}\right) w_{tur}^3 = K_{opt} w_{tur}^3 \tag{5}$$

Eq. (5) indicates that P_{max} is proportional to the cube of turbine speed, hence the mechanical torque T_{mmax} is:

$$T_{m\max} = K_{opt} w_{tur}^2 \tag{6}$$

2.2 Electrical modeling

A wounded-rotor induction motor can operate as a double-fed induction motor (DFIM) with the stator side windings openly attached to the three-phase power grid/load and the rotor side windings attached to a side-by-side moderately measured (20–30) % rating power converter as shown in **Figure 1**. As shown in **Figure 2**, an

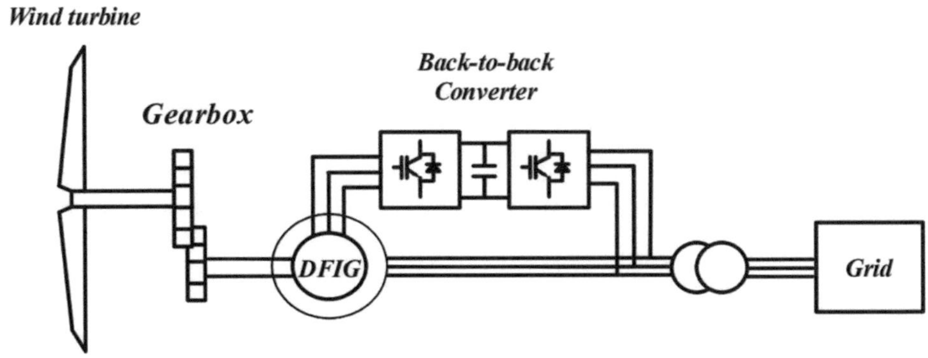

Figure 1.
Grid-connected DFIG [20].

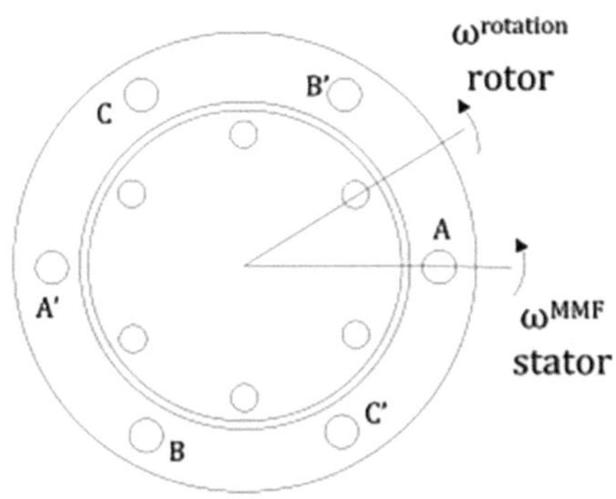

Figure 2.
Electrical modeling of an induction machine winding layout [21].

induction motor works on the interface principle between the stator and rotor magnetomotive forces (MMF). The stator side windings current produce an MMF revolving at power grid side frequency especially including an MMF in the rotor side windings. The rotor speed does not compliment the stator side MMF. This induced rotor MMF will rotate at the so-called slip frequency which possesses the subsequent value [21]:

$$\omega_{slip} = \omega_{mmf}^{rotor} = \omega_{mmf}^{stator} - \omega_{rotor} \tag{7}$$

Where, ω_{slip} is the slip frequency, corresponding to the frequency of rotor current and voltage, ω_{mmf}^{stator} is the stator or the grid frequency in (rad/sec), ω_{rotor} is the rotor rotating frequency (rad/sec). In both sub-synchronous and super-synchronous operations, the DFIM machine can be operated either as a motor ($0 < \text{slip} < 1$) with positive rotor torque or a generator ($\text{slip} < 0$) with negative rotor torque. The Park and Clark transform allowing the transformation of time-dependent variables into constant values. The per-unit electromagnetic torque equation expressed in d-q park reference is given by [22]:

$$T_e = \varphi_{ds} I_{qs} - \varphi_{qs} I_{ds} = \varphi_{qr} I_{dr} - \varphi_{dr} I_{qr} = L_m \left(I_{qs} I_{dr} - I_{ds} I_{qr} \right) \tag{8}$$

Neglecting the power losses associated with the stator and rotor resistances, the active and reactive stator powers for the DFIG are [22]:

$$P_s = \left(\frac{3}{2}\right)(V_{ds}I_{ds} + V_{qs}I_{qs}) \tag{9}$$

$$Q_s = \left(\frac{3}{2}\right)(V_{qs}I_{ds} - V_{ds}I_{qs}) \tag{10}$$

And the active and reactive rotor powers are given by:

$$P_r = \left(\frac{3}{2}\right)(V_{dr}I_{dr} + V_{qr}I_{qr}) \tag{11}$$

$$Q_r = \left(\frac{3}{2}\right)(V_{qr}I_{dr} - V_{dr}I_{qr}) \tag{12}$$

The overall system equations can also be re-written with relation to the rotating frames [21]:

$$P_T = P_s + P_r = \frac{3}{2}\left(V'_{qr}I'_{qr} + V'_{dr}I'_{dr} + V_{ds}I_{ds} + V_{qs}I_{qs}\right) \tag{13}$$

$$Q_T = Q_s + Q_r = \frac{3}{2}\left(V'_{qr}I'_{qr} - V'_{dr}I'_{dr} + V_{ds}I_{ds} - V_{qs}I_{qs}\right) \tag{14}$$

The torque expression and the stator reactive power, which are the control objectives of the rotor-side converter control, are shown in Eqs. 17 and 18. Where, p is the number of pole pairs of the generator, I_{qs} and I_{qr} are the q component of the stator and rotor current, I_{ds} and I_{dr} are the d component of the stator and rotor current, V_{qs} and V_{ds} are the q and d components of the stator voltage. The stator and rotor flux linkages in the synchronous reference frame are expressed as [23]:

$$\psi_s = L_sI_s + L_mI_r \tag{15}$$

$$\psi_r = L_mI_s + L_rI_r \tag{16}$$

The electromagnetic torque can be expressed using the $d-q$ components as following [23].

$$T_m = \frac{3}{2}p\frac{L_m}{L_s}\left(\psi_{qs}I_{dr} - \psi_{ds}I_{qr}\right) \tag{17}$$

$$Q_s = \frac{3}{2}(V_{qs}I_{ds} - V_{ds}I_{qs}) \tag{18}$$

2.3 Drive train model

Considering the mechanical aspect of the wind turbine, the mechanical representation of the drive train of the entire wind turbine is complex. Following four types of the drive train in wind turbine models are generally used [24].

1. Six mass drive train model.

2. Three mass drive train model.

3. Two mass drive train model.

4. one-mass drive train model.

Of the above four types of drive train models, the one that was modeled and implemented is the simplified form of the two mass-shaft model power train systems as shown in **Figure 3** consisting of a shaft and gearbox. As per the two-mass model of the drive train system described in [24], all masses are grouped into low and high-speed shafts. The inertia of the low-speed shaft comes mainly from the rotating blades and the inertia of the high-speed shaft. The input to the model for a two-mass system is established as torque T_A, which is gained by the aerodynamics methodology and the generator response torque T_e. The target is the deviations in the rotor speed ω_r and the generator speed ω_g. The deviations in the mechanically compelled torque T_m, the generator torque response T_e, and torque loss owing to friction T_{fric}, causes the variation of angular velocity $\omega_g^{\bullet}$ [24].

$$T_m - T_e - T_{fric} = j_g \times \dot{\omega}_g \tag{19}$$

The change in the angular speed $\dot{\omega}_r$ is caused by the difference between the aerodynamic torque T_A and shaft torque T_s at a low speed shaft [24].

$$T_A - T_s = j_r \times \dot{\omega}_r \tag{20}$$

$$\dot{\omega}_g = \varphi_g'' \text{ and } \dot{\omega}_r = \varphi_r'' \tag{21}$$

T_m and T_s are connected by the gear ration, as $T_s = nT_m$

$$T_s = K_s \cdot \Delta\varphi + D_s \cdot \varphi = K_s . \Delta\varphi + D_s(\omega_r - \omega_g/n) \tag{22}$$

Where K_s is the stiffness constant and D_s is the damping constant of the shaft. Considering a two-mass free-swinging system the Eigen frequency is as follows:

$$\dot{\omega}_r = \frac{1}{J_r}\left(T_A - D_S \cdot \omega_r + \frac{D_S}{n}\dot{\omega}_g - K_S \int\left(\omega_r - \frac{\omega_g}{n}\right)dt\right) \tag{23}$$

$$\dot{\omega}_g = \frac{1}{J_g}\left(-T_e - \left(D_g + \frac{D_S}{n^2}\right)\omega_g + \frac{D_S}{n}\dot{\omega}_r - \frac{K_s}{n}\int\left(\omega_r - \frac{\omega_g}{n}\right)dt\right) \tag{24}$$

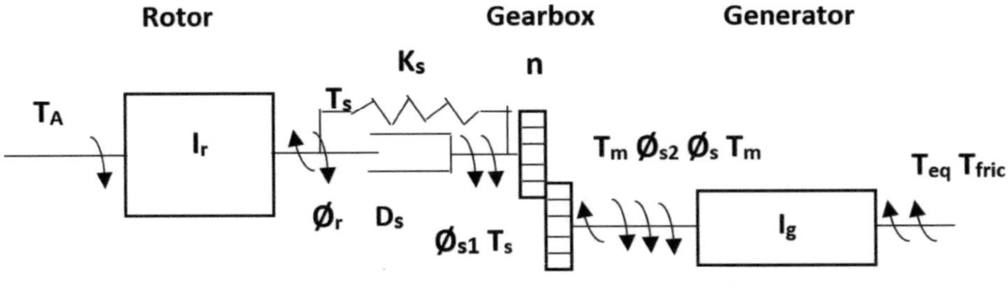

Figure 3.
Schematic drawing of the two mass shaft drive train model [24].

Where, K_s is the stiffness constant and D_s is the damping constant of the shaft. Considering a two-mass free-swinging system, the Eigen frequency is given as:

$$\omega_{os} = 2\pi f_{os} = \sqrt{\frac{K_s}{J_{ges}}} \tag{25}$$

The total inertia of the free-swinging system on the low-speed is calculated by:

$$J_{ges} = \frac{J_r \cdot J_g \cdot n^2}{J_r + J_g \cdot n^2} \tag{26}$$

So, the stiffness constant K_s and the damping constant D_s of the low-speed shaft, with ξ_s as logarithmic decrement is:

$$K_s = J_{ges} \cdot (2\pi f_{os})^2 \tag{27}$$

$$D_s = 2\xi_s \cdot \sqrt{\frac{K_s J_{ges}}{\xi_s^2 + 4\pi^2}} \tag{28}$$

3. Control system

3.1 Conventional wind power control scheme

Owing to the alterations in the timing measures of the mechanical and electrical gestures, the DFIM-centered wind power control system has a multiple-layer control arrangement, with unified sub-systems. At the uppermost developed control stage, a maximum power point tracking procedure is applied to compute the generator speed set-point Ω_m^* to produce the DFIG target power set point T_{em}^*. The other control stage oversees the turbine pitch control scheme. The third and final control stage standardizes the generator torque, the real and reactive power, and the DC linkage voltage. The key limitations of the traditional control approach are explained in [25]. In the fractional load control process, the PI controller does not permit calibration of the commutation between the energy intensification and the momentary load depreciation, and at the full load control stage, the pitch-controlled generator speed directive can source acute power variations. Therefore, this study emphasizes the generator controller utilizing an adaptable tracing, self-adjustable PI controller framework. The suggested control system is disintegrated into various sub-categories entailing the fréchet derivative, the proportional, integral, and the derivative control. The PID control variables, K_p, K_i, and K_d are observed as the adaptable interfaces among the above sub-categories to self-adjusting these variables.

A reference current calculation and current control loop are presented as shown in **Figure 4** [26] and both the reference reactive power Q_{sref} and Q_{gref} are usually set to zero and can be modified depending on the grid requirements. The DC link reference voltage has a fixed value while the reference torque is determined by the maximum power point tracking control system. The vectorial control system of a grid-integrated DFIM is very similar to the traditional vectorial control system of a

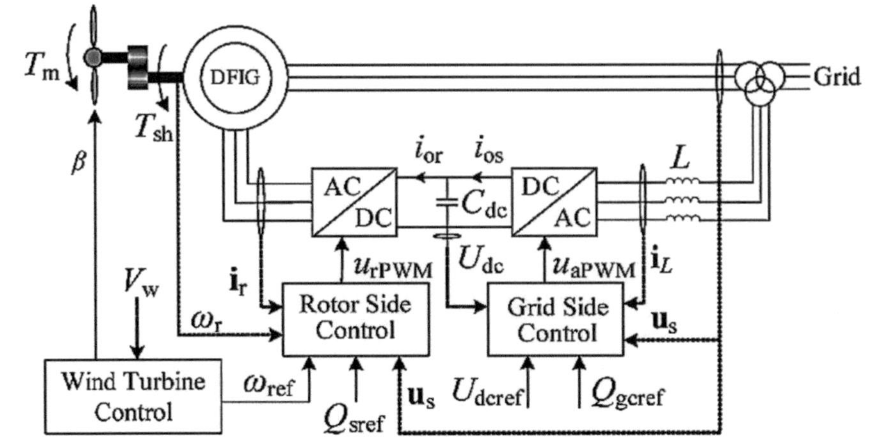

Figure 4.
Reference values entered in DFIG back to back converter [26].

squirrel cage machine. DFIM is controlled in a synchronously revolving *dq* orientation structure, with the *d* axis adapted to the rotor flux space vector locus. The direct current is therefore related to the rotor side magnetic flux linkage although the quadrature current is related to the electromagnetic torque. By regulating autonomously, the two current modules, a disintegrated control between the torque and the rotor excitation current is achieved. Likewise, in the vectorial control system of a DFIM, the *d* and *q* axis components of the rotor current are controlled.

3.2 Maximum power point control

The most commonly used wind turbine control strategy is illustrated in **Figure 5**, and consists of four operation zones, this shows the wind speed as a function of the wind speed [27]. This resembles an operation at full load condition. Here, the

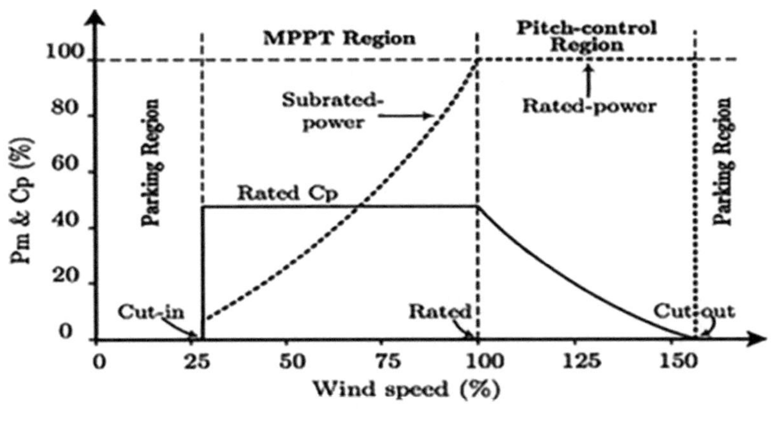

Figure 5.
The operation zones for power point tracking for wind turbine [27].

mechanical power can be restricted moreover by changing the pitch or using torque control. Usually, the electromagnetic torque is retained at an insignificant value and regulates the pitch angle to retain the wind turbine at extreme speed to maintain power output at a higher than rated wind speed.

The maximum power deviation with the rotational speed of DFIM is pre-established for every individual wind turbine. Owing to the intermittent character of the wind, it is vital to comprise a control unit to be able to follow maximum peak irrespective of the wind speediness. Due to the adaptive tracking and self-tuning capabilities, the two best MPC control methods are described in [27] as indirect speed control and direct speed control. The direct speed controller (DSC) as shown in **Figure 6** follow the maximum power curve more narrowly with rapid dynamics. Observing the description of the tip speed ratio, the optimum VSWT rotating speed Ω_{topt} may be established from the wind speed V_w, whereas T_{em} is the turbine electromagnetic torque, Ω_m is rotating speed, T_{t_est} is assessed turbine aerodynamics torque, and $P_{\max}$ is the maximum powering point.

$$\Omega_m^* = N\sqrt{\frac{T_{t\_est}}{k_{opt\_t}}} \tag{29}$$

$$P_{\max} = \frac{1}{2}\rho\ \pi\ R^5 \frac{\left(C_{p\_\max}\right)}{\left(\lambda_{opt}^3\right)} \left(\Omega_m^*\right)^3 \tag{30}$$

$$P_{MPPT} = K_{opt}\left(\Omega^*{}_m\right)^3 \tag{31}$$

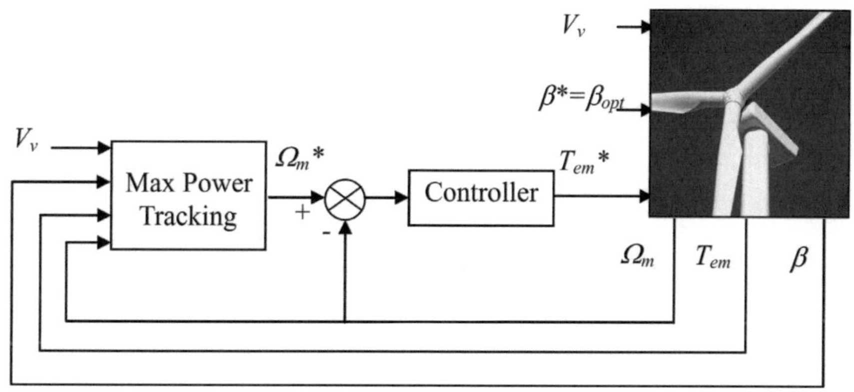

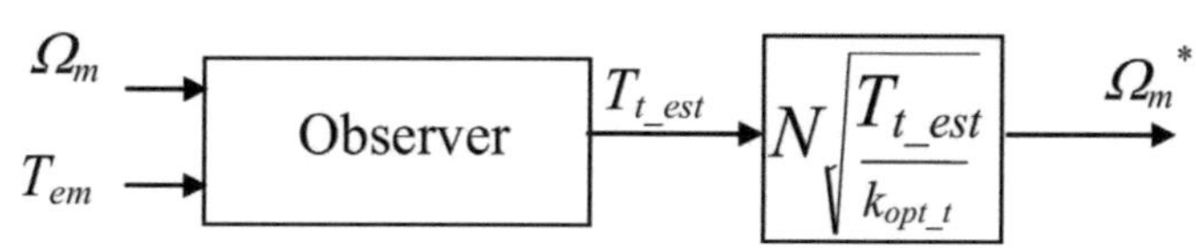

Figure 6.
Direct speed control [27].

3.3 Rotor side control

The DFIG rotor variable's orientation must follow the orientation of the selected orientation parameter. Here, two algorithms are implemented, stator side voltage aligned control and stator side flux aligned control. Once this parameter vector is computed in the rotor side orientation structure, its comparative angle (δ) to the rotor side orientation structure is designed and the rotor parameters are altered into the novel control-reference structure, as shown in **Figure** 7 [21].

Vector control is applied to the rotor side converter to control the stator's active and reactive power. The direct axis loop is used to control reactive power whereas the quadrature axis is for active power control. The rotor converter obtains evaluations of rotor circuit parameters and is accountable for handling the reactive power flow between the stator and the power grid as well as regulating the generator torque. Its input parameters are not associated with the stator orientation structure. Though, it is exactly the stator target that must be measured and controlled. For the RSC to calculate an output stable with the stator's parameters, the rotor parameters articulated in the rotor d-q orientation structure must be revolved to be oriented with the control orientation structure. The RSC controller MATLAB block diagram is shown in **Figure 8.**

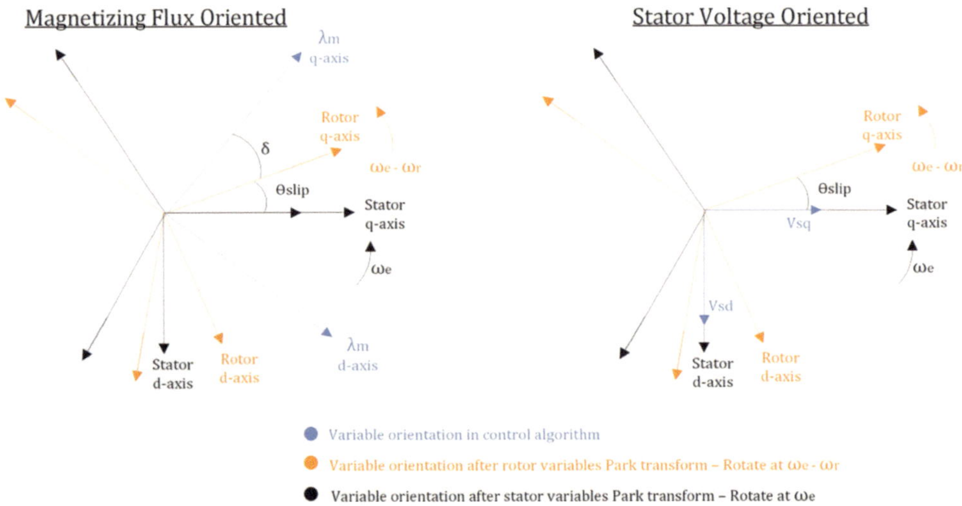

Figure 7.
Reference frames used in park transform [21].

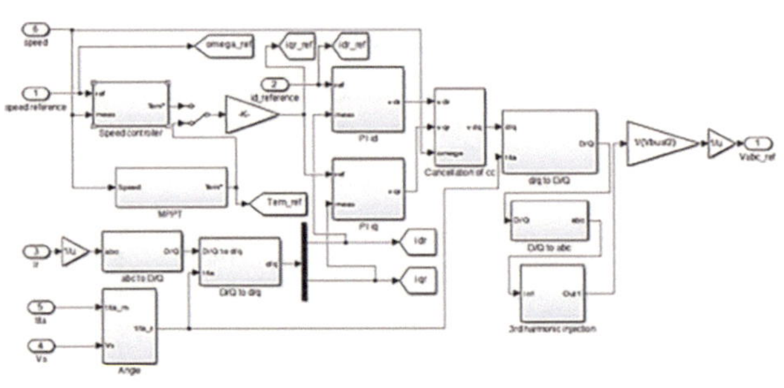

Figure 8.
RSC controller MATLAB block diagram.

The equations used in the orientation process are [21]:

$$S_{r\ d} = S'_{r\ q}\sin(\delta) + S'_{r\ d}\cos(\delta) \tag{32}$$

$$S_{r\ q} = S'_{r\ q}\cos(\delta) - S'_{r\ d}\sin(\delta) \tag{33}$$

Where:

- S' represents a d-q rotor variable expressed in the rotor reference frame rotating at slip frequency.
- S represents the same rotor variable oriented along with the control reference frame.
- δ represents the alignment angle by which the rotor reference frame must be rotated.

3.4 Grid side control

The main objective of the grid side converter control model with ideal bidirectional switches as shown in **Figure 9** is to focus on the active and reactive powers delivered to the grid, keeping a constant DC-link voltage independent of the value and direction of the rotor power flow, and grid synchronization control. The grid side of the wind turbine system is composed of the grid side converter, the grid side filter, and the grid voltage. It converts voltage and currents from DC to AC, while the exchange of power can be in both directions from AC to DC (rectifier mode) and from DC to AC (inverter mode). The $d - q$ axis voltage V_{gd} and V_{gq} of the grid side converter from the original three phases V_{ga}, V_{gb}, V_{gc} is as below [27]:

$$V_{g\ d} = V_{g\ d} + R_g I_{g\ d} + L_g \frac{dI_{g\ d}}{dt} - \omega_s L_g I_{g\ q} \tag{34}$$

$$V_{g\ q} = V_{g\ q} + R_g I_{g\ q} + L_g \frac{dI_{g\ q}}{dt} + \omega_s L_g I_{g\ d} \tag{35}$$

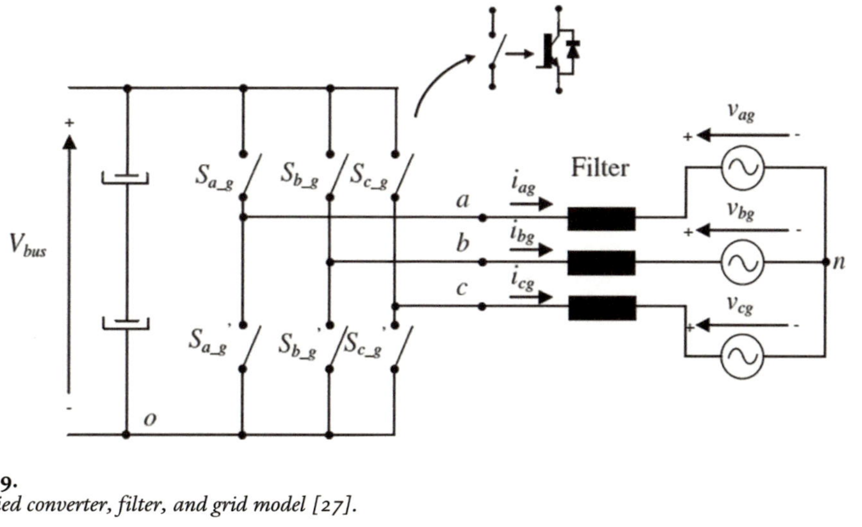

Figure 9.
Simplified converter, filter, and grid model [27].

4. Simulation and discussion

4.1 Steady-state simulation

The system under study is shown in **Figure 10**. The 2 MW, 1500 rpm, 50 Hz, 690 V, 1760 A, and 12,732 Nm torque DFIG model was used to model and simulate the systems. The simulation was used to analyze the challenges with power system stability of integrating the WTG into the grid, considering intermittent wind characteristics and the problem of slip convergence. This task was executed by creating a steady-state Matlab function, to calculate the steady-state operation points and reveals how the rotor speed of the modeled DFIG involves the power flow of the studied system. Speed array and torque array were considered as inputs into the stimulated three-blade wind turbine connected with DFIG. Two different work frames of generation strategies $Q_s = 0$ and $I_{dr} = 0$ were considered here.

4.1.1 Steady-state simulation results and analysis

Simulations were carried out for variable wind speeds ranging from 5 m/s (cut-in speed) to 25 m/s (cut-off speed) in progressive steps of 2 m/s, with reactive power $Q_s = 0$, (red plot) and $I_{dr} = 0$ (green plot) as a control strategy separately. The results for the DFIG voltage, torque, generated real power, efficiency, and consumed reactive power in both the methods are plotted in **Figures 11–18**. The influence of two different generation strategies does not make big differences for variables such as T_{em}, P_t, P_s and P_r, however, some other variables such as I_s, I_r, Q_s, and Q_r, (**Figures 14–17**) was found to have some big differences in amplitudes, concerning rotor speed.

Figure 11 shows the DFIG's torque vs. speed characteristics, which stimulate the three-blade wind turbine with a minimum speed of 900 rpm and a maximum speed of 1800 rpm. The DFIG can perform above and under the synchronization speed for power generation. The generation model of DFIG matching negative torque values

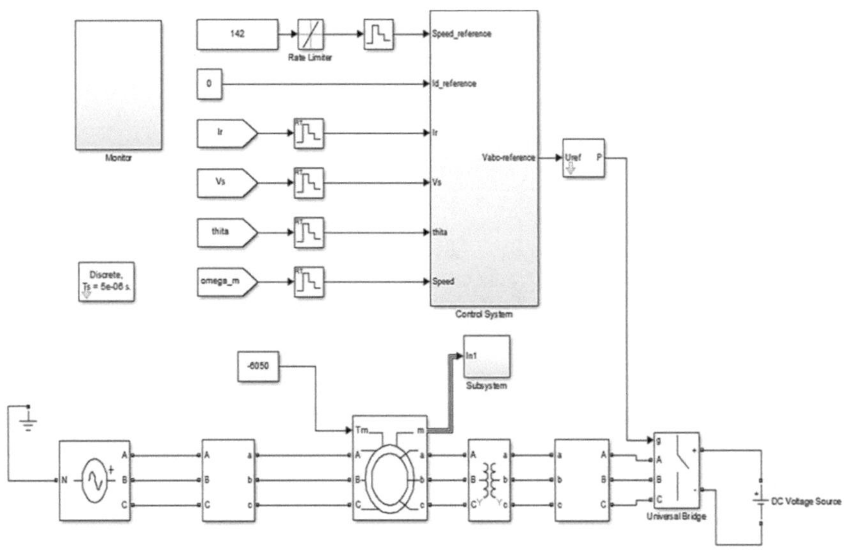

Figure 10.
Steady-state simulation program block model.

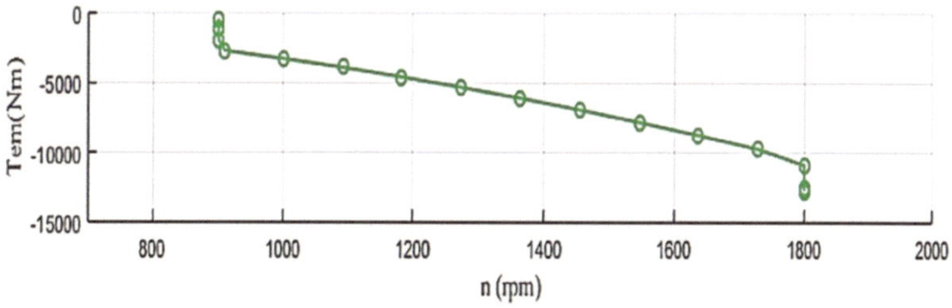

Figure 11.
The graph of torque (T_{em}) vs. rotor speed (n).

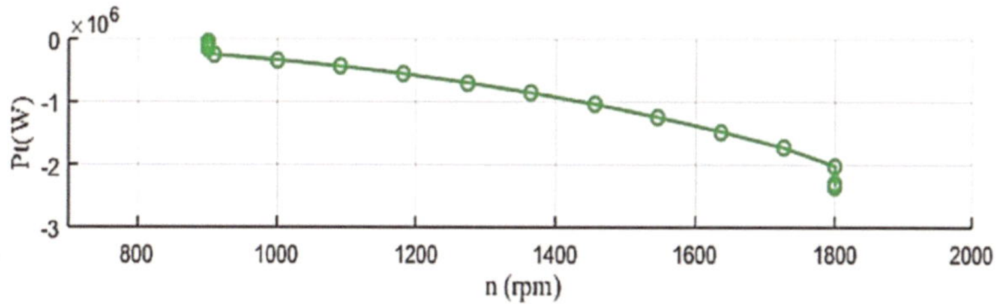

Figure 12.
DFIG's active power Pt (W) vs. speed n (rpm).

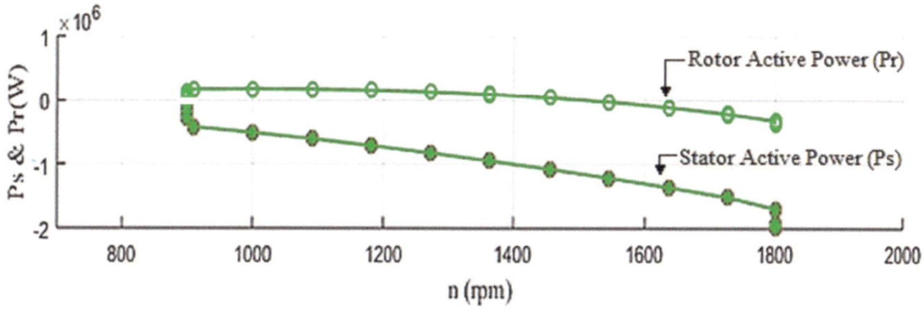

Figure 13.
DFIG stator and rotor active power Ps & Pr (W) vs. speed n (rpm).

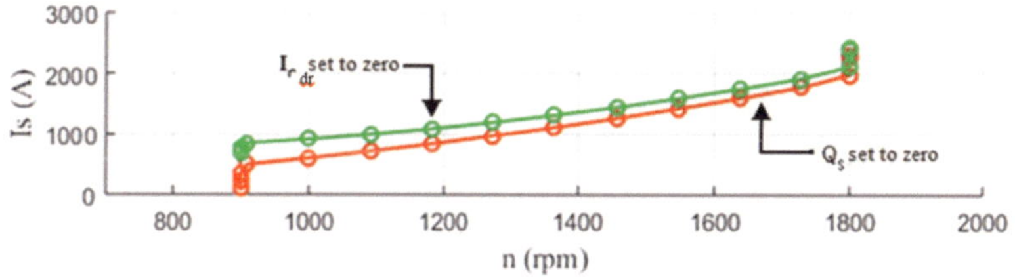

Figure 14.
DFIG I_s (A) vs. n (rpm), red plot: Qs = 0, green plot: I_{dr} = 0.

covers from the negative slip to the positive slip state. Therefore, the turbine target power and electromagnetic torque features of variable speed DFIGs are unlike the customized constant-speed induction machine. **Figure 12** shows the plotting for the total mechanical power of the turbine shaft, which is the product of torque and speed, from sub synchronous to super synchronous speed, with a maximum power value of

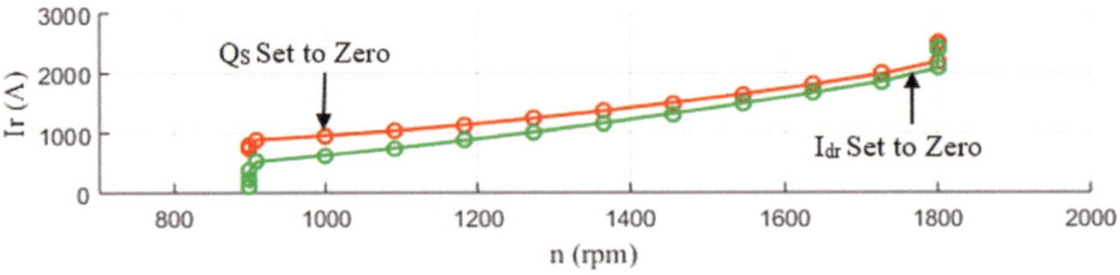

Figure 15.
DFIG I_r (A) vs. n (rpm), red plot: Qs = 0, green plot: I_{dr} = 0.

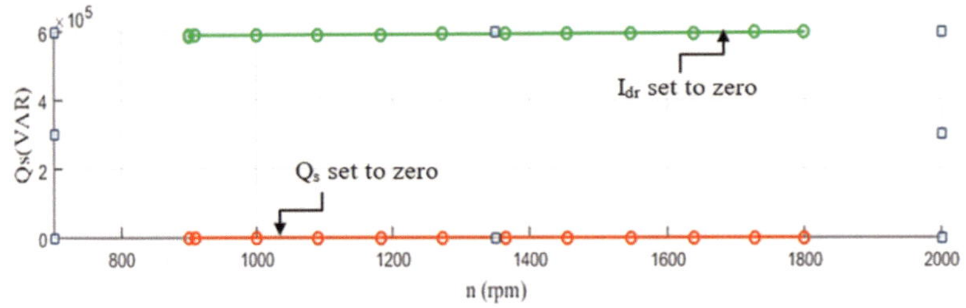

Figure 16.
DFIG Qs (VAR) vs. n (rpm), red plot: Qs = 0, green plot: I_{dr} = 0.

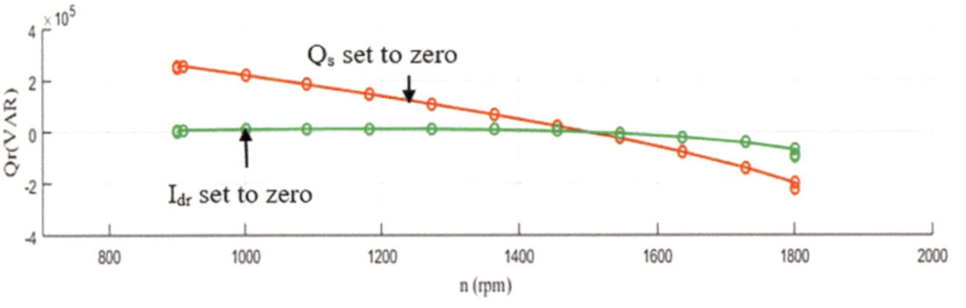

Figure 17.
DFIG Qr (VAR) vs. n (rpm), red plot: Qs = 0, green plot: I_{dr} = 0.

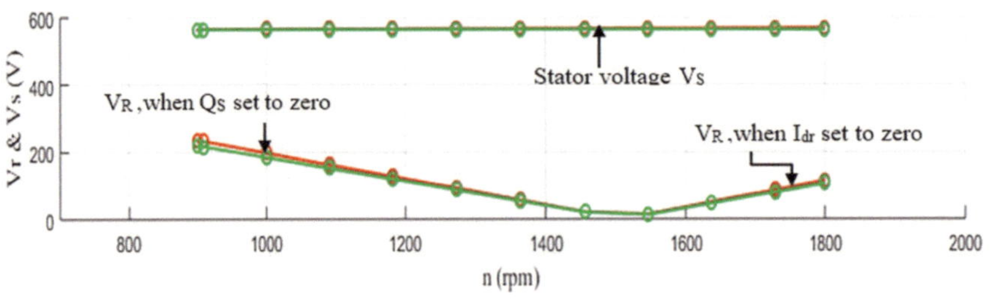

Figure 18.
DFIG Vr & vs vs. speed n (rpm), red plot: Qs = 0, green plot: I_{dr} = 0.

−2.54 MW at 1800 rpm. **Figure 12** shows, the rotor's active power P_r is absorbed by the induction machine at below synchronous speed, and the active power is supplied above the synchronous speed from the induction generator to the grid. **Figure 13**, shows, with $Q_s = 0$ (red plot) as an adopted control strategy the stator current value I_s is on the lower side. **Figures 14** and **15** shows during $I_{dr} = 0$ (green plot) as a control

Sn	Speed		Torque (N-m)		I_r		I_s		V_r	
	Steady-state	Simulation Model	Steady-state	Simulation Model	Steady-state	Simulation Model	Steady-state	Simulation Model	Steady State	Simulation Model
1	1182	1182	−4550	−4550	875	920	1090	1100	110	100
2	1356	1356	−6050	−6000	1160	1200	1325	1350	60	60
3	1691	1691	−9450	−9455	1770	1790	1850	1860	73	75

Table 1
WT simulated parameters at defined DFIG speeds.

strategy the rotor current I_r is on the lower side and Q_s is on the higher side. **Figure 16** shows the rotor reactive power $Q_r = 0$ at synchronous speed 1500 rpm, with both control strategy $Q_s = 0$ and $I_{dr} = 0$, indicating the reactive power varies according to the wind turbine speed. **Figure 17** shows, a constant stator voltage V_s amplitude throughout the variable speed range, while the variable rotor voltage V_r amplitude is very low at synchronous speed 1500 rpm, with two peak voltage amplitudes at a minimum and maximum rotor speeds.

4.1.2 Evaluation of simulation modeling at defined speeds

Table 1 shows the specified wind turbine DFIG speeds that are compared and used to evaluate the simulation model parameters with the steady-state model parameters as obtained from **Figures 11–18**. The simulation graphs shown in **Figures 19** and **20** represent the torque vs. time and rotor current vs. time characteristics at 1356 rotor rpm and steady-state simulation period of 1.5 sec and 1691 rotor rpm at a steady-state simulation period of 2.0 sec for the entire modeling period of 3.0 sec. Simulated torque values −6050 Nm and − 9450 Nm, rotor current values 1200 amps and 1790 amps, and stator current values 1325 amps and 1850 amps at pre-defined speeds are close to steady-state parameter values as shown in **Figures 11–18**.

4.1.3 Simulation model of DFIG using wind turbine MPPT block

In this section, a 2 MW stator power DFIG model and a three-blade wind turbine model with gear ratio $n = 100$, blade radius 42 m, $C_p = 0.42$, and $\lambda_{opt} = 7.2$ were used for the wind turbine maximum power point tracking simulation control as shown in **Figure 21**. **Figures 22** and **23** show the wind turbine MPPT simulation model characteristics at 8 m/sec and 10 m/sec of wind speed. Observed in **Figure 11**, the torque response for the wind speed and i_q current indicates that more oscillations occur at the low torque due to the fact reduced mechanical inertia. The more the mechanical inertia, the more the torque oscillations. On achieving the steady-state condition at

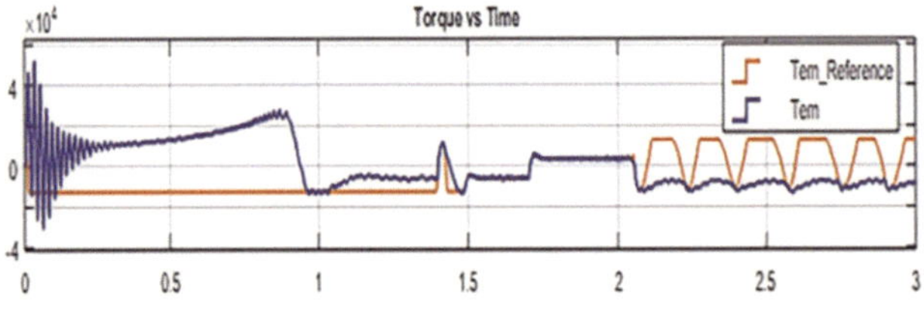

Figure 19.
Torque vs. time graph @ 1356 rotor rpm and @ 1691 rotor rpm.

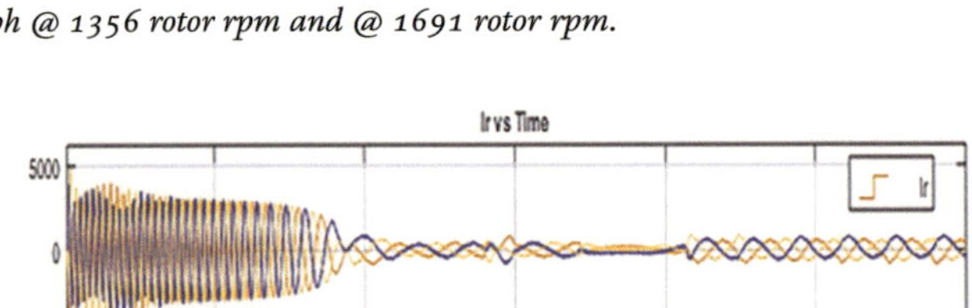

Figure 20.
I_r vs. time graph @ 1356 rotor rpm and @ 1691 rotor rpm.

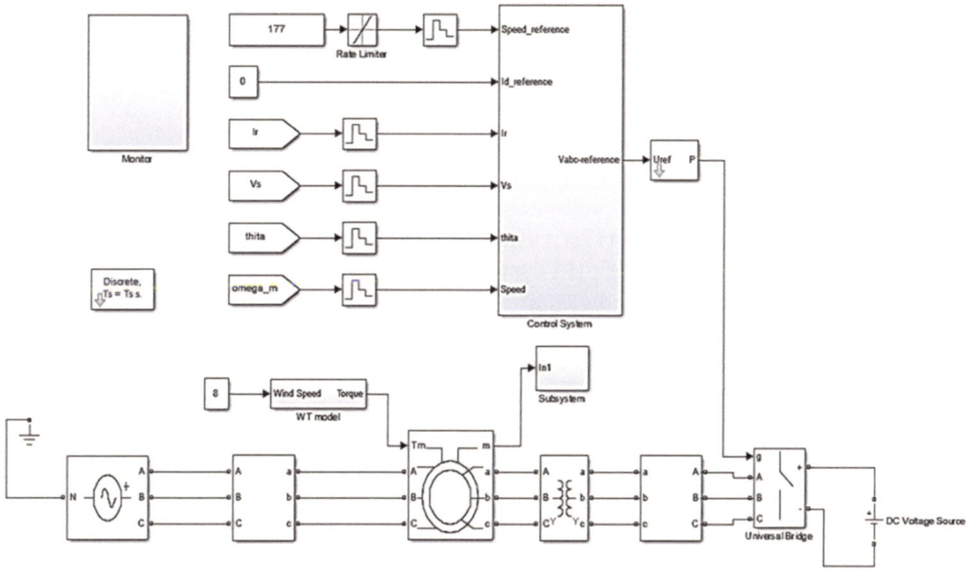

Figure 21.
WT MPPT control model.

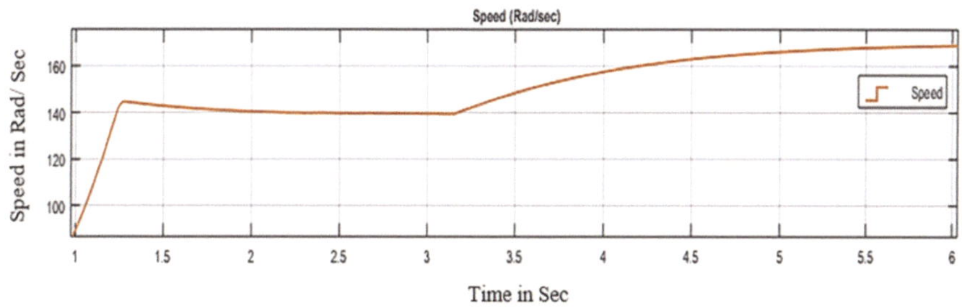

Figure 22.
The dynamic state WT MPPT graph of speed vs. time (sec).

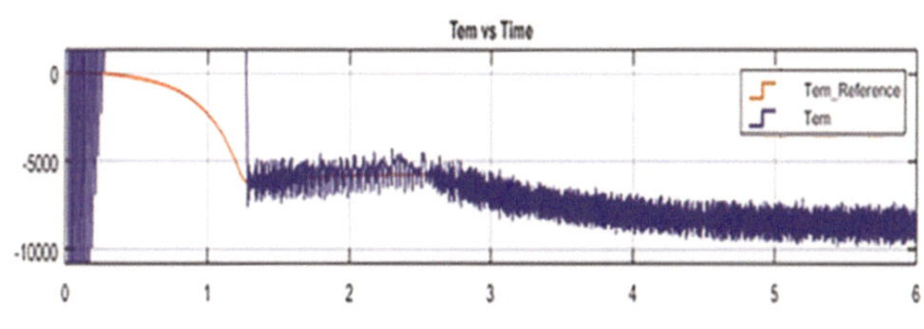

Figure 23.
WT MPPT torque vs. time characteristic curve.

2 sec of modeling time, wind turbine speed and correspondence torque values were tabled for angular speed of 140 rad/sec, at a torque of −5500 Nm, and a mechanical wind turbine power output approximately equal to 770 kW was obtained. Further, the wind speed was increased from 8 m/sec to 10 m/sec and the steady-state simulation at 6 sec of modeling time was observed. On achieving the steady-state at 6 sec of modeling time, wind turbine speed and correspondence torque values were tabulated for 170 rad/sec, at −8800 Nm respectively and a mathematical wind turbine power

outcome equivalent to 1.49 MW was recorded. These two simulated outcomes are very close to the steady-state characteristics graph numerical values as shown in **Figures 11** and **12**.

5. Conclusions

This study was to focus on investigating the influences of the integration of wind power generators into the power grid systems. The rotor side converter control unit is utilized for, real and reactive power control by regulating the rotor current and the speed of the DFIG. With the computed stator voltage, stator current, rotor current, and the rotor location by encoder response signal the active PI measured and controlled procedure results in a considerable enhancement in control system sturdiness and advances its indemnity to produced system noise. The engaged PI control unit attests to the grid side converter control by sustaining the stable generated power frequency and voltage with the grid frequency and voltage. The controller scheme and the simulation mode controller employed for the study assure the wind generator supplying into the grid at varying wind speeds behaves like a synchronous generator, at a zero Hz rotor frequency.

Acknowledgements

I wish to thank the almighty God for giving me life and enabling me to reach the heights that I have reached.

I wish to thank my parents and my siblings, for their tireless and relentless love, continuous support, and the countless sacrifices they have made on my behalf. To my family, for being great inspirations and believing in us even when we have stopped believing in ourselves. This would not have been possible without the family's help.

Finally, I wish to everyone not mentioned above but directly or indirectly contributed to our work, your input is much acknowledged.

Abbreviations

DFIGs	Double Fed Induction Generators
DSC	Direct Speed Control
ESO	Extended State Observer
GSC	Grid Side Converter
IEA	International Energy Agency
IRENA	International Renewable Energy Agency
MMF	Magnetomotive Force
MPPT	Maximum Power Point Tracking
PCC	Point of Common Coupling
RSC	Rotor Side Converter
SARGEC	The South African Renewable Energy Grid Code
VSCF	Variable Sped Constant Frequency
WECS	wind energy control System
WT	Wind Turbines
WTGCS	Wind turbine generators control system

C_f	Turbine blade density constant
C_p	Coeficient of performance
E	Kinetic energy of air mass of m kg
K_s	Stiffness constant
D_s	Damping constant
P_s	Stator active power in watts
P_r	Rotor active power in watts
Q_s	Stator reactive power in watts
Q_r	Rotor reactive power in watts
R	Turbine blade radius in meters
T_e	Per-unit electromagnetic torque in d-q park reference newton/meter
Tf	Turbine friction torque in newton/meter
T_{mmax}	Turbine maximum torque in Newton/meter
$V_m(t)$	Wind Speed in meter/sec
$V_w(t)$	Wind Speed of air mass of m Kg in meter/sec
$V_{wt}(t)$	Wind Speed turbulence component speed in meter/sec
$V_{wa}(t)$	Wind Speed constant component speed in meter/sec
$V_{gt}(t)$	Wind Speed gust component in meter/sec
ρ	Air density in Kg/m^3
β	Turbine blade pitch angle
λ	Turbine blade tip speed ratio
ω_{tur}	Turbine rotational angular speed in mechanical radian/sec
ω_{slip}	Slip frequency
ω_{mmf}^{stator}	Slip frequency corresponding to the frequency of rotor current and voltage in radian/sec
ω_{rotor}	Rotor rotating frequency in radian/sec
ωr	Rotor speed in radian/sec
ωg	Generator speed in radian/sec
Ψ_s	Stator flux linkage in synchronous frame weber
Ψ_r	Rotor flux linkage in synchronous frame weber

Author details

Ramesh Kumar Behara[1*] and Kavita Behara[2]

1 University of KwaZulu Natal, Durban, Republic of South Africa

2 Mangosuthu University of Technology, Durban, Republic of South Africa

*Address all correspondence to: bramesh250770@gmail.com

References

[1] India Energy Outlook. India Energy Outlook 2021 – Analysis - IEA. India Energy Outlook, 2021

[2] Renewable I and Agency E. Renewable Energy Statistics 2021. 2021

[3] European Wind Energy Association and Greenpeace. WIND FORCE 12 a blueprint to achieve 12% of the world's electricity from wind power by 2020. Wind Force. 2005;**12**(June 2005):1-52

[4] Dr P, Li M. World energy 2017-2050: Annual report. World Energy Reports. 2017;(June):1-29

[5] Scholtz L, Muluadzi K, Kritzinger K, Mabaso M & Forder S. Renewable Energy: Facts and Futures - The energy future we want. 2017

[6] AbdElhafez AA, Alruways SH, Alsaif YA, Althobaiti MF, AlOtaibi AB & Alotaibi NA. Reactive Power Problem and Solutions: An Overview. Journal of Power and Energy Engineering. 2017;**5** (5):40–54

[7] Robert LMT, Gabler E, Petersen JF. Atmospheric pressure, winds, and circulation patterns. Weather. 2008: 113-137

[8] Wang T. A brief review on wind turbine aerodynamics. Theoretical and Applied Mechanics Letters. 2012;**2**(6):1-8

[9] Ragheb M, Ragheb AM. Wind turbines theory - the Betz equation and optimal rotor tip speed ratio. Intechopen. 2011;**1**(1):1-22

[10] Mcdonald JD, Wojszczyk B, Flynn B, Voloh I. Distribution systems, substations, and integration of distributed generation. In: Encyclopedia of Sustainability Science and Technology. Springer Science+Business Media New York; 2013. pp. 1-63

[11] Sewchurran S, Davidson IE. Introduction to the south African renewable energy grid code version 2.9 requirements (part III - discussions and conclusions). IEEE AFRICON: Science, Technology and Innovation for Africa. 2017;**17**(4):1231-1235

[12] Jabal Laafou A, Ait Madi A, Addaim A, Intidam A. Dynamic modeling and improved control of a grid-connected DFIG used in wind energy conversion systems. Mathematical Problems in Engineering. 2020;**2020**(10):1-15

[13] Monteiro CP and Da Costa JS. Simulation and Control of a Dfig-based Wind Turbine. Res. gate NECEC. 2020. no. December, p. 9

[14] Boukili Y, Aguiar AP, Carvalho A. A DFIG-based wind turbine operation under balanced and unbalanced grid voltage conditions. IFAC-PapersOnLine. 2020;**53**(2):12835-12840

[15] Akrama K, Xiao Ming H, Mohamed AK, and Paul B. Doubly Fed Induction Generator Open Stator Synchronized Control during Unbalanced Grid Voltage Condition. 2020;**13**(3155):1-13

[16] Eltamaly AM, Al-Saud M, Sayed K, Abo-Khalil AG. Sensorless active and reactive control for DFIG wind turbines using opposition-based learning technique. Sustain. 2020;**12**(9): 1-14

[17] Slootweg JG, Polinder H, Kling W. Dynamic modelling of a wind turbine with doubly fed induction generator. IEEE Power Engineering Society

Transmission and Distribution Conference. 2001;**01**(9):644-649

[18] Singh M. Dynamic Models for Wind Turbines and Wind Power Plants. 2011

[19] Zhang Y, Zhang L, Liu Y. Implementation of maximum power point tracking based on variable speed forecasting for wind. PRO. 2019;**158**(7): 1-18

[20] Abo-Khalil AG. Modeling and Control Aspects of Wind Power Systems. Impacts of wind farms on power system stability. Wind Power In Power System. 2013:133-151

[21] Jonathan Fournier. Modeling, Control and Experimental Validation of a DFIG-based wind Turbine test bench. Universitat Politecnica de Catalunya, Barcelona, Catalonia Institute for Energy Research (IREC). 2013

[22] Badreldien M, Usama R, El-wakeel A, Abdelaziz AY. Modeling, Analysis, and Control of Doubly Fed Induction Generators for Wind Turbines. ICEENG Conference ICEENG-9. 2014:1-18

[23] Arnaltes S, Rodriguez-Amenedo JL, Montilla-D Jesus ME. Control of variable speed wind turbines with doubly fed asynchronous generators for stand-alone applications. Energies. 2018;**26**(11):1-16

[24] Soriano LA, Yu W, Rubio JDJ. Modeling and control of wind turbine. Mathematical Problems in Engineering. 2013;**2013**:1-13

[25] Prabakar K. Gain Tuning of Proportional-Integral Controller Based on Multiobjective Optimization and Controller Hardware-in-loop Microgrid set up. 2015

[26] Michalke G, Hansen AD, Hartkopf T. Control strategy of a variable speed wind turbine with multipole permanent magnet synchronous generator. European Wind Energy Conference and Exhibition. 2007:1371-1378

[27] Abad G, Lopez J, Rodriguez M, Marroyo L, Iwanski G. Doubly Fed Induction Machine. 2011th ed. Hoboken, New Jersey, Wiley Publication: John Wiley & Sons, Inc.; 2011

Chapter 11

Wind Turbine Aerodynamics and Flow Control

Karthik Jayanarasimhan and Vignesh Subramani-Mahalakshmi

Abstract

Aerodynamics is one of the prime topics in wind turbine research. In aerodynamics, the design of a flow control mechanism lays the foundation for an efficient power output. Lift generation in the airfoil section leading to rotary motion of blade and transfer of mechanical to electrical power generation through gearbox assembly. The primary objective of a flow control mechanism in wind turbine blades is to delay the stall and increase the lift, thereby an efficient power generation. Flow control is classified into active and passive flow control mechanisms. Active flow control works on an actuation mechanism that comes into action when required during varied operating conditions. Passive flow control devices are designed, developed, and fixed on the surface to extract the required output through effective flow control. Vortex generators are the simplest, most cost-effective and efficient passive flow control devices. These devices influence the power of wind turbine blades in various ways, such as placement of generator along the chord, distance between pairs of a generator, angle of inclination of a generator with the blade surface, the height of generator. Flow control device needs to be optimized with the aforementioned parameters for efficient stall delay and power generation.

Keywords: aerodynamics, wind turbine, boundary layer, flow control, vortex generator

1. Introduction

Wind is an abundant resource available in the earth's atmosphere, and the need for renewable energy is demanding due to climate change and the energy crisis. Wind energy is low carbon footage leads to importance in research increasing the efficiency and use of the wind resource even in low wind speed. In the case of renewable and carbon-free emission energy production; firstly, solar power gains less attraction due to the less efficient and cannot produce energy on a night or cloudy days. Secondly, hydropower depends on rainfall; has a high impact on river ecosystems and forest environments. Additionally, tidal power and geothermal energy are far away from mass energy production. At last, carbon-free energy production can be achieved in nuclear energy but gain a vast life risk during a disaster and handling nuclear waste is a big challenge. Therefore, wind energy gains significance in the technological and political community in fighting climate change without compromising the modern depend and national economy.

In Denmark, 28% of wind energy is generated at total consumption in 2018 [1], and wind energy capacity almost doubled in 2020, where China had a major part of 72 GW [2]. The wind farm located in China (Gansu Wind Farm) with a capacity of 7965 MW is the world's largest and the second-largest is located in India (Muppandal Wind Farm) with a capacity of 1500 MW [3]. Wind energy production is increasing globally by installing wind turbines in large offshore farms located in agricultural lands, valleys and hills. In addition, onshore wind turbines on the sea bed and new initiatives for installing wind turbines in urban areas (University Campus or highway street lights) [4].

The power extraction from the wind is by converting the wind energy into useful mechanical energy by rotating the turbine or through vibration. The latest research trends in wind energy are in the construction of horizontal wind turbines (liftbased rotation), vertical wind turbines (drag-based rotation) and bladeless wind turbines (aero-elastic-based vibration). The other significant research focused on the pattern of sitting wind to gain more aerodynamic efficiency to get more power output in farm and urban areas, the aerofoil and flow control mechanism in the blade increase the power output efficiency and decrease the cutoff wind velocity. The major challenges in wind energy are turbine transportation and installation, especially in the hilly area, bird's attack in turbines, the need for extensive land acquisition and recycling of retired wind turbines.

1.1 Wind and wind turbine history

The wind played an important role in ancient civilization in developing sailing boats, kites, agriculture, and metrology. In the ancient period, there are a lot of myths about wind being raised and a hole in the sky which blew it from the sky to earth. In Greek mythology, the God of the Sea, Aeolus, is a guardian of the wind. Feng Po (Wind God) had a sack with an opening that controlled the wind in China. In 3500 BC, Egyptians used wind power to sail the boat in the Nile river, and in Persia, 500 BC millstone, the water pump is driven using wind power. In 1300–1850 AD windmill was designed for water pumping and large-scale milling, which is similar to modern wind turbine design [5]. In 1887, a wind turbine was firstly used to generate electricity built by Prof. James Blyth in Scotland. In 1900, 30 MW of power was generated with around 2500 windmills in Denmark. A Smith Putnam 75—feet wind turbine blade generated 1.25 MW of power for local energy needs gained colossal importance and possibilities in the wind energy sector. In 1975, a wind turbine was developed by NASA—with a composite material blade with pitch control, steel tube tower installed with aerodynamics and structural design ignited more possibilities in the max power output and led to building large wind turbines for energy production [6]. Today, the Sea Titan three-bladed wind turbine can generate 10 MW of power with a rotor diameter of 190 m.

2. Aerodynamics basics

Aerodynamics is a branch of fluid dynamics, the study of the motion of air with forces and moments that act on the body. Aerodynamics plays a vital role in the flight

DOI: http://dx.doi.org/10.5772/intechopen.103930

of the aeroplane and helicopter, rocket technology, designing high speed and fuel-efficient cars, reducing the drag on the athlete in sports events and a lot more engineering applications. For example, the aerodynamics of the wind turbine is an important area to increase power output and design a large turbine blade.

2.1 Aerodynamics forces and moments

The forces and moments on the body are due to pressure and shear stress distribution (**Figure 1**). The pressure acts perpendicular to the surface, which acts as a load on the wind turbine and shear stress is the frictional force tangential to the surface. The pressure difference between the bottom of the blade and the top of the blade generates the lift force (Eq. (1)) (perpendicular to freestream velocity) the wind turbine blade generates the power by rotating the generator.

$$\mathbf{L} = \frac{1}{2}\rho V^2 S C_L \tag{1}$$

Where,
L = lift force (N)
ρ = density of air (Kg/m^3)
V = wind velocity (m/s)
S = blade span area (m^2)
C_L = coefficient of lift

The lift force on the wind turbine blade is proportional to the square of the wind velocity gains essential parameters in the wind energy generation. The blade span area depends on the length and width of the blade throughout the cross-section and C_L depends on the shape (aerofoil selection) and orientation (pitch angle) of the blade. Aerofoil is a streamlined, cross-sectional blade shape that produces high lift compared to other shapes. The aerofoils were first used in the aeroplane wings to generate lift and are now widely used for energy production. Terminologies define the shape of the

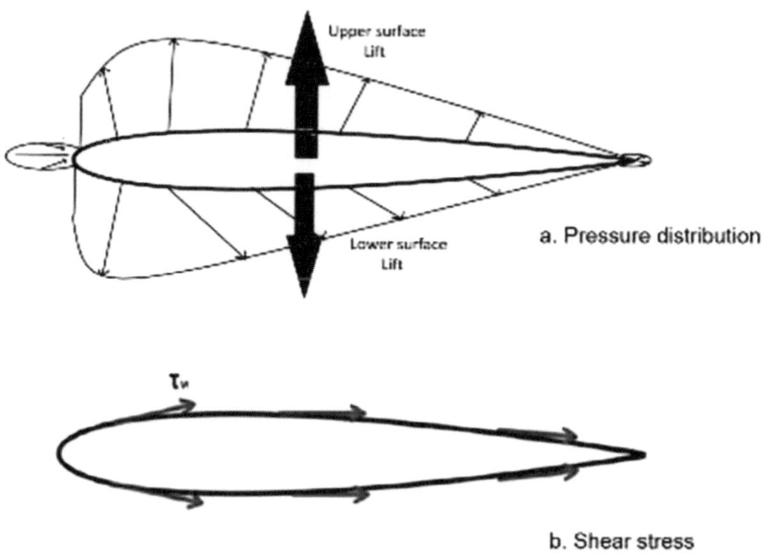

Figure 1.
Aerodynamic forces in the aerofoil.

aerofoil, the frontal part is the leading edge and rearward part is the trailing edge, and the chord line is the straight line connecting the leading edge to the trailing edge. A mean locus between the upper and bottom of the aerofoil is the mean camber line and the maximum distance between the chord and camber line is called camber. The well-known and basic aerofoil series is NACA aerofoil. NACA stands for National Advisory Committee for Aeronautics, which designed aerofoil with a series to understand the shape [7] easily. For example, in a NACA-4 series aerofoil, the first digit represents maximum camber at 0–9.5% chord, the second digit represents the location of maximum camber at 0–90% chord and the last two-digit represents the thickness of the aerofoil at 1–40% chord. NACA-0012 is a symmetrical aerofoil with zero camber at 12% chord thickness. NACA-2412 is asymmetrical aerofoil with 2% chord of maximum camber, location of camber at 40% of chord, and 12% thickness of chord.

2.2 Reynold's number

Reynolds number (Re) is a non-dimensional number used to predict the behavior of the fluid at varying environments and used to model the scale-down model [8]. The Reynolds number is named after Irish-born Osborne Reynolds, who predicted the different flow patterns by inducing die in the pipe flow. Reynolds number (Re) is the ratio of inertial force to viscous force (Eq. (2)).

$$\mathbf{Re} = \frac{\rho V D}{\eta} \tag{2}$$

Where,
Re = Reynolds number
ρ = density of air (Kg/m^3)
V = wind velocity (m/s)
D = characteristic length or diameter (m)
ή = dynamic viscosity of air (Pa.s/Kg m^{-1} s^{-1})

The flow pattern is differentiated into laminar flow and turbulent flow. Both possess different characteristics in nature. Laminar flow is a smooth and regular streamline pattern, whereas turbulent flow is a random and irregular flow pattern. The critical Reynolds number is 5×10^5 transition between the laminar to turbulent flow over a flat plate.

2.3 Boundary layer

In 1904, Ludwig Prandtl developed the theory boundary layer [9], the flow field around the body had two areas where flow is frictional and non-frictional. The boundary layer is the area where the friction of the flow is considered due to viscous characteristics. The thickness of the boundary layer is a distance between the surface to freestream velocity of flow, the velocity at the surface is zero (V = 0) due to shear stress with the surface and air. The velocity will increase with the increase in thickness (**Figure 2**) in the boundary layer and attain the freestream velocity on point and outside the boundary layer; the flow is considered a non-frictional flow regime. The laminar boundary layer is less thick than the turbulent boundary layer and the turbulent boundary layer will have high kinetic energy and mixing rate.

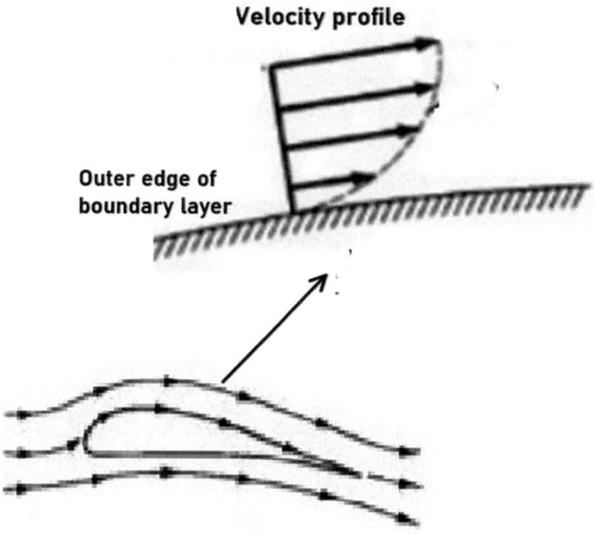

Figure 2.
Velocity profile in boundary layer.

2.4 Pressure coefficient

Pressure is a dimensional quantity (Eq. (3)) (SI unit N/m^2) and important variable to express the force that acts on the body. The pressure must be expressed in the dimensionless quantity pressure coefficient (C_p) like Reynolds number for similarity in aerodynamics.

$$C_p = \frac{P - P_\infty}{q_\infty} \tag{3}$$

To measure the pressure coefficient, the pressure tapping is distributed around the model's surface in the wind tunnel. To measure the pressure coefficient, on the surface of the model in the wind tunnel the pressure tapping will be distributed around the surface. The tubes will be connected to multi-tube manometer or pressure sensors to measure the pressure difference at the tappings (p) in the surface and the free stream pressure (p_∞). The dynamic pressure is measured through freestream quantity ($(q_\infty = 1/2\rho_\infty V_\infty^2)$, where ρ_∞ is freestream or sea-level density and V_∞ is freestream velocity.

2.5 Generation of lift

Aerodynamics lift is a complex topic for understanding, the lift generated by wings made the heavier than air flight possible. There is much debate on how the wing or turbine creates lift with aerofoil cross-sections. When the fluid flow over an object, the force exited due to the fluid motion where the lift is perpendicular to the freestream and drag is parallel to the freestream. Concentrating on the lift produces a high lift with minimum drag on the streamlined body like an aerofoil.

The aerofoil shape is used in aeroplane wings, wind turbines and propellers to generate the lift and based on the application and need the different aerofoil profiles are used. Consider a wind turbine aerofoil where the wind flow over it causes a pressure distribution with high pressure in the bottom and low pressure on the top cause a lift generation on the turbine to rotate the generator to produce electricity. The shape of the aerofoil creates an uneven pressure when fluid moves over it to generate the lift, but how is the uneven pressure distribution formed on the aerofoil? It is a tricky question to answer. We discuss two widely accepted explanations of lift generation in the aerofoil. The following explanation is based on Newton's third law of motion, where the fluid nature is considered in lift generation. When fluid flows over an aerofoil, the fluid will suddenly experience the aerofoil where the flow moves upward, called upwash and downward called downwash. Due to the large fluid volume displacement, every action has an equal and opposite reaction, the aerofoil creates lift as a reaction force by turning down the incoming air. In conclusion, the lift is created due to uneven pressure distribution, but the pressure distribution is complex and has a different explanation based on the approach.

We will now discuss how the aerofoil shape and orientation affect lift generation. At freestream velocity V (relative wind) over an aerofoil will generate a lift, drag and moment due to pressure and shear stress distribution. The angle of attack (ρ) is the angle between chord (c) and the relative wind velocity (V_{∞}) of the aerofoil. The coefficient of lift (C_L) will increase with an increase in the angle of attack till C_{Lmax} and stalls (lift decrease) due to flow separation on the upper surface. The symmetrical aerofoil (NACA 0012) has a similar shape on both sides of the chord line and $C_L = 0$ when the angle of attack is zero because pressure distribution will same on both but asymmetrical aerofoil (NACA 4412) will generate lift even at a zero angle of attack (**Figure 3a**). There are two kinds of stalls based on the aerofoil thickness: leading-edge and trailing-edge stall. Let us compare NACA 4412 and NACA 4421 (**Figure 3b**), both aerofoils have the same mean camber line and camber location but the thickness varies, NACA 4421 have 10% extra thickness to NACA 4412. In both cases, the aerofoil has the same lift slope, C_L increases with increasing angle of attack but C_{Lmax} various. In NACA 4412, the flow separation occurs at the leading edge of the aerofoil where the stall will be sudden and cover the entire upper surface, the phenomenon is known as a leading-edge stall. On the other hand, NACA 4421, where the aerofoil thickness is high enough to make the separation occur in the trailing edge and stall will be gradual. The lift curve evident that the NACA 4412 C_{Lmax} is increased little compared to NACA 4421 where the curve bend over at C_{Lmax} means that stall was soft and gradual at maximum lift (**Figure 3b**).

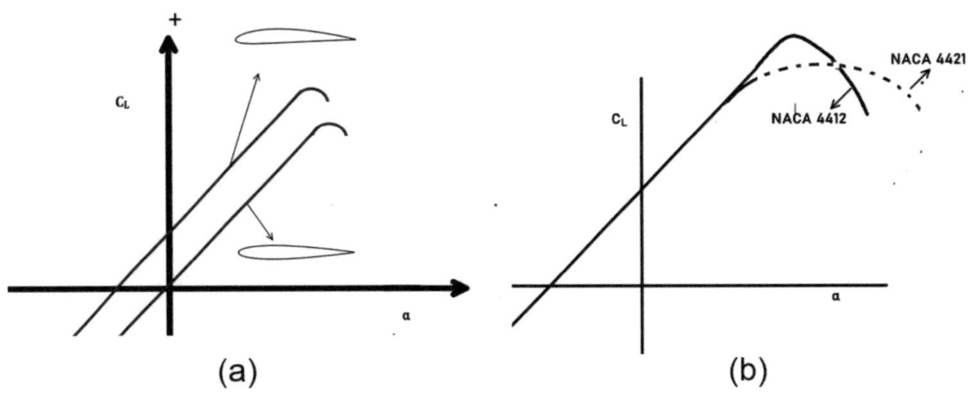

Figure 3.
(a) C_L vs. α for symmetrical and asymmetrical aerofoil. (b) C_L vs. α for thin aerofoil (NACA 4412) and thick aerofoil (NACA 4421).

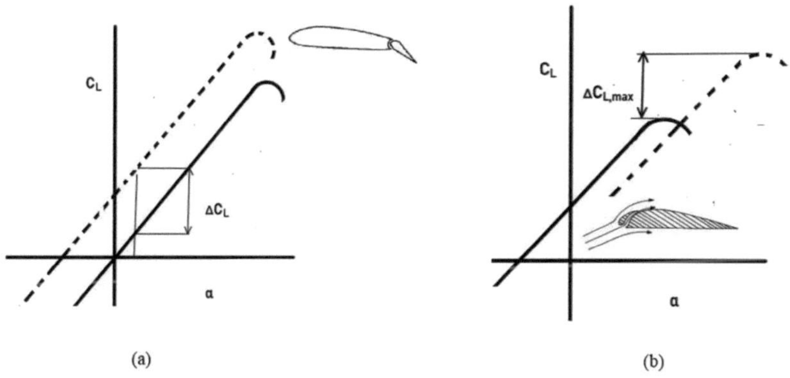

Figure 4.
(a) C_L *vs.* α *tailing edge flap effect in* C_L*. (b)* C_L *vs.* α *for leading edge slat effect in* C_{Lmax}.

The flow control technique (flaps and slats) alters the lift slope and increases the C_{Lmax}. The flaps and slats are also known as high lifting devices, which increase the lift higher than the actual aerofoil lifting capacity. The flap (**Figure 4a**) is a moving element in the trailing edge that moves up and down when the flap deflects downward camber of the aerofoil increase to shift the lift curve upward to increase C_{Lmax}. The slat (**Figure 4b**) will be fixed in the leading edge of the aerofoil which moves front to allow the airflow between the slat and aerofoil to the upper surface to delay the flow separation thereby increasing the C_{Lmax} in lift slope which is evident that the stall is delayed in high angle of attack.

3. Wind turbine

A wind turbine is a mechanical device that converts the kinetic energy of the incoming airflow striking the blade surface, producing considerable lift on the airfoils; thereby, rotation of blades is effected and successfully converted to electrical power through gearbox assembly. According to the mode of operation, wind turbines can be classified as follows.

Each type of wind turbine mentioned in (**Figure 5**) above can be summarized as:

1. Horizontal axis wind turbine: It is a type of wind turbine in which the rotor's axis of rotation is parallel to wind flow.

   a. Dutch type grain grinding windmill: It operates at the thrust exerted by wind, and the number of blades in a turbine is four. Wooden slats have been used for making the blades of the turbine.

   b. Multiblade water pumping windmill: Blades of this type of turbine are made of metal or wood and the selection of a site depends on the water availability of the area. It operates at low velocities and is also called a fan mill.

   c. High-speed propeller-type wind machines: The working of this turbine is only dependent on the aerodynamic force generated when wind flows on the airfoil surface of the blade section. They find their applications in the

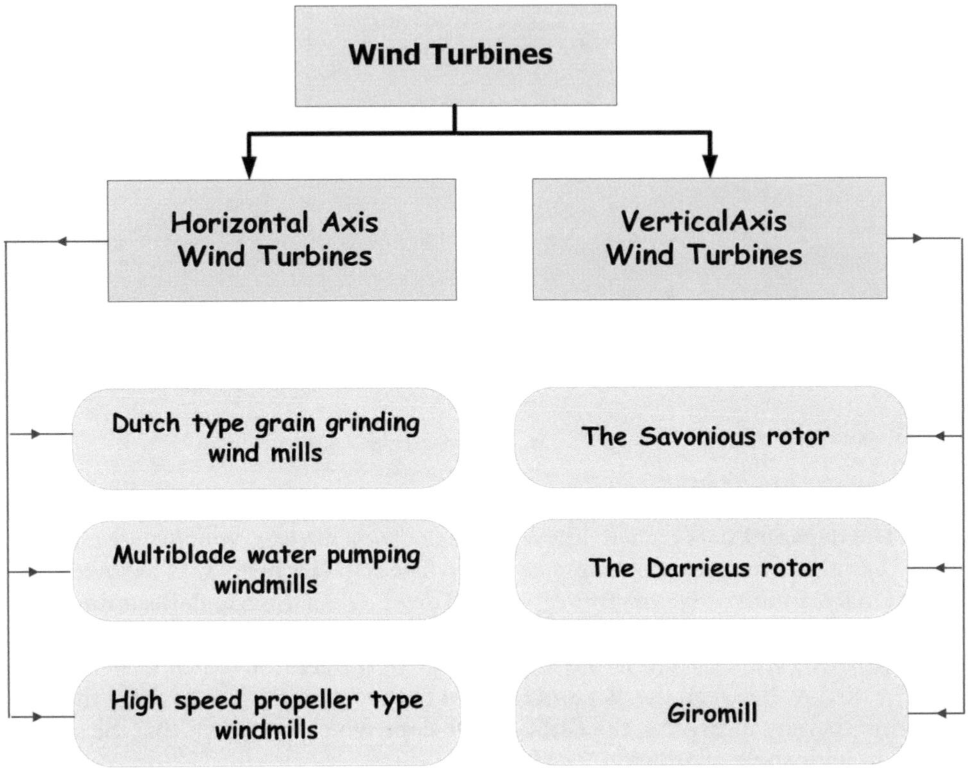

Figure 5.
Classification of wind turbines.

electricity generation of our modern era. The selection of the airfoil section forms the core of the blade design of modern wind turbines.

2. Vertical axis wind turbine: It is a type of wind turbine in which the rotation axis is placed vertical or perpendicular to the ground.

    a. The Savonius rotor: This wind turbine consists of a drum cut into two halves and attached opposite to the vertical shaft. The rotor torque is generated due to wind flow on concave and convex surfaces.

    b. The Darrieus turbine: This type of wind turbine has two or more blades made flexible and attached in the shape of a bow to the vertical shaft. The rolling action of blades generates the torque.

3.1 Components of wind turbine

- Rotor: Rotor blades of wind turbines work under the principle of an aircraft wing. The airflow on their surface creates pressure difference; blades rotate to produce electrical power.

- Nacelle: It forms the housing, which contains gearbox, generator, drive train, brakes, etc.

- Blades: Blade is a critical part of any wind turbine design as they are responsible for lift and power by rotation. The blade section close to the rotor is the hub, whereas the section away from the rotor is the tip of the blade. Hub is designed thicker, and the blade's tip is thinner to facilitate the airflow.

- Tower: It is designed to hold the rotor blades and whole assembly off the ground. Usually, a tower is constructed 50–100 m above the ground surface or water (in the case of offshore wind turbines).

3.1.1 Wind turbine aerodynamics and flow control

- Brake: The braking system is specifically designed to stop the whole machine when there is a flaw or damage in a component of the turbine. The braking system demands higher cycle rates and reliability. The brake pad of the modern turbine is coated with Kevlar to ensure longevity and robustness.

- Gearbox: The gearbox is used to is to increase the rotational velocity of the low-speed rotor to an electrical generator by gearing arrangement. The gearbox ratio varies from 15:1 to 30:1, depending on the power output of turbines.

- Anemometer: Instrument used to measure the velocity of incoming wind flow, and it transmits the wind speed to the controller.

- Controller: A wind turbine controller is a series of systems connected to monitor the operation of the wind turbine and adjacent turbines (wind farm). It is responsible for the initiation and shutdown of the system in adverse conditions.

- Yaw system: The orientation of the wind turbine towards the incoming wind is done by the yaw system. It has two systems; active and passive yaw systems and comprises mainly of yaw drive, yaw brake, and yaw bearing.

3.2 Design of horizontal axis wind turbine blade

Horizontal axis wind turbine (**Figure 6**) blades demand a pre-requisite of specific terminologies and mathematical formulas, which converge to a critical section called blade element momentum theory [10]. The preliminary step in blade element momentum theory is dividing the blade into equal sections and let each sectional element has a radius "*r*."

The output power (*P*) of the blade is determined by (Eq. (4)):

$$\mathbf{P} = \frac{1}{2}\rho A V^3 \tag{4}$$

Where,

$A = \pi R^2$ is the rotor's surface area (m_2)

V = velocity of incoming wind flow (m/s)

Betz law states that "The power extracted from the wind is independent of wind turbine design in the open flow. Therefore, it is impossible to capture more than 59.3% of Kinetic energy from the wind." From the Betz law, power is validated from the above equation.

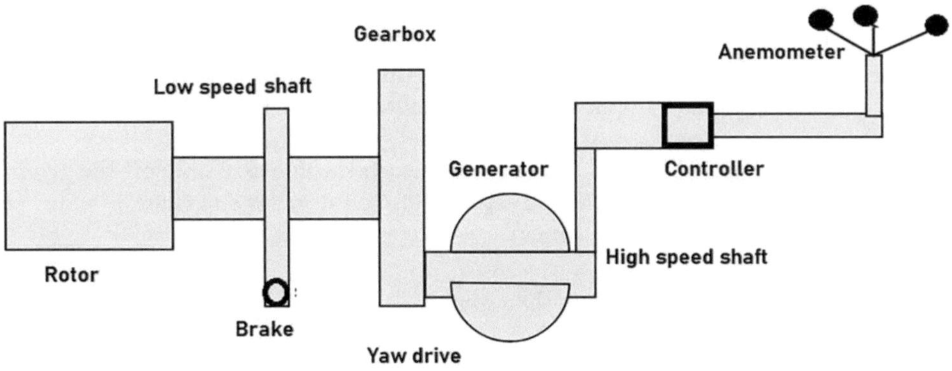

Figure 6.
Horizontal axis wind turbine mechanism.

- The angle of attack (α) is defined as the angle between the chord line and incoming wind. The optimal angle of attack of a wind turbine falls in the range of 25°–35°.
- Tip speed ratio: Tip speed ratio of the wind turbine is defined as the ratio of blade tip velocity to the wind velocity as mentioned in (Eq. (5)).
- The tip speed ratio of wind turbines should be greater than 4 for electrical power generation applications. The optimum value for TSR is 6 for a horizontal axis wind turbine blade.

$$\wedge = \frac{V_{tip}}{V} \tag{5}$$

- The number of blades (B) is an essential criterion in the power performance of blades. In horizontal axis wind turbines, the number of blades is chosen to be three as it is 40% more efficient when blades are reduced (wobbling) or increased (high drag). In the case of vertical axis wind turbines, blade number varies from 2, 3, or 4 depending on the operating conditions.
- Once the number of blades is fixed, the immediate next step in blade design is evaluating the relative wind angle (r). It is done by:

$$\psi = \frac{2}{3}\tan^{-1}\frac{1}{\wedge_r}\wedge = \frac{V_{tip}}{V} \tag{6}$$

In Eq. (6), $\wedge r = \wedge_* \left(\frac{R}{r}\right)$ where,
R = rotor radius, r = radius of element (refer (**Figure** 7)).

The design lift coefficient is measured from the properties of airfoil used in a wind turbine blade. For example, if the analyst uses NACA 4418 airfoil [10] for the wind turbine analysis, the aerodynamic properties of an airfoil can be extracted from the lift curve and lift-drag curve.

- Maximum lift coefficient, (C_{Lmax}) =1.797

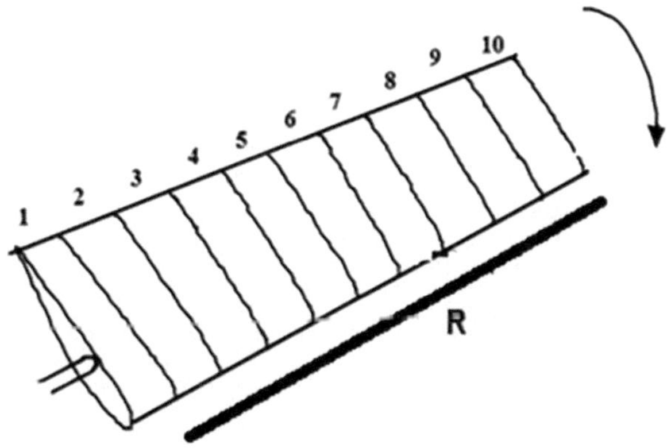

Figure 7.
Schematic representation of blade elements.

- Critical angle of attack, ($\alpha_{critical}$) =15°
- Zero lift angle, $\alpha_{L\,=\,0} = -4°$
- $\left(\frac{C_L}{C_D}\right)_{max}$ = 44.44, which occurs at an angle of attack α = 6.5°
- Design lift coefficient, $C_{L,design}$ = 1.209

The next step in the design process is the evaluating the chord length of airfoil sections in the blade by using (Eq. (7)) below:

$$\mathbf{c} = \frac{(8\pi r)(1 - \cos \psi r)}{BC_{L,\text{design}}} \tag{7}$$

Pitch angle (β) is measured theoretically from 0°, and they form the crucial part in the design of blades. For example, the optimum pitch angle for low velocity such as 15 m/s is 20°, and it varies depending on the conditions.

Mathematically pitch angle is calculated using (Eq. (8)) by the difference between blade angle and angle of attack.

$$\beta = \psi_r - \alpha \tag{8}$$

The twist angle at each section of the blade is calculated using (Eq. (9)) by subtracting the blade pitch with the pitch at the tip:

$$\theta_\tau = \beta - \beta_0 \tag{9}$$

In this expression, β_0 is the blade pitch angle at the tip.

The twist angle reduces from the hub to zero at the tip. From the above data, we can create a table for the geometric design of the horizontal axis wind turbine blade, as shown in **Table 1**. The geometrical modeling of the blade can be done using commercial software ANSYS (or) SOLIDWORKS.

S. No.	Radius of element (r) [mm]	Chord length (c) [mm]	Twist angle (θ_{T_0})
1	$r1$	$c1$	$\theta_{T_1}(Hub)$
2	$r2$	$c2$	$\theta_{T_2}(Hub)$
…	…	…	…
10	$r10$	$c10$	$\theta_{T_{10}}(Tip)$

Table 1.
Blade geometry.

3.3 Computational analysis

Computational analysis (3D) of the blade is a tedious process as modeling of the blade is a complex process to the core. The computational domain involves a stationary element and a rotational element to perform the moving reference frame approach, as shown in (**Figure 8**). Moving reference frame involves varied translation and rotational velocities of individual cell zones of the mesh. Stationary equations are generated and solved for stationary element. The rotating element is solved by moving reference frame equations such as centripetal acceleration and Coriolis acceleration in the momentum equation. The flow variables in one zone are extracted to calculate the adjacent zone by transforming the local reference frame in the interface between the cell zones.

Usually, the computational domain for horizontal axis wind turbine blade is designed as follows.

- Diameter of inner cylinder = 1.5 D
- Length of inner cylinder = 0.5 D
- Diameter of outer cylinder = 5 D
- Length of outer cylinder = 20 D
- Distance between the cylinder and upstream domain = x = 5 D where "D" is the diameter of the rotor.

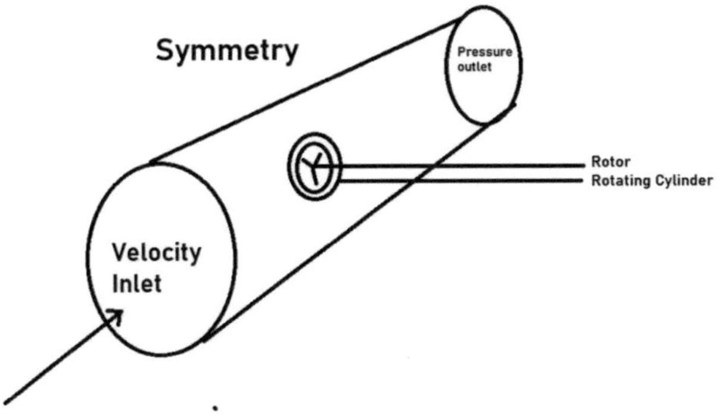

Figure 8.
Computational domain of wind turbine blade.

The meshing of domain involves creating unstructured mesh [11] around the domain with tetrahedral elements as they give good results during the simulation. The exploded view of mesh and meshing elements around the blade (**Figure 9**).

Simulation of the turbine blade is done using commercial software such as ANSYS-FLUENT/CFX. The turbulence model suitable for external flows [12] such as wind turbine flows is the k-ω SST (shear stress transport). In this model, "k" denotes turbulent kinetic energy, and "ω" denotes a specific dissipation rate. This turbulence model is widely used for wind turbine blades as it predicts the flow separation more efficiently than other RANS (Reynolds average numerical solution) models. It also performs well in adverse pressure gradients as it takes the principal shear stress transport into account while solving the equations.

Experimental analysis of wind turbine blades involves modeling and fabrication of blade setup as its preliminary step. Fabrication of blade is done using 3D printing of reinforced composite material.

The velocity profile of the rotor is extracted by fixing a pitot tube with equal holes in the X and Y-axis along the surface. Then, the pressure difference readings can calculate the velocity using (Eq. (10)) derived from (Eq. (3)).

$$V = \sqrt{\frac{2(p - p_0)}{(C_p)\rho}} \tag{10}$$

3.4 Flow control

Flow control [12] is one of the essential phenomena to be addressed in aerodynamics. As the name says, the flow control mechanism aims to control the flow of wind, thereby delaying the flow separation leading to the generation of lift and power output. Flow control is primarily classified into two types: active flow control and passive flow control mechanism.

Active flow control mechanism involves an instantaneous change in the design of the installation the installed device to increase the $\frac{\text{lift}}{\text{power}}$, whereas passive flow control mechanism [13] involves a fixed surface to influence the flow purely by its

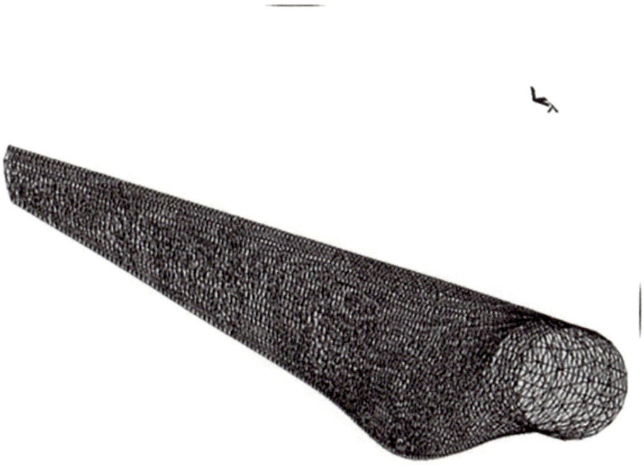

Figure 9.
Mesh elements of wind turbine blade.

geometrical characteristics. The passive flow control mechanism requires a more efficient design process as it does not have the luxury of displacement. This book will discuss one of the simplest and most effective passive flow control devices called vortex generator [14]. The design methodology [15], parameters influencing the design of vortex generators [16] and the aerodynamic effects of the vortex generator [17] are discussed in detail, taking a sample analysis for reference.

Vortex generator was introduced by Taylor [18] during 1947 as thin plates arranged in a spanwise manner projecting on the airfoil surface. Intensive research in Vortex generators had its roots in the 1970s when Kuethe [19] performed analysis on wave-type vortex generators with (h/δ) of 0.27 and 0.42 using the Taylor-Gortler mechanism [20] to create a vortex stream when a concave surface experiences and incoming flow. NASA performed much qualitative research on the design, development and testing of vortex generators [21] and preliminary analysis results suggested vortex generators (**Figure 10**) as a passive add-on control for Carter model airfoils resulting in efficient momentum transfer. The installed vortex generators on the surface have to be at the height of 1–2% of chord length and length should be approximately 2–3% of chord length with the angle of attack (α) varying from 150 to 200. The vortex generators are placed on the inboard span, outboard span, midspan, and whole span along the surface and the resulting power output is compared as shown in (**Figure 11**).

The performance comparison is shown in (**Figure 12**) depicts the increment in output power due to the addition of vortex generators. The vortex generators placed in the airfoil surface's whole span predominantly produce a 6% increase in power output with a mean wind speed of 7.15 m with a counter-rotating arrangement. The optimum dimensions suggested are pair width of vortex generators should be 0.1 c and pair spacing between generators is 0.15 c where "c" is chord length of airfoil. It

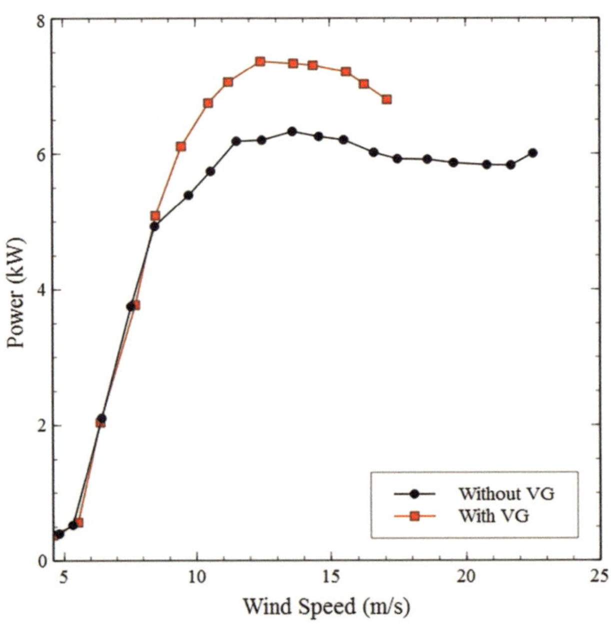

Figure 10.
Effect of leading-edge VG on the power curve.

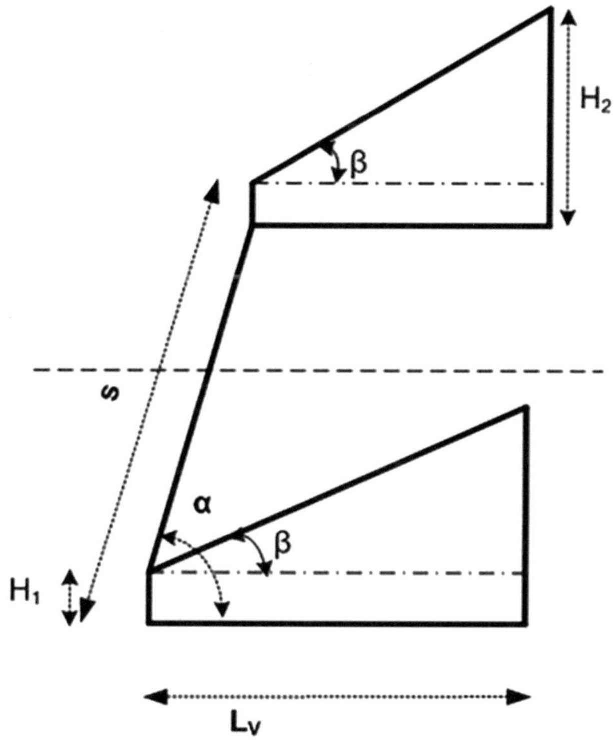

Figure 11.
Triangular Vortex generator.

also deduces vortex generators used to suppress the sensitivity of the blade to dirt accumulation on the leading edge. The following research step is optimizing the design and performance prediction of turbines [22] installing vortex generators [23]. Integrating vortex generators in wind turbines is the next giant leap in aerodynamic research.

Design risks and modifications in the vortex generators are studied [24] thoroughly for different radius as tabulated in **Table 2**.

The design of the vortex generator depends on parameters such as:

1. Height of vortex generator: In most analyses, the boundary layer thickness (δ) of the flow is taken as the height of the generator as it proves to be in good accordance with the results.

2. Spacing between generators: The spacing between a pair of vortex generators depends on the chord length of the airfoil element of the surface and flow characteristics.

3. Position of vortex generator: The position of the vortex generator is fixed by the prediction of flow separation point in the blade surface extracted from the CFD analysis of the blade.

A triangular vortex generator [25] is designed for a wind turbine blade as a sample analysis as it is simple and effective under varied operating conditions.

Radius	Radius	Modification
0–30 m	Laminar flow is observed at 25% radius. Forward placement of VG leads to early transition and increased drag penalty.	VGs are placed aft outboard of the blade.
0–45 m	Vortex generators are positioned to stall at a velocity range of 14.3–15.6 m/s where a portion of the blade is installed sharply, leading to adverse effects.	The slope of the chordwise VG locations is increased, leading to the smooth progression of the stall.
5–60 m	A stall angle closer to maximum peak rotor power may lead to an unwanted increase in the rotor power. As a result, outboard sections are less significant and sometimes lead to additional drag.	Removal of unwanted outboard vortex generators will compromise the drag penalty.

Table 2.
Design risk and modification for varied dimension.

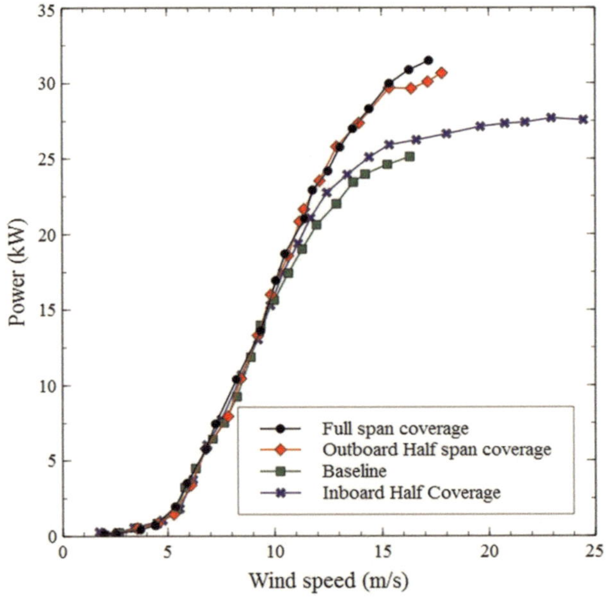

Figure 12.
Influence of span-wise location of vortex generator on power output.

In a preliminary analysis, one of the airfoil elements in BEM analysis is taken, and the vortex generator is placed at different locations in the chordwise direction. The meshing of an airfoil with VG involves special near-wall mesh. The flow can be captured on the surface without any jumps in this mesh type.

From the wall shear analysis, we can predict the flow separation point, forming the underlying basics for consequent 3-dimensional analysis.

The flow separation point is decided by fixing the vortex generator in different positions on the elemental surface and it is evident from CL vs. angle of attack (**Figure 13**) and recirculation zone (**Figure 14**) that the highest lift is obtained when the vortex generator is placed on the flow separation point [26]. The experimental analysis is validated from the CFD analysis to get qualitative results [27].

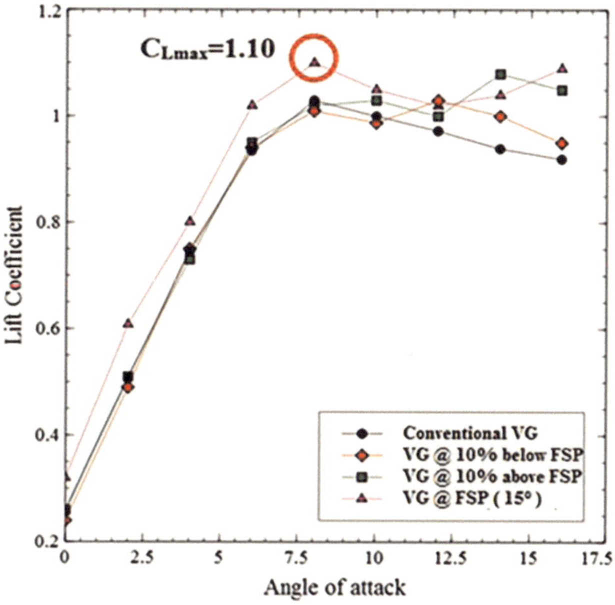

Figure 13.
Lift coefficient vs. angle of attack.

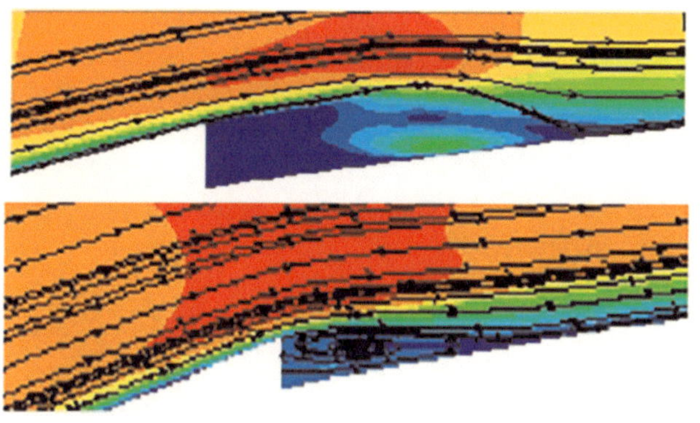

Figure 14.
Recirculation zone behind the vortex generator.

4. Conclusion

Wind turbine aerodynamics is one of the intriguing sections in the field of aerodynamics with much varied scope in the future years. Wind turbine blade analysis is practically a tedious and challenging area as the design parameters are vast, and each of them has a specified impact on the turbine performance either directly or indirectly. Effects of climatic change, terrain location, the wind rose of a particular area, environmental effects of the wind turbine, impact of blade materials in performance, height of tower and impact of the surrounding environment on the turbine's performance. Research on offshore turbines and bladeless turbines has started and improvement of performance with considerable cost will be the key objective.

The effect of ocean currents, ecosystem, and airflow in the ocean are exciting areas to ponder as energy conservation will be the prime focus for the future. Wind energy, the cheapest energy source, will be looked upon in the immediate future. The chapter gives a preface to the concept of aerodynamics and explains wind turbine terminologies to briefly explain the design and analysis of turbines to form a formidable and appealing pre-requisite for researchers to begin their work on wind turbine analysis.

Author details

Karthik Jayanarasimhan[1*†] and Vignesh Subramani-Mahalakshmi[2†]

1 Vellore Institute of Technology, Chennai, India

2 Tamil Nadu Physical Education and Sports University, Chennai, India

*Address all correspondence to: karthik.j2014phd1170@vit.ac.in

† These authors contributed equally.

References

[1] The International Renewable Energy Agency (IRENA) [Internet]. Available from: https://www.irena.org/Statistics/ [Accessed: 5 January 2022]

[2] World Adds Record New Renewable Energy Capacity in 2020 [Internet]. Available from: https://www.irena.org/newsroom/pressreleases/ [Accessed: 8 January 2022]

[3] National Institute of Wind Energy (NIWE) [Internet]. Available from: https://niwe.res.in/aboutusdgm.php [Accessed: 8 January 2022]

[4] Vignesh SM, Karthik J, Jaganraj R. Investigation of siting the wind turbine on Vel Tech University Campus. In: Bajpai R, Chandrasekhar U, Arankalle A, editors. Innovative Design, Analysis and Development Practices in Aerospace and Automotive Engineering. Lecture Notes in Mechanical Engineering. Vol. 1. New Delhi: Springer; 2019. pp. 97-110

[5] Solari G. Wind Science and Engineering Origins, Developments, Fundamentals and Advancements (Springer Tracts in Civil Engineering: XV). 1st ed. Switzerland: Springer Nature; 2021. p. 944. DOI: 10.1007/978-3-030-18815-3 [Accessed: 20 December 2021]

[6] History of Wind Turbines [Internet]. 2014. Available from: https://www.renewableenergyworld.com/storage/history-of-wind-turbines/#gref [Accessed: 6 January 2022]

[7] NACA Airfoils [Internet]. 2017. Available from: https://www.nasa.gov/image-feature/langley/100/naca-airfoils [Accessed: 2 January 2022]

[8] Jackson D, Launder B. Osborne Reynolds and the publication of his papers on turbulent flow. Annual Review of Fluid Mechanics. 2007;**39**:19-35. DOI: 10.1146/annurev.fluid.39.050905.110241 [Accessed: 12 January 2022]

[9] Eckert M. Ludwig Prandtl and the growth of fluid mechanics in Germany. Comptes Rendus Mécanique. 2017;**345**(7): 467-476. DOI: 10.1016/j.crme.2017.05.005 [Accessed: 12 January 2022]

[10] Keerthana M, Sriramkrishnan M, Velayutham T, Abraham A, Selvi Rajan S, Parammasivam KM. Aerodynamic analysis of a small horizontal axis wind turbine using CFD. Journal of Wind Engineering. 2012;**9**(2):14-28

[11] Diniz M. Unstructured mesh parameterization for a small wind turbine simulation. Journal of Wind Engineering and Industrial Aerodynamics. 2014;**125**:102-110

[12] Costa Rochaa PA, Barbosa Rochaa HH, Moura Carneiroa FO, Vieira da Silva ME, Valente Buenoc A. K-ω SST (shear stress transport) turbulence model calibration: A case study on a small scale horizontal axis wind turbine. Energy. 2014;**65**(1):412-418

[13] Zhen TK, Zubair M, Ahmad KA. Experimental and numerical investigation of the effects of passive vortex generators on Aludra UAV performance. Chinese Journal of Aeronautics. 2011;**24**:577-583

[14] Viswam R, Sankar S. Efficiency improvement of wind turbine generator by introducing vortex generator. International Research Journal of Engineering and Technology (IRJET). 2015;**2**(3):2271-2274

[15] Li B, Cao H, Deng S. Vortex generator design and application on the

flow control of top-mounted subsonic intake at high angle of attack. Journal of Vibroengineering. 2014;**16**(2):808-817

[16] Fouatih OM, Medale M, Imine O, Imine B. Design optimization of the aerodynamic passive flow control on NACA 4415 airfoil using vortex generators. European Journal of Mechanics - B/Fluids. 2015;**56**:82-96

[17] Li X, Yang K, Wang X. Experimental and numerical analysis of the effect of vortex generator height on vortex characteristics and airfoil aerodynamic performance. Energies. 2019;**12**:959-979

[18] Taylor HD. The elimination of diffuser separation by vortex generators. United Aircraft Corporation Report No. R-4012-3. June 1947

[19] Kuethe AM. Effect of stream wise vortices on wake properties associated with sound generation. Journal of Aircraft. 1972;**9**(10):715-719

[20] Kobayashi M, Maekawa H. Turbulent flow accompanied by Taylor-Goertler vortices in a two-dimensional curved channel. Flow Measurement and Instrumentation. 1995;**6**(2):93-100

[21] Gyatt GW. Development and testing of vortex generators for small horizontal axis wind turbines. Technical Report. NASA CR-179514. 1986

[22] Pesmajoglou SD, Graham JMR. Prediction of aerodynamic forces on horizontal axis wind turbines in free yaw and turbulence. Journal of Wind Engineering and Industrial Aerodynamics. 2000;**86**:1-14

[23] Manolesos M, Voutsinas SG. Experimental investigation of the flow past passive vortex generators on an airfoil experiencing three-dimensional separation. Journal of Wind Engineering and Industrial Aerodynamics. 2015;**142**: 130-148

[24] Brandner PA, Walker GJ. Hydrodynamic performance of a vortex generator. Experimental Thermal and Fluid Science. 2003;**27**:573-582

[25] Wendt BJ. The modelling of symmetrical airfoil vortex generators. NASA Report AIAA-96-0807

[26] Szwaba R. Comparison of the influence of different air-jet vortex generators on the separation region. Aerospace Science and Technology. 2011;**15**:45-52

[27] Senthil Kumar S, Ganesh RPV, Karthikeyan S, Karthik J. Numerical analysis of suppression of laminar bubble at low Reynolds number using different protrusions. In: Bajpai R, Chandrasekhar U, Arankalle A, editors. Innovative Design, Analysis and Development Practices in Aerospace and Automotive Engineering. Lecture Notes in Mechanical Engineering. New Delhi: Springer; 2014. DOI: 10.1007/978-81-322-1871-522

Chapter 12

On the Design and Manufacture of Wind Turbine Blades

Mohamed Mahran Kasem

Abstract

Wind turbines become extremely important worldwide along with the need for clear energy sources. The concept of wind turbines is based on using the wind energy to produce lift that turns into toque, which rotates the wind turbine blades and subsequently produces electric power using a proper generator. However, the wide use of wind turbines and their design and manufacturing process are a challenge. Therefore, much research has been conducted to improve and develop new methods for the design and manufacturing of wind turbines. In this chapter, the author discusses some techniques for wind turbine design and manufacturing, including airfoil appropriate selection, design optimization methods, and manufacturing techniques. One of the manufacturing techniques that are found to be superior is the use of chordwise and spanwise stiffeners to increase the stiffness of the skin of carbon fiber wind turbine blades. Those stiffeners are not bonded externally to the skin; otherwise, they are layers of carbon fibers that are buried inside the skin of the wind turbine blades.

Keywords: wind turbine blades, blade manufacturing, blade design methods

1. Introduction

The design of wind turbine blades has two objectives: (1) to determine the blade geometry that can produce an optimum power and (2) to determine the optimum structure required to create the wind turbine blade. The objective of the former is to obtain the wind turbine blade geometry that maximizes the power generated at different tip speed ratios. **Figure 1** illustrates the variation of the power coefficient (C_p) with the tip speed ratio (λ) for two different blade designs. Design 1 has the maximum C_p but with large drop with small and high λ. Design 2 has smaller C_p, but performs better over the range of λ. Therefore, design 2 seems to be better than design 1; however, it has smaller maximum power coefficient C_p.

The aerodynamic design also includes the selection of optimum chord and twist distribution for the wind turbine blade. The objective of the latter is to create a wind turbine blade structure that satisfies the aerodynamic requirements. A typical blade cross section is shown in **Figure 2**. A blade structure is usually constructed from external skin and internal spar.

This chapter summarizes the key steps required to perform an appropriate aerodynamic and structural designs for wind turbine blades. This includes the design

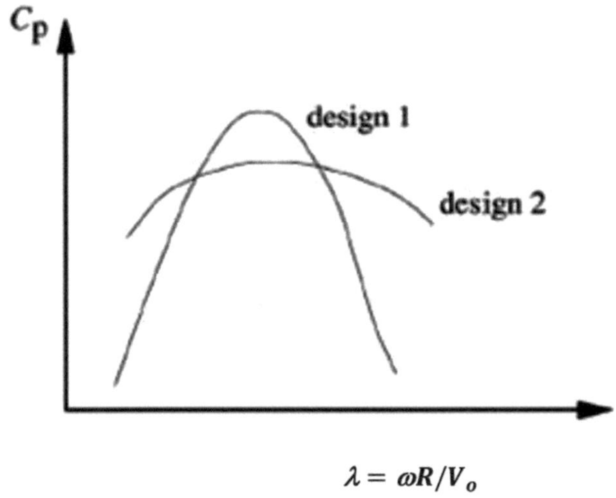

Figure 1.
Variation of power coefficient with tip speed ratio for two different blade designs [1].

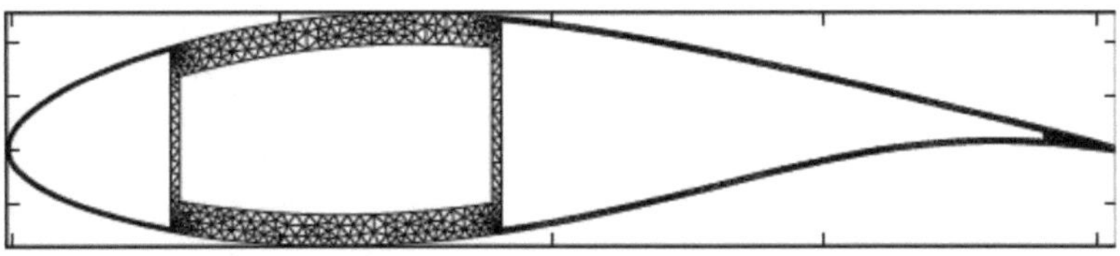

Figure 2.
Typical cross section of a wind turbine blade [2].

process, unsteady aerodynamic analysis, design optimization, and structural design of the blade.

2. Design of wind turbine blades

A major objective of wind turbine design is to maximize the output power and improve its performance. This objective can be accomplished by maximizing the aerodynamic lift and minimizing the drag. The process of designing a wind turbine blade starts by the airfoil selection in addition to selecting the appropriate wind turbine geometries according to the required performance. **Figure 3** shows the main variables in a typical wind turbine blade. **Figure 4** shows the relation between the wind turbine power and diameter.

2.1 Airfoil selection

Wind turbine blades are usually constructed with high taper ratio and twisting angle. Small wind turbines usually have one airfoil type, whereas large-scale wind turbines need different airfoils along the blade radius. An airfoil should be selected with maximum lift-to-drag ratio and minimum pitching moment coefficient. Most optimization models concentrate on improving wind turbine blade performance by enhancing the taper ratio, aerodynamic twist, and geometric twist of the blade;

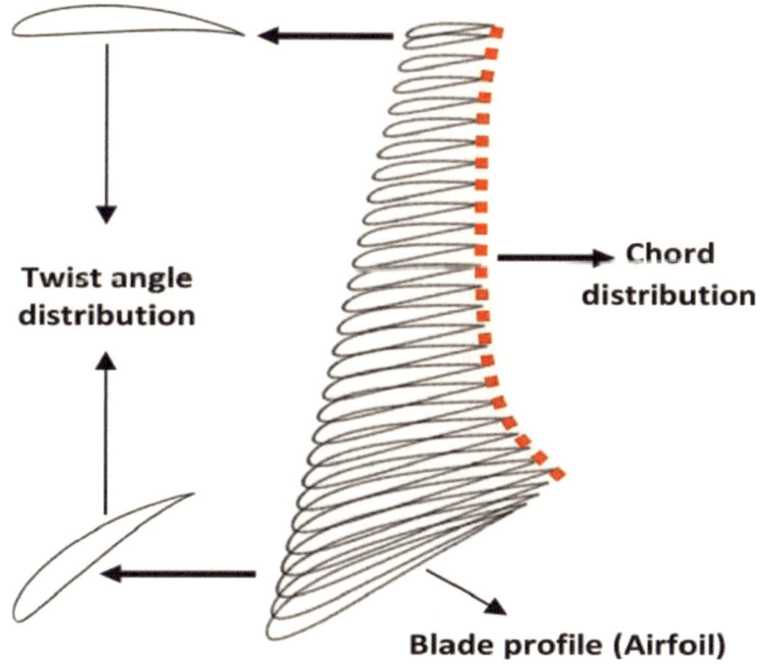

Figure 3.
Wind turbine blade variables [3].

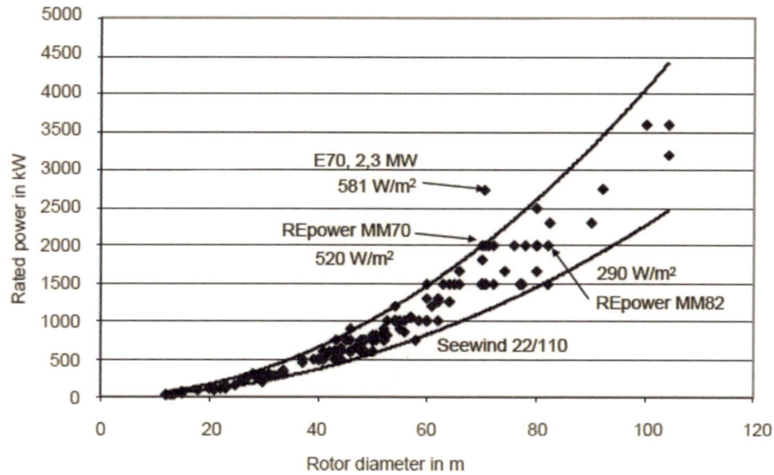

Figure 4.
Relation between wind turbine power and diameter.

however, some optimization models improve the wind turbine performance by changing the airfoil shape. The latter can be conducted either by considering different airfoil shapes in the optimization problem or by defining control points over the airfoil and change its shape during the optimization process (**Figure 5**).

Sometimes, special types of airfoils are required for the wind turbine based on its characteristics. For instance, low-speed wind turbines require special types of airfoils to generate the torque required to rotate the blades [5]. In most cases, it is required to

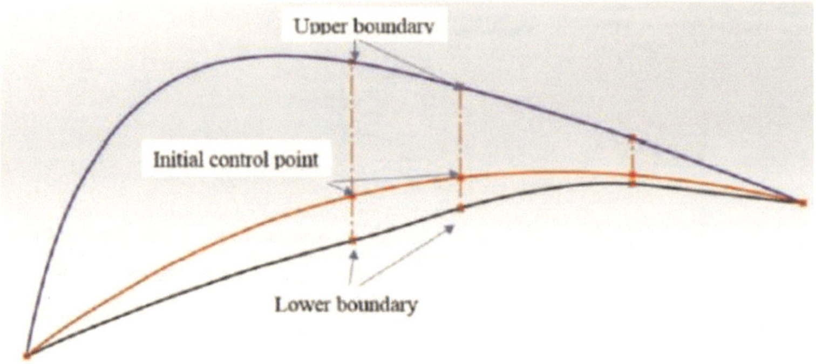

Figure 5.
The control point motion [4].

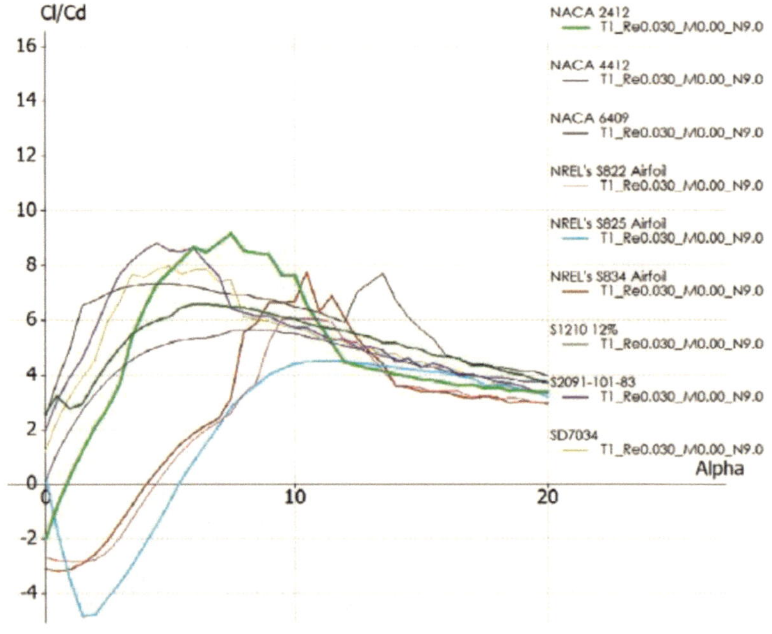

Figure 6.
Comparison between different airfoils at Re = 3×10^4.

compare between different airfoil types and select the best airfoil to be integrated with a certain wind turbine. The airfoils are evaluated based on their $\frac{C_l}{C_d}$ ratio. The maximum is $\frac{C_l}{C_d}$; the airfoil can produce more lift and smaller drag.

There are several airfoils' families suitable for wind turbine blades, such as the NACA family and the S series. **Figure 6** shows a comparison between nine airfoils from different series at Reynold's number 3×10^4. The performance of each airfoil is different in relative to the angle of attack.

Large wind turbines are usually constructed from more than one airfoil. It could have two or three different airfoils along its radial position. In this case, a linear or higher order chord variation can be assumed between the airfoils.

2.2 Unsteady aerodynamic

Before starting the design optimization process, steady and unsteady aerodynamic analyses should be conducted. These analyses can be accomplished either using the blade element momentum (BEM) or by applying computational fluid dynamics (CFD). A detailed comparison between the two methods can be found in [6]. The two methods are used to solve the blade mathematical model, which is usually has a form of differential equation. The mathematical model should provide the relation between the different variables and parameters of a typical wind turbine blade. One of the most popular and widely used mathematical models in wind turbine analysis and design is the blade element momentum (BEM) method. A general procedure for applying the BEM method can be summarized as follows [1]:

1. Define the geometry.

2. Discretize the blade into elements.

3. Initialize the induced and relative blade velocities.

4. Determine the airfoil data including the lift.

5. Compute new values of wind velocities.

6. Compare between the old and new values till convergence is achieved.

More details about the BEM method can be found in [6].

2.3 Design optimization

The design of wind turbine blades is twofold: first, the correct selection of the optimization method, and, second, the proper definition of design variables and other optimization parameters. A genetic algorithm (GA) is one of the popular methods that are widely used in design optimization [7]. The GA is based on the process of natural selection, in which the new generation is selected based on the fitness of the parents. Thus, the parents with high fitness supposed to produce offspring better than those with low fitness score in the optimization process. The process is keeping in iteration until the best design variables are selected through mutation, crossover, and selection steps. **Figures 7** and **8** show details of the GA optimization process.

One of the advantages of using the GA in optimization is that it can be applied to both discrete and continuous optimization. In the GA, the population is generated randomly, and a candidate solution is defined for the design variables. The best solutions are selected based on their fitness defined from the objective function. Those solutions define the parents. Their children are produced by crossover operation. Then, to ensure global optimization, a mutation operation is applied [3].

Wind turbine optimization requires the definition of an objective function. The objective function differs based on the purpose of the optimization. If the purpose is to improve the wind turbine aerodynamic performance, the objective function may be to increase the wind turbine lift and/or decrease the drag. If the optimization purpose is to improve the wind turbine structure performance, then the design objective could be to maximize the wind turbine stiffness and/or minimize its weight. If one objective

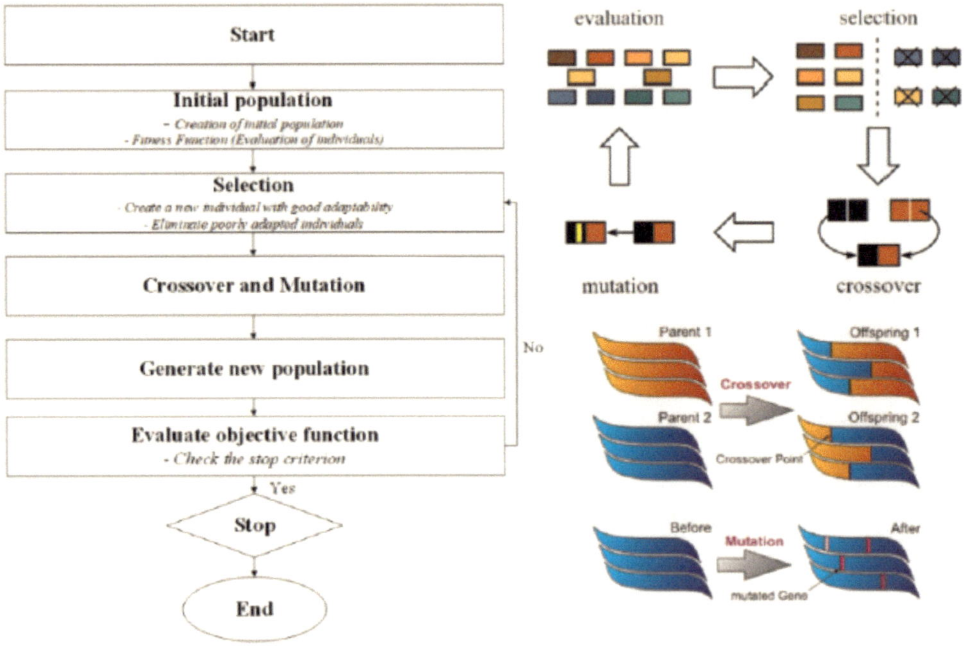

Figure 7.
GA process [4].

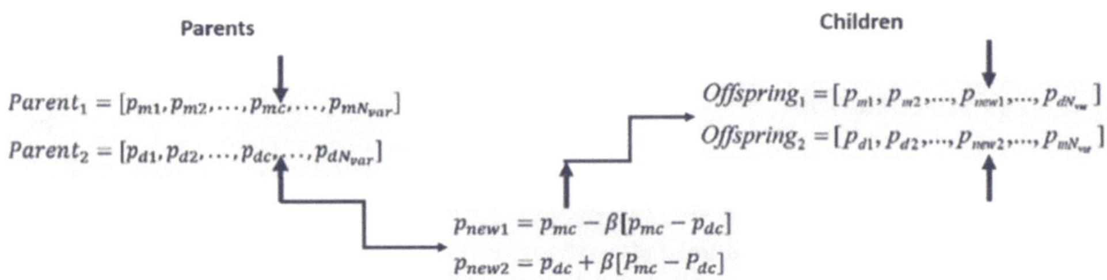

Figure 8.
Single-point example for the crossover process [3].

function is defined, we called it single-objective optimization. In case of more than one objective function, the optimization is called multiobjective optimization.

3. Structural design

After determining the wind turbine geometry, a structure design should be accomplished to create the wind turbine. The structure design includes several analyses, such as static analysis, modal analysis, dynamic analysis, and aeroelastic analysis. There are two schools in wind turbine structure analysis and design. The first school suggests the design of the wind turbine structure by approximating the wind turbine blade into a beam model [2]. This method is very efficient in computational time and cost, but it cannot provide a detailed solution for deformation, strain, and

	Beam analysis	3D simulation
Time cost	Low	High
Money cost	Low	High
Accuracy	Applies 1D approximation to blades using beam theory	Applies numerical approximation to the governing equation using variational methods
Solution	Based on analytical or numerical methods	Based on numerical methods
Results	Provides 1D solution	Provides detailed 3D solution

Table 1.
Comparison between 1D beam and 3D numerical solutions.

stress distribution. The second school prefers to analyze the wind turbine blade as a full 3D model using numerical software. The most popular method to conduct these analyses is the finite-element method. However, the 3D simulation can provide a detailed solution to the wind turbine blade structure; it costs lots of time and money in comparison to the 1D beam analysis. **Table 1** provides a detailed comparison between the beam and numerical solutions.

Usually, four solutions should be studied to make sure that the wind turbine structure is safe and stiff enough:

1. Static analysis—by which steady loads are applied to the structure and the static displacement and stresses are determined.

2. Modal analysis—by which the blades' natural frequencies and mode shapes are calculated.

3. Dynamic analysis—by which the dynamic displacement and stresses are determined in response to unsteady aerodynamic loads.

4. Aeroelastic analysis—by which the divergence and flutter speeds are calculated to make sure that the blade is safe from any aeroelastic instability.

4. Manufacturing of wind turbine blades

The wind turbine blade structure usually consists of upper skin, lower skin, and spar (**Figure 9**). These structural elements help in resisting the direct and shear stresses applied to the blade.

Spar is the main structural element in the wind turbine blade. It transforms all the blade loads to the wind turbine hub. Thus, the selection of appropriate spar shape is a corner stone in structural design. In the following figures, a comparison between the most common spar cross-sectional shapes is provided. **Figure 10** shows a comparison between the rectangular shape, circular shape, I section, double I shape, and C section spar elements. **Table 2** provides the mathematical equations for a detailed comparison. Two performance parameters are defined to measure the stiffness of the spars (m and m_d). In terms of the performance parameters, the I and C sections are found to have the best bending and torsional stiffnesses in comparison to the other candidates.

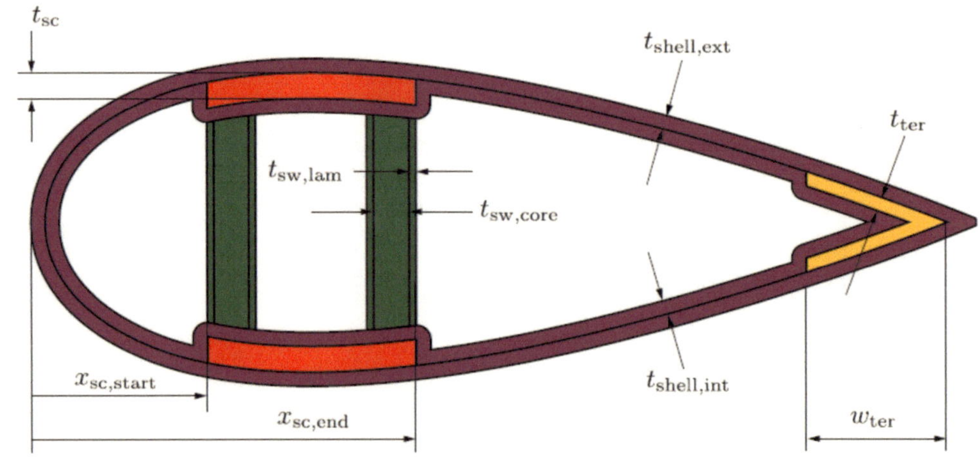

Figure 9.
Wind turbine construction [8].

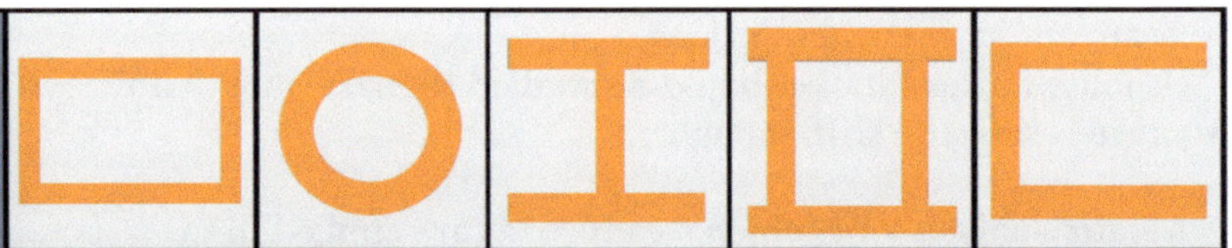

Shape	Box	Circular	I-beam	Double I	C-section
Advantages	Balances between bending and torsional stiffness	Best in torsional stiffness	Best in bending stiffness	Balances between bending and torsional stiffness	Best in bending stiffness
Disadvantages	Relatively heavy	Lower bending Stiffness	Bad in torsional stiffness	Relatively heavy	Shear center not coincident with centroid may resulting in more twist deformation

Figure 10.
Comparison between different spar shapes [9].

Figure 10 and **Table 2** can help in selecting the appropriate spar cross section for bending and torsional applications.

In **Table 2**, b_w is the web height (the section height), b_f is the flange width (the section width), D_i is the inlet diameter, t is the thickness, t_w is the web thickness, $\hat{EI}$ is the equivalent bending stiffness, $\hat{GI}$ is the equivalent torsional rigidity, d, a_{66}, d_{66} are composite stiffness coefficients, y_{tip} is the tip displacement, ψ_{tip} is the tip rotation, p is the applied load, T_{max} is the maximum torque, and L is the beam length.

In small wind turbines, it is difficult to add a spar inside the wind turbine blade because the blade thickness is small. In this case, a lateral and longitudinal stiffener can be bonded inside the wind turbine skin to stiffen the skin. In composite manufacturing, those stiffeners can be inherent inside the skin during the manufacturing process. We found this technique efficient in increasing the wind turbine blade stiffness.

DOI: http://dx.doi.org/10.5772/intechopen.104490

Shape					
Area (64π)	$2(b_w + b_f)t$	$\frac{\pi}{4}\left((D_i + 2t)^2 - D_i^2\right)t$	$2b_f t + b_w t$	$2(b_f t) + 2(b_w t)$	$2b_f t + b_w t_w$
Dimension	$b_w = 30$ $t = 2$ $b_f = 16\pi - 30$ $d_f = b_f - t$	$D_i = 30$ $t = 2$	$b_w = 30$ $t = 2$ $b_f = 16\pi - 15$	$b_w = 30$ $t = 2$ $b_f = 16\pi - 30$ $d_f = \frac{b_f}{2}$	$b_w = 30$ $t = 2$ $b_f = 16\pi - 15$
$\hat{EI}$	$\frac{b_f}{(a_{11})_f}\frac{d^2}{2} + \frac{2b_f}{(d_{11})_f} + \frac{2b_w^3}{12(a_{11})_w}$	$\pi\left(\frac{\left(\frac{D_i+t}{2}\right)^3}{a_{11}} + \frac{\left(\frac{D_i+t}{2}\right)}{d_{11}}\right)$	$\frac{b_f}{(a_{11})_f}\frac{d^2}{2} + \frac{2b_f}{(d_{11})_f} + \frac{b_w^3}{12(a_{11})_w}$	$\frac{b_f}{(a_{11})_f}\frac{d^2}{2} + \frac{2b_f}{(d_{11})_f} + \frac{2b_w^3}{12(a_{11})_w}$	$\frac{b_f}{(a_{11})_f}\frac{d^2}{2} + \frac{2b_f}{(d_{11})_f} + \frac{b_w^3}{12(a_{11})_w}$
	2.8711e6	2.4840e6	3.9196e6	2.8711e6	3.9196e6
$\hat{GI}$	$\frac{2d_f^2 d^2}{(a_{66})_f d_f + (a_{66})_w d}$	$2\pi\frac{\left(\frac{D_i+t}{2}\right)^3}{a_{66}}$	$4\left(\frac{2b_f}{(d_{66})_f} + \frac{b_w}{(d_{66})_w}\right)$	$\frac{2d_f^2 d^2}{(a_{66})_f d_f + (a_{66})_w d}$	$4\left(\frac{2b_f}{(d_{66})_f} + \frac{b_w}{(d_{66})_w}\right)$
	4.8951e5	9.2678e5	8.2639e3	1.7972e5	8.2639e3
$m = 0.75\frac{\hat{EI}}{\hat{EI}_{max}} + 0.25\frac{\hat{GI}}{\hat{GI}_{max}}$					
max	0.681	0.725	0.752	0.598	0.752
$y_{tip} \times 10^7$	$3.4830\ \frac{pL^4}{8}$	$4.0258\ \frac{pL^4}{8}$	$2.5513\ \frac{pL^4}{8}$	$3.4830\ \frac{pL^4}{8}$	$2.5513\ \frac{pL^4}{8}$
$\psi_{tip} \times 10^6$	$2.0429\ \frac{T_{max}L}{2}$	$1.0790\ \frac{T_{max}L}{2}$	$121.01\ \frac{T_{max}L}{2}$	$5.5642\ \frac{T_{max}L}{2}$	$121.01\ \frac{T_{max}L}{2}$
$m_d = 0.75\frac{y_{tip_{min}}}{y_{tip}} + 0.25\frac{\psi_{tip_{min}}}{\psi_{tip}}$					
max	0.681	0.725	0.752	0.598	0.752

Table 2.
Detailed comparison between different spar cross sections [9].

5. Conclusion

This chapter summarizes the methods and techniques usually used in the design and manufacturing of wind turbine blades.

Author details

Mohamed Mahran Kasem[1,2]

1 Aerospace Engineering Department, Cairo University, Giza, Egypt

2 School of Engineering and Applied Science, Nile University, Shaikh Zayed City, Egypt

*Address all correspondence to: mohamed.kasem@cu.edu.eg; mkasem@nu.edu.eg

References

[1] Hansen MO. Aerodynamics of Wind Turbines: Rotors, Loads and Structure. Vol. 17. London, United Kingdom: Earthscan; 2000. Available from: http://books.google.com/books?hl=en&lr=&id=d_REBCyBuKkC&oi=fnd&pg=PR7&dq=%22to+run+for+a+minimum+of+20+years+and+thus+performs+in+the+order+of%22+%22the+main+loads+on+a+horizontal-axis+wind+turbine+is+given.+To%22+&ots=KFaPhqpZpf&sig=qdF-2Qs_w8WEPlst8nSkIo5lDQU [Accessed: April 11, 2017]

[2] Otero AD, Ponta FL. Structural analysis of wind-turbine blades by a generalized Timoshenko beam model. The Journal of Solar Energy Engineering. 2010;**132**(1):011015. DOI: 10.1115/1.4000596

[3] Pourrajabian A, Dehghan M, Rahgozar S. Genetic algorithms for the design and optimization of horizontal axis wind turbine (HAWT) blades: A continuous approach or a binary one? Sustainable Energy Technologies and Assessments. 2021;**44**:101022. DOI: 10.1016/j.seta.2021.101022

[4] Hoang Quan NN, Lam PV, Long LV. Wind turbine blade design optimization using OpenFOAM and DAKOTA software. Transportation Research Procedia. 2021;**56**:71-78. DOI: 10.1016/j.trpro.2021.09.009

[5] Gray A, Singh B, Singh S. Low wind speed airfoil design for horizontal axis wind turbine. Materials Today: Proceedings. 2021;**45**:3000-3004. DOI: 10.1016/j.matpr.2020.11.999

[6] Abdou Mahran Kasem M. Aerodynamic structural and aeroelastic design of wind turbine blades. In: Maalawi KY, editor. Design Optimization of Wind Energy Conversion Systems with Applications. London: IntechOpen; 2020. DOI: 10.5772/intechopen.89761

[7] Kasem MM, Maalawi KY. Optimization techniques and models for structural design. Journal of Mechanical Engineering Science. 2021;**15**(3):26

[8] Chiu PK, Roth-Johnson P, Wirz RE. Optimal structural design of biplane wind turbine blades. Renewable Energy. 2020;**147**:2440-2452. DOI: 10.1016/j.renene.2019.08.143

[9] Kasem MM, El-Sayed H, Halaka J, Morcos M, Shaker P, Kasem Y. Design and Manufacturing of Full Composite Unmanned Flying Wing. Egypt: Cairo University; 2021

Chapter 13

Wind Turbine Bearing Failure: A Personal View

John Campbell

Abstract

The writer describes his experience of a lifetime of casting metals, and how the casting technique controls the quality of the metal and offers answers to engineering failures. In view of the wide denial of this aspect of process metallurgy, the author takes the opportunity to present a personal view, backed up by additional evidence in a bibliography. It is a concern that the failure of wind turbine bearings continues, on occasions, to defy substantial metallurgical efforts. It is proposed here that there is good reason to identify the casting process as the generator of pervasive defects, which the writer calls bifilms. These defects originate from the casting process during the pouring of the liquid steel. They are simply doubled-over oxide films originating from the surface of the melt. They are inherited by the solidified steel and are resistant to bonding by mechanical working. They, therefore, exist in finished steel components as a substantial population of cracks. These pre-existing cracks are usually the initiators of fatigue failure, as well as other failure modes. Techniques to eliminate bifilm cracks during the casting of steel are now known and require to be implemented to produce steels that will naturally eliminate failure. We shall have, for the first time, steels we can trust.

Keywords: ingot casting, inclusions, defects, oxides, Bifilms, contact pour

1. Introduction

Wind turbines are typically designed for a minimum 20-year life. However, failure of the main bearing of the turbine after only a few years, perhaps 5 years, can involve the immense expense of dismantling, lowering, transporting away for repair or replacement, raising, and re-installing the new bearing. These costs are enhanced for off-shore turbines and threaten the economic case for wind energy.

It is important therefore that bearings are reliable. The engineering involved in the modern bearing designs, optimised by computer simulation, and manufacture involving precision machining ensure an extremely high standard of 'designed in' reliability. The steel is also held to within close limits of its chemical specification, generally based on the composition 1C–1.5Cr, the typical 'carbon chrome' steel which is widely used for ball and roller bearings. The use of this bearing steel for over 70 years or more has generated an optimised material backed up by an immense volume of development and production experience.

It is a source of surprise and disappointment, therefore, despite all this vast accumulation of knowledge and experience, bearing failures of large turbines still occur prematurely.

A higher strength steel, with an unusual structure of lower bainite, has more recently been available; it is hoped that this improved material will provide greater reliability and longer life. Experience with its longevity should be emerging over the next few years, so it is too early to include the new bearings in this report. This chapter concentrates, therefore, on the known behaviour of the carbon-chrome steel bearings. Even so, it seems likely that the proposals in this chapter will also benefit the low bainitic steels in due course.

This chapter draws attention to the universally neglected role of the casting process in the behaviour of the steels. Unfortunately, defects are introduced by the casting process which can be sufficiently serious to dominate the failure mechanism of the steel. This widely overlooked effect is considered in detail. The mechanisms cited in this chapter are described in more detail by the author elsewhere [1–3].

2. The background to inclusion creation and the cracking of the liquid metal

In the liquid state, all metals oxidise to some degree in the air, forming a thin surface oxide film. This surface oxide is not a problem while on the surface, and is valuable, limiting the rate of further reaction with oxygen, conferring a kind of pseudo-inertness of the liquid, preventing the liquid completely oxidising away or even, some metals, bursting into flames.

The structure of the surface oxide on the liquid is significant. Its lower surface is completely 'wetted', being in atomic contact with the liquid, from which it has grown, atom by atom. In contrast, the top surface of the oxide is completely dry, and when viewed under the microscope has an irregular surface like sandpaper, or sometimes a microscopic downtown Manhattan.

When filling a mould, if the metal proceeds upwardly, the surface oxide trapped between the liquid and the mould wall cannot rise with the metal. Consequently, the oxide splits at the crown of the upward progressing meniscus, to allow the metal to rise. Instantly, new oxide forms at the newly revealed liquid surface. The newly forming oxide splits and moves sideways to become trapped as the skin of the casting against the mould. Importantly, the oxide does not become entrained in the matrix. In this way, the filling of the mould is achieved without the formation of defects, and the presence of the surface oxide, thickened during its travel across the meniscus and down the walls, protects the metal from the ingress of contaminants such as gases and other solutes by reaction with the mould, and mechanically supports the surface, bridging the gaps between asperities to confer a smooth 'averaged' cast surface. When the mould filling is completed only by this mechanism, it is known as Counter-Gravity casting (**Figure 1**). It is described in more detail below.

However, as everyone knows, most castings are not made only by this upward counter-gravity progress of the liquid. Most castings are poured, using gravity, which accelerates the metal as it falls into the mould. The result is vigorous turbulence and mixing of the bulk liquid and its surface so that its surface oxide film becomes entrained into the bulk liquid metal.

The entrainment of the surface film happens by a number of interesting, related mechanisms. The surface can (i) fold, or (ii) form droplets and splashes, or (iii) form

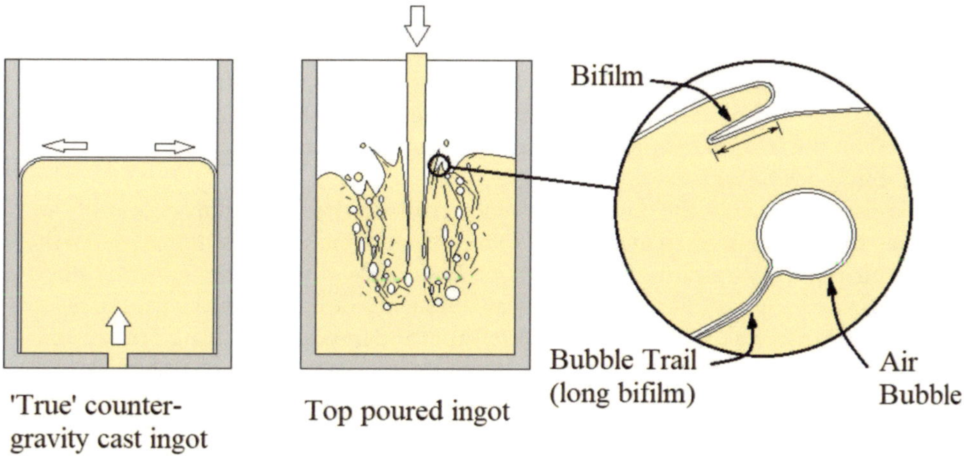

Figure 1.
Contrasting casting techniques.

bubbles which during their motion through the liquid, slough off their surface oxide to form oxide tubes sometimes meters long, which collapse to form lengthy bifilms (**Figure 1**). Naturally, the bubble trails are generally shredded to shorter lengths in the violent turbulence, so that the final bifilms population consists of a fragmented collection of largely indistinguishable defects, although some bubble trails survive to form leak paths through 100 mm thick walls in shaped castings (if well-fed to avoid any shrinkage porosity, how otherwise could leaks through thick steel walls be explained?). In all cases, the mutual impingement of surfaces causes the oxides on the opposing surfaces to come together as a dry side to dry side. Only the high spots meet, so the films make little contact on a microscale, and little bonding can take place. This double film now takes on a life of its own. I call it a 'bifilm'. Each bifilm finds itself now immersed in the bulk liquid, but has practically no bonding between its two opposed films. It, therefore, acts as a crack in the liquid. Very severe turbulence, as in the pouring of a steel ingot, results in a dense population of cracks in suspension, akin to a snowstorm [4].

For liquid steel, the snowstorm takes time to clear; although the oxides are less dense than the steel and therefore expected to float out, (i) they have nearly zero volume because of their extreme thinness and so can exert only minimal buoyancy force, plus (ii) their relatively large area is characterised by high drag—a parachute action to slow progress.

Thus, the liquid steel, now damaged by a semi-permanent population of cracks only slowly recovers its integrity: the larger bifilms separated by flotation within minutes, forming the observed layer of oxide slag on the surface of the steel. However, a large population of smaller bifilms remains to be trapped by solidification.

Because bifilms are not 'clean' cracks but faced with highly stable oxides such as alumina (Al_2O_3) and chromia (Cr_2O_3), the cracks tend to survive plastic working such as forging and rolling. After plastic working, most of our steels remain impaired by a dense population of cracks introduced by the casting process. This is particularly common during the pouring of ingots for special purposes, such as large bearing rings for wind turbines, because top pouring of the ingot is the cheapest casting technique. As will be explained below, even if the ingots are bottom gated (uphill poured) the current technology only improves the surface of the ingots a little but makes little change to the internal integrity.

For a particularly large ingot, required for the largest bearing rings, as the steel freezes in a direction away from the mould wall, the advancing dendrites 'push' the bifilms, tending to concentrate them in the centre of the ingot. When the ingot is pierced and opened by forging to form a ring, the bifilm defects are naturally concentrated on the inner working surface of the ring. Regrettably, the distribution of defects could not be worse.

This chapter reviews the damaging mechanisms of the casting process and proposes alternative casting techniques to reduce or avoid damage. It is noteworthy that the current carbon-chrome steel is usually capable of providing good service life of main bearings, and most probably suffers early failure from the presence of occasional material defects. It is expected that the implementation of processes to eliminate the larger bifilm cracks, the major defects in the material, should significantly assist to eliminate failure.

3. Steel casting processes

Bulk steels are now commonly cast by the continuous casting process. This is an efficient process in which a ladle of steel spends a significant length of time delivering metal from a bottom nozzle into a launder (a trough, or channel) which continuously supplies the water-cooled mould. The metal in the ladle, therefore, has a lengthy period in which its bifilm population can be reduced by upward flotation (despite huge research and development efforts, relatively little can be separated in the launders). Fortuitously, the metal is delivered to the mould from the base of the ladle, providing the cleanest metal. The result is that continuous cast steels are usually significantly cleaner and more fatigue resistant than ingot cast steels.

However, for limited batches of special steels, and especially those required in large pieces, ingot casting remains the only practical production route. The ingots are filled from a ladle which the crane positions above the mould. The metal is poured from the base of the ladle via a nozzle sealed with a stopper, or by a sliding gate.

A common casting technique for large ingots is top pouring, in which a ladle opens its bottom nozzle above the mould (**Figure 1**). The metal jets from the nozzle at speeds of the order of 5 m/s. The speed of the falling jet increases during the remainder of its fall into the ingot mould, reaching up to approximately 10 m/s. This high-energy liquid fragments and churns, introducing masses of bifilm cracks. It is a casting technique to be avoided for steel bearings.

The alternative technique for introducing the liquid metal at the base of the mould is widely used in the belief that it constitutes a kind of counter-gravity filling. It certainly improves the surface finish on the outside of the cast ingot because of the reduced amount of splashing, slopping and surging (**Figure 2**). However, it is not generally realised that the interior of the metal suffers a disaster. The reason for this is the universal use of a conical intake (the trumpet) at the entrance to the running system. The slightly ragged edges of the high-velocity falling jet of metal accelerate the closely surrounding air and take this down into the filling system. The conical entrance acts as a venturi pump, concentrating air into the system. A resulting 50/50 mix of air and steel is now known to be formed and is clearly observed in water model experiments and computer simulations. An observer looking down into the mould during the filling process observes the melt to 'boil' as it rises. Once again, the steel is severely degraded. Some restoration of the properties

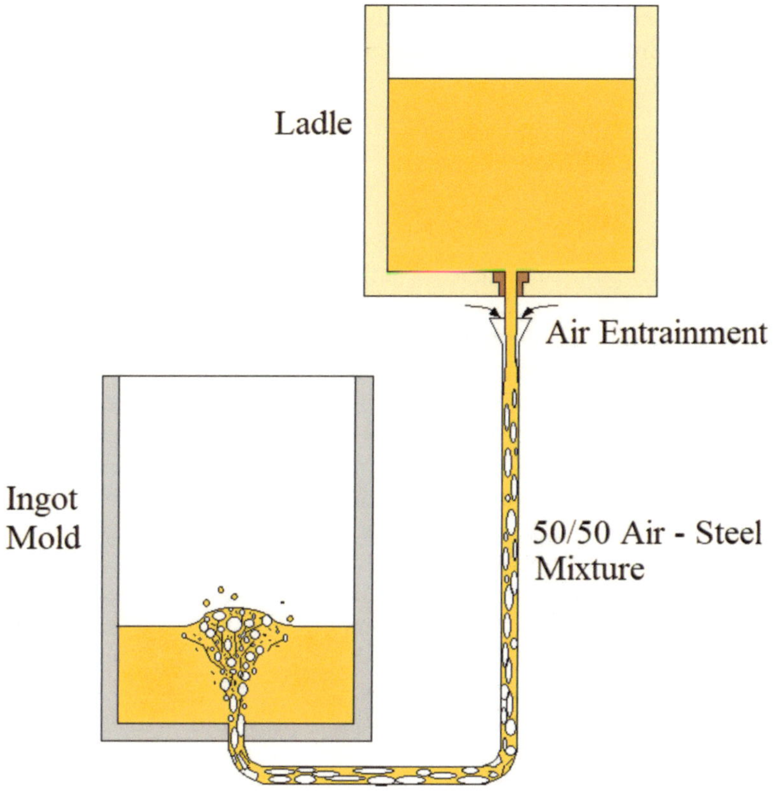

Figure 2.
Bottom-gated or uphill teemed ingot.

will occur by flotation before freezing, but an immense population of defects will remain and be trapped by the freezing process. A viewing of a water model of the filling process should convince even the most sceptical traditional steelmaker that the 50/50 emulsion of steel and air will be impossible to convert into a reasonably clean steel (Video 1, https://acrobat.adobe.com/link/track?uri=urn:aaid:scds:US:8ce23dc8-fbc5-4892-9934-439f732f3e57). We need to agree that this is not the way to treat liquid steel.

The problem is the conical intake which acts as an air pump. During the casting process, it is necessary to eliminate the massive ingestion of air at the entrance to the filling system.

Shrouds are widely used to reduce air entrainment by the trumpet. The trumpet is surrounded by an enclosure (the 'shroud') filled with argon. In this way, the percentage of oxygen from air entrained into the filling system is reduced from 20% down to perhaps 5%. However, when it is considered that pouring in a vacuum, in which the oxygen levels may be less than 0.01%, the amount of oxygen is clearly still massively over-sufficient to form very effective bifilm cracks throughout most steels. Thus, even though the use of shrouds reduces the content of oxide inclusion particles, and reduces oxygen in solution in the steel, the amount of turbulence is largely unaltered, so the area of bifilms is unchanged. However, of course, the bifilms are now thinner, so they are now much more difficult to see even though they continue to act as effective cracks.

It is necessary to conclude that the pouring of steel is *the* most damaging feature of the steelmaking process. It has to be improved if ultra-low defect steels are to be achieved.

4. Contact pour

The procedure which is practically 100% effective to reduce air entrainment during pouring is 'contact pour' in which the ladle nozzle is directly put into contact with the entrance to the filling system pipework. After the first few seconds of eliminating the air already in the filling system, the system runs clear of bubbles: entrained air is excluded (Video 2, https://acrobat.adobe.com/link/track?uri=urn:aaid:scds:US:4a0df8eb-5ec4-480a-b908-34a98165cef4). This simple solution has been demonstrated with success for those ladles with stopper/nozzle systems for steel castings over the size range of 1–50 metric tons. If contact pour can be implemented, no massive investment in new equipment would be required to revolutionise steel quality. There are additional helpful developments in which the few seconds of damaged metal and bubbles originating from the priming process can also be diverted from entering the ingot [1, 2]. Thus, it is possible for metal totally free from contamination with air to be cast into the ingot. The metal will be free from its normal population of cracks.

Even so, there are several issues that many steel casting shops would need to solve to implement the system successfully, and success may not be easy. Although the x-y positioning of the ladle is a challenge, this is not expected to be insuperable in view of modern laser triangulation and computer control of the crane. The z-control of the crane is more significant to avoid the hundred or more tons of ladle crushing the mould filling system. Most difficult is the need for some casting systems to use oxygen lancing to initiate pour. The latter may be a non-trivial requirement.

If contact pour cannot be implemented, other solutions to avoid the damage of pouring are needed. The known outstanding solution is described below.

5. Counter-gravity

Counter-gravity casting is *the* optimum solution. In **Figure 1**, the metal is contained in a vessel below the mould and is preferably treated to be of high quality. Zero entrainments of the surface oxide film occur if the process is carried out correctly. The process is widely used in the aluminium casting industry, where uphill displacement by differential pressure or pump is relatively easy and is widely used. It is not so common for the casting of steel. However, there are some notable success stories for smaller steel castings of up to 100 kg which have been routinely made in their millions with excellent freedom from inclusions. A certain amount of scaling up seems feasible, so that steel ingots of a ton or more might be possible. In particular, for large bearing rings, the ingot could be cast as a ring, saving the cost of reheating and forging.

The standard objections to the elimination of forging are the loss of so-called 'densification' and the possible loss of a so-called 'favourable grain flow', or texture. Neither of these conventional benefits of forging are to be expected to be required for an ingot without bifilms. The oxide cracks are not present to pin grain boundaries and so the alignment to develop texture which occurs in traditionally cast steels is not to be expected. Similarly, the oxide cracks are the usual initiators of porosity of various kinds, which once again will be absent, so the natural soundness of counter-gravity cast steel should be near-perfect and not improvable by plastic working. The counter-gravity cast structure is expected to have excellent, and substantially unimprovable, homogeneous properties which cannot be enhanced by forging.

6. Fatigue failure

Turning now to a practical example. **Figure 3** shows an image of the microstructure of a main outer bearing ring of carbon-chrome steel which had failed by rolling contact fatigue [5]. It seems to have been generally assumed that the large array of cracks had been formed by fatigue. This is not true.

The large array of cracks is a typical tangle of bifilms, introduced into the steel as a result of the turbulence during pouring of the originating ingot. This was probably top poured for the economy, but even if uphill teemed (bottom gated) the turbulence and air entrainment issues are immensely damaging and certainly capable of creating such extensive defects. The bifilm population in the solidifying ingot will tend to be segregated into the ingot centre because of the 'pushing' action of the advancing solidification front (advancing dendrites cannot grow through the 'air layer' in the bifilm). Probably, the solidified ingot is now forged, opening it into the shape of a ring. The inner working surface of the bearing will naturally contain the highest density of bifilm defects from the centre of the original ingot, typical of those seen in **Figure 3**.

The enlarge detail provided in **Figure 3b** shows a fractured inclusion together with light etching cracks and 'wings' on either side, as in a classical fatigue structure. The 'fractured' inclusion appears fractured because of its growth either side of a bifilm (it is worth emphasising that the 'fracture' of inclusions is not normally the result of stress, but of growth on bifilms). However, this diminutive region constitutes the real fatigue failure. One can imagine that among the massive bifilm array, of the order of millimetres in size, large blocks of metal will be stressed by the passing of the rollers, and the stress will be concentrated in those small remaining regions which connect the block to the main mass of the bearing. The gradual failure of these ligaments by fatigue will eventually release the block into the rollers, causing catastrophic failure. The size of the 'butterfly wings' is of the order of 10 μm—only 1% of the size of the pre-existing bifilm cracks, but, of course, necessary for releasing the final failure.

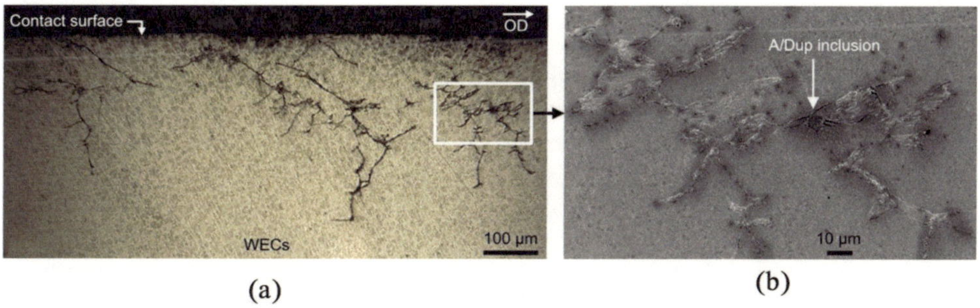

Figure 3.
(a) Array of bifilm cracks under the bearing surface; (b) inclusion with butterfly wings and adjacent white-etching cracks (courtesy ref. [5]).

In summary, extensive pre-cracks (bifilms) provide major weakening of wind turbine bearings, but final failure is by the fatigue of microscopic ligaments in which the rolling stresses are concentrated. The ligaments may or may not contain inclusions. It seems that extensive bifilm pre-cracks and microscopic fatigue cracks may be expected to be common conditions for failure. Work on the newer bainitic steels [6] is expected to *reduce* the fatigue failures of wind turbine bearings which is, of course, welcome. However, the *complete* elimination of failures is only to be expected if casting techniques can be improved [3].

7. Conclusions

Any improvement to the casting technique will be of immense value to a wide range of designers and fabricators worldwide. Without the initiation sites for failure which bifilms provide, steels (and other metals and alloys) should become free from the normal modes of failure such as tensile fracture, creep, fatigue, stress corrosion cracking and hydrogen embrittlement [3]. For the first time, the world would have metals that it could trust.

Author details

John Campbell
Faculty of Engineering, School of Metallurgy and Materials, University of Birmingham, Birmingham, UK

*Address all correspondence to: jc@campbelltech.co.uk

References

[1] Campbell J. Complete Casting Handbook. 2nd ed. Oxford OX5 1GB, UK: Netherlands Elsevier; 2015

[2] Campbell J. Mini Casting Handbook. 2nd ed. Aspect Design, Malvern: UK; 2018

[3] Campbell J. The Origin of Fracture – The Mechanisms of Metallurgical Failure. Oxford OX5 1GB, UK: Netherlands Elsevier; 2020

[4] Fox S, Campbell J. Visualisation of oxide film defects during solidification of aluminium alloys. Scripta Materialia. 2000;**43**(10):881-886

[5] Evans M-H. Literature review: White structure flaking (WSF) in wind turbine gearbox bearings; effects of 'butterflies' and white etching cracks (WECs). Materials Science and Technology. 2012;**28**(1):3-22

[6] Bhadeshia HKDH. Steels for bearings. Programs in Materials Science. 2012;**57**:268-435

Chapter 14

Wind Turbine and Synchronous Reluctance Modeling for Wind Energy Application

Tefera Kitaba

Abstract

The Chapter discusses the turbine characteristics to design low-power rating generators. The low-power machine results in small wind turbines, hence distribution of power generators have attracted a growing interest from the demand, for remote and rural electrification. In renewable energy generation the design of the generator from the wind turbine is the most challenging part of the design. The generator specifications have been obtained from wind turbine models such as torque, speed and power. Based on these specifications the design of the generator with rating of 1 kW has been achieved. The turbine characteristics have been studied and various parameters of the designed machine are analyzed through analytical model and finite element analysis.

Keywords: synchronous reluctance generator, wind turbine, finite element analysis, resistive load, inductance

1. Introduction

In recent years, the evolution of renewable energy sources such as solar, hydroelectric, wind energy, biogas and geothermal energies have gained global attention. Rahim et al. [1] reported that an isolated self-stand generating strategy is desired to achieve electrification in distant areas. The conventional stand-alone energy generation system utilizes a synchronous generator that requires a direct current field excitation. Nagria et al. [2] identified that the convectional energy generating methods are not suitable for rural electrification. Instead, a self-stand and self-excited generating system will be more convenient for such applications, as depicted in **Figure 1(b)**. For machines like synchronous reluctance machines and induction machine, self-excitation can be realized by the interconnection of excitation capacitors in star or delta across the stator end terminals, allowing them to be used as a self-stand generator [3–5]. In self excited induction generator, SEIG offers certain merits over a synchronous generator (convectional) as a source of separated to supply electric power, such as rugged structure, reduced size, low cost, low maintenance requirements and lack of DC source for excitation [5–7]. In spite of that, in SEIG the generated voltage frequency is influced by the loads and the capacitor bank. A self-excited or self-stand synchronous reluctance generator (SRG) has almost all the advantages of a

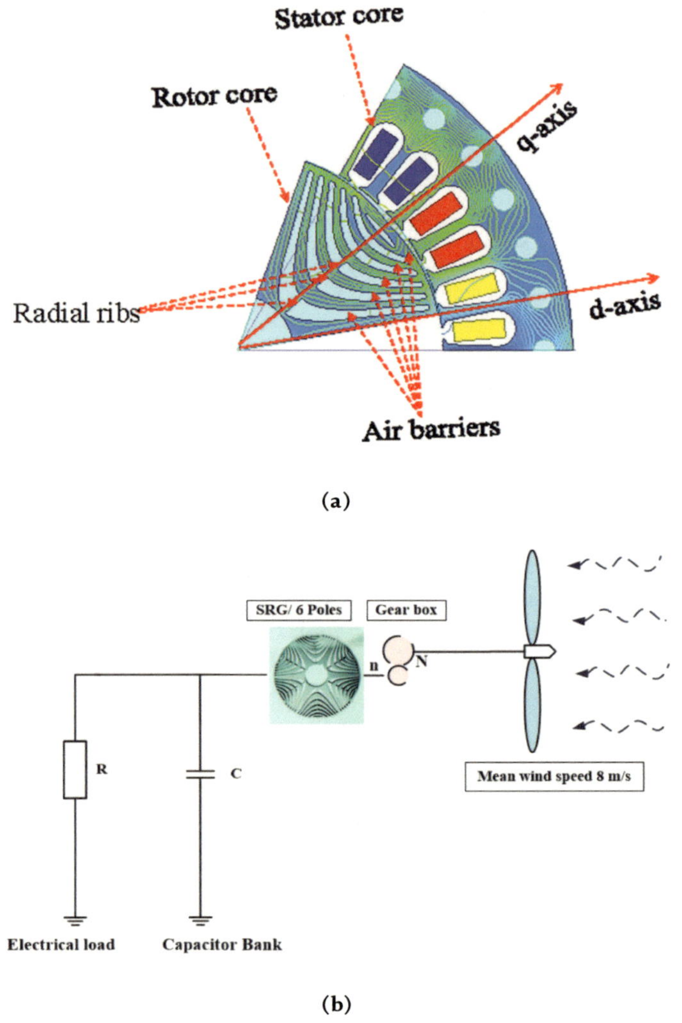

Figure 1.
(a) Dq axes reference frame of the SRG. (b) Power generating system.

self-excited induction generator. In addition, its rotor copper losses and the output frequency are not much influenced by the load, i.e., the load variations do not significantly affect the rotor speed and frequency of output voltage [8, 9].

In literature, only few mathematical models have been built to determine the performance of a self-excited synchronous reluctance generator. Abdel-Kader et al. [10] attempted to develop an equivalent circuit for the SRG in the same manner as the SEIG. Yawei Wang et al. [5] also attempted analysis and modeling of self-excited synchronous reluctance generators. But, the load is limited to no load and resistive load conditions. Moreover, the effects of core losses and saturation effects are neglected, although the losses and saturation have significant effect on the performance of the machine. Rahim H. et al. [11–14], investigated dq axes transformation based model and demonstrated its validity. T. F. Chan. et al. [12], develops a two-axis theory to model and analyze a three-phase self-excited reluctance generator which supplies to an isolated inductive load. However, in this survey, simple method to estimate the excitation capacitor is not included. Nevertheless, these research papers are mostly based on conventional salient rotor synchronous reluctance generators, i.e., no magnetic material or bridges in the rotor.

In this Chapter, wind turbine modeling, design and an analytical model in the dq rotor reference frame is developed as shown in **Figure 1(a)**. In addition, the resistive load is considered to estimate the performance of the SRG. A new and simple method to estimate the minimum capacitance requirement for the resistive load is applied.

2. Wind turbine model

Among renewable energy sources in the world, the wind energy is more environmental and economical to generate electricity. In recent days, there is remarkable expansion in the use of wind energy generation. This result in the importance in developing in turbine and generators of maga power rating. For this reason the recent work is more based on the development of high efficiency, and low cost generators for remote area applications. The rating of the turbine is in the range of kilo watt. The turbine performance coefficient in terms of blade pitch angle and tip speed ratio (TSR) is given in Eq. (1) [15]:

$$C_p(\lambda, \alpha) = C_1\left(\frac{C_2}{\lambda_1} - C_3\alpha - C_4\right)\exp\frac{-C_5}{\lambda_1} + C_6\lambda \tag{1}$$

Letting,

$$\lambda_1 = \frac{(\alpha^3+1)(\lambda+0.08\alpha)}{\alpha^3 - 0.028\alpha - 0.035\lambda + 1} \tag{2}$$

For various types of wind turbine, the values of C1 $\sim$ C6 are different. **Figure 2** depicts the group power coefficient curves and tip speed ratio for C1, C2, C3, C4, C5 and C6 as 0.200, 119, 0.4, 5.5, 12.5, and 0, respectively. The turbine output power is given in (watts)

$$P_m = \frac{1}{2}R^3\pi\rho_a C_p(\lambda, \alpha)V_w^3 \tag{3}$$

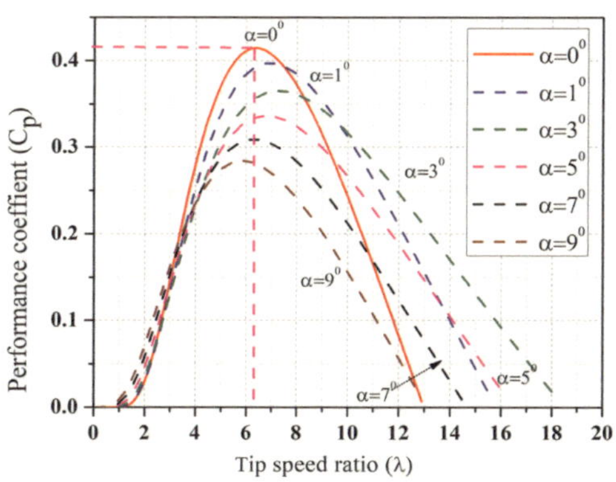

Figure 2.
Performance coefficient at pitch angle α = 0.

The value of λ has been selected for optimal point of C_p at α = 0 as shown in **Figure 2**. Once the values of λ and C_p are found, the turbine speed and the radius are calculated from (3) and (4). The tip speed ratio can be determine from:

$$\Omega = {}^{\lambda v_w}\!/_{R} \tag{4}$$

The torque causing the rotation of the wind turbine shafts depends on the turbine rated power output and angular velocity. It can be expressed as:

$$T_m = {}^{P_m}\!/_{\Omega} = \frac{1}{2\lambda} R^4 \pi \rho_a C_p(\lambda, \alpha) \mathrm{V}_w^2 \tag{5}$$

The gearbox is utilized to transfer torque to the generator shaft rotating at a higher speed in the wind turbine. The wind turbine side to generator side gear ratio can be expressed as:

$$R_G = {}^{\omega}\!/_{\Omega} \tag{6}$$

Here, Ω and ω are the turbine and rotational speed of the generator, respectively. Eqs. (1) to (6) are used to design a wind turbine that can produce an output shaft power of 1 kW at a rated mean wind speed of 8 meter per second.

The designed parameters of the wind turbine is summarized in **Table 1**. The variation of power coefficient of the wind turbine for different blade pitch angle α is shown in **Figure 2. Figure 3**, shows the generated mechanical power at different wind velocity. From **Figure 3**, it is observed that as the wind speed increases, the rotational speed of the turbine also needs to be increased to extract maximum power out of the turbine. It is also observed that for the designed wind turbine, the maximum power attains near 32 rad/s for the average wind speed of 8 m/s.

Figure 4, shows the torque causes mechanical rotation at different wind velocity. From **Figure 4**, it can be observed that as the wind speed increases, the rotational speed of the turbine also needs to be increased to extract maximum rotational torque of the turbine.

Parameters	**Values *(Units)***
TSR(λ)	6.4
Mean wind speed	8 m/s
Gearing ratio	4.9
Blade speed	32 rad/s
Performance coefficient limit	0.414
Mechanical transmission efficiency	90%
Mechanical power	1 kW
Mechanical torque of turbine rotor	31.36 Nm
Mechanical torque of generator shaft	6.4 Nm
Generator rated speed	157.08 rad/s
Turbine blade radius	1.6 m

Table 1.
Initial data and calculated values of the wind turbine parameters.

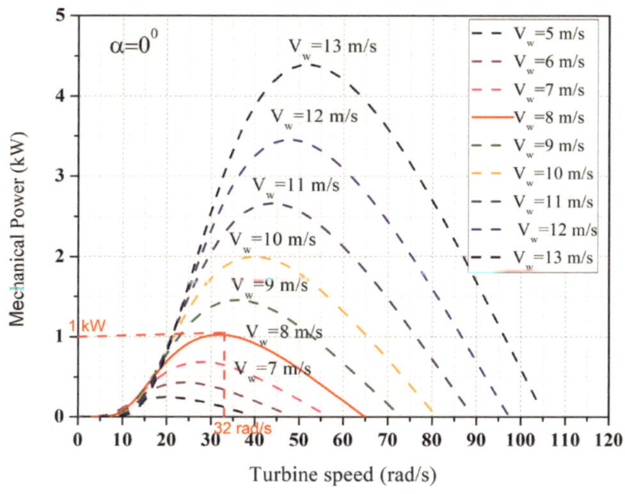

Figure 3.
Turbine power characteristics at pitch angle α = 0.

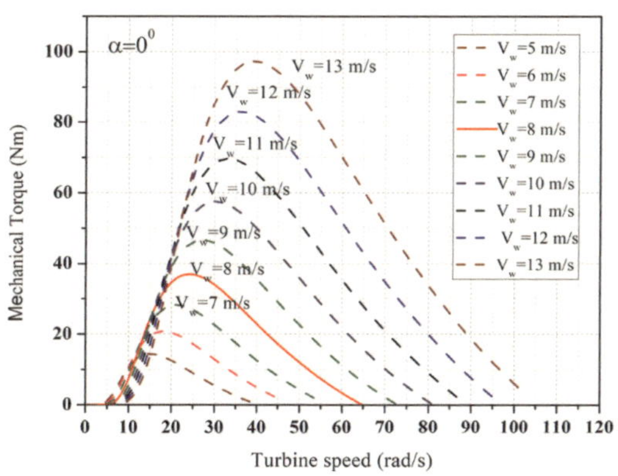

Figure 4.
Turbine torque characteristics at pitch angle α = 0.

3. SRG design algorithm

The sizing procedure of the proposed generator starts with assigning the initial key parameters of wind turbine such as, speed and maximum torque. These assigned parameters are used in the calculations of magnetic, geometric and electric parameters in together with the analytical model of the generator.

The design of the generator starts with the precondition design output parameters such as stator geometry, rotor outer diameter etc. A new repetition with revised assigned parameters will be done, if the geometries such as stack length and stator outer diameters cannot satisfy the design parameter requirements. Otherwise, the estimated magnetic, electric and geometrical parameters are used as the design parameters. In other words, stator geometry, inductances, winding specifications, maximum torque, saliency ratio, etc., are determined if the required conditions are satisfied. Finite element (Ansys Electronic Desktop) software is used to analyze the generator's performance related to the output functions such as torque quality, maximum developed torque, and the magnetic properties such as, magnetic field H and magnetic field B. The process ends if the Ansys Electronic Desktop software results satisfy the design requirements. Otherwise, the process is repeated by updating the assigned parameters such as tip speed ratio, efficiency coefficient, air mass density, current and magnetic loading, pole pitch to air gap ratio and stack aspect ratio to obtain the proper size (**Figure 5** and **Table 2**).

The design constraint need to be satisfied are:

The power rated 1 kW, with electromagnetic torque ≥6.4 Nm, maximum torque ripple ≤9%, maximum back-emf ≥ 100 V.

3.1 Analytical modeling of SRG

For the developed analytical model, the following simplifying assumptions are considered:

- The time harmonics in current and space harmonics in air gap flux are neglected
- The core loss resistance is assumed to be constant and has no effect on excitation.

Pole pitch τ, as given in the equation below, is the prime parameter to obtain the outer rotor diameter of the generator.

$$\tau = \sqrt{\frac{T_e k_d^2 \mu_o}{B_m^2 P^2 (k_d - k_q)\sqrt{L_d/L_q} k_c (1 + k_s) l_g L/\tau}} \tag{7}$$

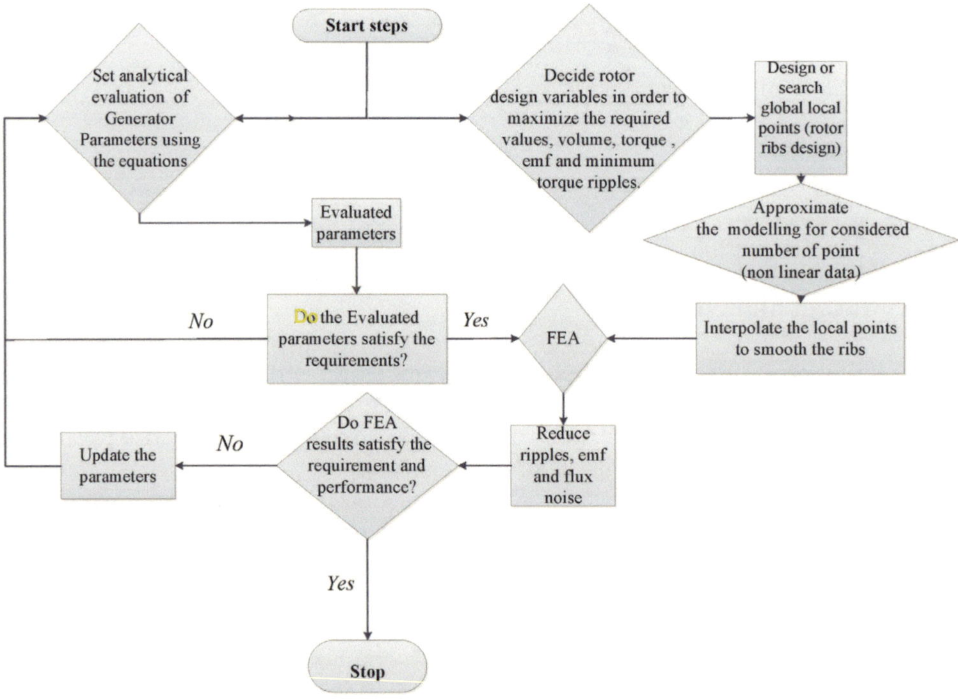

Figure 5.
Flow chart of SRG design process.

Dimension of the Generator	Values
B_{in}	[0.5, 1] T
k_d	[0.6, 1]
k_q	[0, 0.4]
L/τ	[0.3, 6]
v_w	[5, 25]m/s
k_u	[0.4, 0.7]
J	[4, 9] A/mm^2
S	36
P	3

Table 2.
Assigned parameters.

The outer rotor diameter, D_r and stack length, L could be:

$$\begin{cases} D_r = \dfrac{2P\tau}{\pi} \\ L = \tau\left(\dfrac{L}{\tau}\right) \end{cases} \tag{8}$$

where, T_e, L/τ, $k_c = 1.05$, $ks = 1.4$, P, k_d, and k_q are generator torque from the wind turbine, stack aspect ratio, Carter factor, saturation factor, number of pole pairs, the ratio of $d-$ and $q-$ axes inductance to magnetizing inductance, respectively. The parameters are defined in **Table 2**, while the saliency ratio defined as $L_d/L_q = (k_d - k_q)/2k_q$.

The design of the stator core geometry, i.e., the stator slot dimensions and rotor design in details the structure of rotor ribs and stator have been determined. The distance of the rotor air gap ribs from the shaft radius is designed. It also shows the separation of the edge of the ribs along the inner core radius of the rotor with an angle of ∂_m.

The segments/points are selected and are interpolated to get the structure of the 6-poles rotor [16].

Every flux, and air barrier consists of trapezoid shape segment with a radial thickness and tangential thickness, and the end point angle other air gap, ∂_m. The parameter is designed in such a way that to ensure required electromagnetic and the structure stability at high speed.

The maximum rotor tips mechanical end point angle, ∂m expressed in terms of number of flux barriers (q_i), poles pair (P) and floating angle (β) is given as (9):

$$\partial_m = \left(\frac{\pi}{2P} - \beta\right) / (q_i + 1/2) \tag{9}$$

Here, the floating angle β assumed to be in between 0 to 10° (0 to π/18 rad).

The total slot, $d-$ and $q-$ axes components of ampere turns can be expressed as:

$$nI_m = \sqrt{(nI_d)^2 + (nI_q)^2} \tag{10}$$

Here,

$$nI_d = \frac{B_m \pi (1 + k_s) k_c l_g}{3\sqrt{2}(q k_w k_d \mu_o)} \tag{11}$$

and,

$$nI_q = nI_d \sqrt{L_d / L_q} \tag{12}$$

where, $I_m = I_{max}$, and k_w is stator slot winding factor, $k_w = 0.955$. The conductor per slot turns n is one of the key parameter in the design such as stator resistance, leakage inductance and machine inductances. The resistance per phase R, is:

$$R = \frac{2(L + L_e) P J n q \rho_r}{I_m} \tag{13}$$

where, L_e is end winding length, $L_e = \pi\tau/2$, J is current density, and ρ_r is copper resistivity at temperature of 120°C. The leakage inductance L_{ls} is given as:

$$L_{ls} = (c_s + c_e + c_a) 2 L P q n^2 \mu_o \tag{14}$$

Here, the calculated slot permeance, $c_s = 2.12535$, air gap coefficient, $c_a = 0.2$ and the end winding length coefficient, $c_e = 0.00071817$. The magnetizing inductance L_m, for uniform air gap can be presented as:

$$L_m = \frac{6 \mu_o \pi L P (q k_w n)^2}{\pi^2 l_g k_c (1 + k_s)} \tag{15}$$

Therefore, for double layer winding, the stator resistance, leakage inductance, magnetizing inductance, $d-$, and $q - axes$ inductances are the function of n. By simplifying the calculation, they are expressed as in **Table 3**:

Using the d-q – axes rotor reference frame in **Figure 1(a)**, equations of the SRG in the transient state are written as below.

$$\begin{cases} V_d = R_s I_d + \rho \lambda_d - \omega P \lambda_q \\ V_q = R_s I_q + \rho \lambda_q + \omega P \lambda_d \\ V_{ph} = \sqrt{V_d^2 + V_q^2} \end{cases} \tag{16}$$

Parameters	Magnitude
R_s	$0.92\text{x}10^{-3}\ (n)^2\ \Omega$
L_{ls}	$1.4\text{x}10^{-6}\ (n)^2$ H
L_m	$2.99\text{x}10^{-5}\ (n)^2$ H
L_d	$2.93\text{x}10^{-5}\ (n)^2$ H
L_q	$2.993\text{x}10^{-6}\ (n)^2$ H

Table 3.
Parameters expression.

At steady state, the voltage equation can be represented as:

$$V_{ph} = n\sqrt{10} \tag{17}$$

The above Eqs. (16) are written utilizing the motor convention for the reluctance machine. When synchronous reluctance machine work in generating mode, it converts mechanical energy in to electrical energy, however, it requires capacitor over reactive p to magnetize their magnetic field paths for self-excitation. In motor mode, I_q and I_d are of the same sign, while in generator mode they are of opposite sign. Referring to the current q and d-axes frame, motor operations are in the first and third quadrants, in other words the motoring mode is when I_q and $I_d < 0$ or I_q and $I_d > 0$, $T_e > 0$, generating operations are in the second and fourth quadrants which means $I_d > 0$ and $I_q < 0$ or $I_d < 0$ and $I_q > 0$, $T_e < 0$. **Figures 6** and **8**, show the reluctance torque developed by the SRG. Neglecting the effect of stator resistance, the torque equation is given as:

$$T_e = \frac{3}{2}\mathrm{P}(\mathrm{L_d} - \mathrm{L_q})\mathrm{I_d I_q} = \frac{3}{2}\mathrm{P}(\lambda_d \mathrm{I_q} - \lambda_q \mathrm{I}_d) \tag{18}$$

Since, the wind turbine rotates at low speed, gearbox are utilized to increase the speed of rotation of generator shaft. The swing-equation corresponding to the combined turbine generator system is given as:

$$\rho\omega = \frac{1}{2J_i}(\mathrm{T}_m - T_e - 2\mathrm{B}\omega) \tag{19}$$

3.1.1 Excitation capacitance and load modeling

The inductive load R_L, X_L is connected to a capacitor bank in shunt at the stator terminals.

The equation which relates the stator current, load current, and terminal voltages are presented as follow:

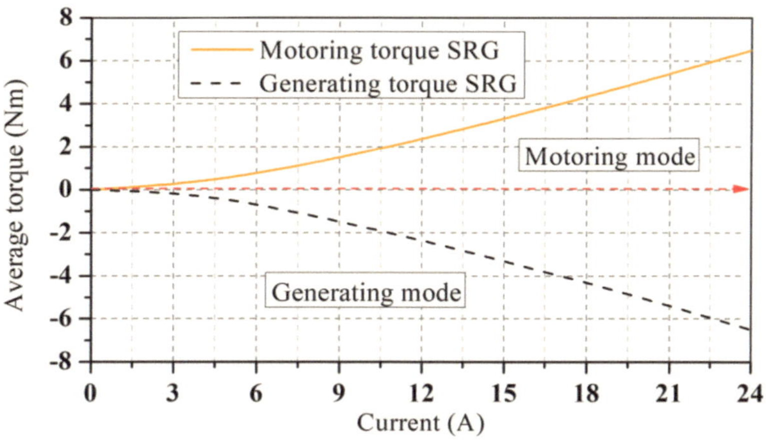

Figure 6.
Torque verses current curves of the machine.

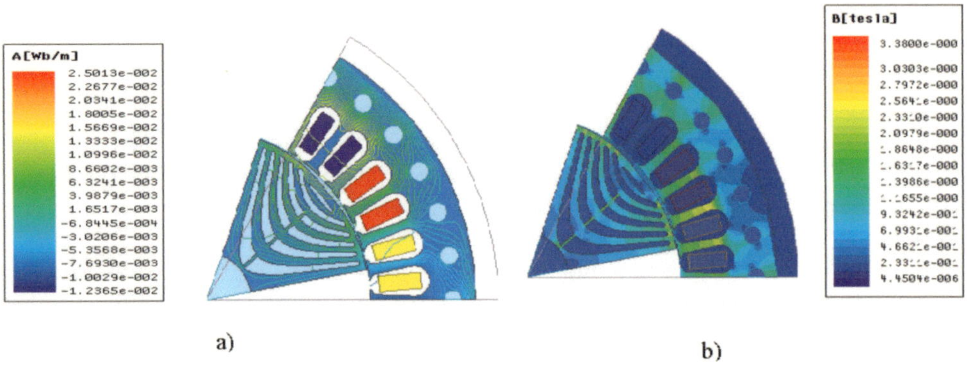

Figure 7.
Flux lines and magnetic flux density distribution of the machines.

$$\begin{cases} I_{dc} = -I_d - I_{dL} \\ I_{qc} = -I_q - I_{qL} \end{cases} \tag{20}$$

The excitation capacitance in rotor reference frame is as follows (21)

$$\begin{bmatrix} I_{dc} \\ I_{qc} \end{bmatrix} = \begin{bmatrix} \rho C & \omega PC \\ -\omega PC & \rho C \end{bmatrix} \begin{bmatrix} V_{dL} \\ V_{qL} \end{bmatrix} \tag{21}$$

Whereas the voltages in rotor reference frame are given as:

$$\begin{cases} \rho V_{qL} = \omega PV_{dL} + \left(-I_q - I_{qL}\right)/C \\ \rho V_{dL} = -\omega PV_{qL} + \left(-I_d - I_{dL}\right)/C \end{cases} \tag{22}$$

The R-L load model are obtained as (23)

$$\begin{cases} I_{qL} = \int \frac{1}{L}\left(V_{qL} - I_{qL}R_L + \omega PLI_{dL}\right)dt \\ I_{dL} = \int \frac{1}{L}\left(V_{dL} - I_{dL}R_L - \omega PLI_{qL}\right)dt \end{cases} \tag{23}$$

Eq. (23) is obtained using a general balanced RL load model ($V = R_L I + LdI/dt$). **Table 4** summarizes the calculated parameters, the designed parameters of the generator, and their performance.

3.1.2 Resistive load condition

For the resistive load case, the capacitor C, and the load, R_L are connected in parallel to the stator terminals, as shown in **Figure 8**. Therefore, the impedance can be determined as:

$$Z = -BjR_L + BX_c \tag{24}$$

Where,

Parameters	Quantity	Parameters	Quantity
L_m	17.24 mH	D_r	104 mm
L_d	16.9 mH	D_{in}	105 mm
L_q	1.73 mH	D_o	180 mm
L_{ls}	0.80 mH	h	13 mm
I_m	20.6 A	n	24
L	40 mm	nI_q	453.6 AT
l_g	0.5 mm	nI_d	195.2 AT
k_u	0.5	E_{ph}	100 V
R_s	0.53 Ω	nI_m	493.8 AT

Table 4.
Evaluated performance parameters and approximated quantity of 1 kW of SRG.

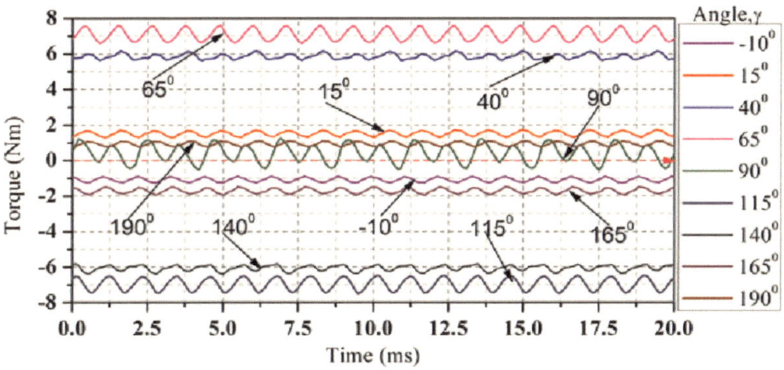

Figure 8.
Variation of SRG electromagnetic torque with power angle, for $R_s \neq 0$.

$$B = \frac{X_c R_L}{{R_L}^2 + {X_c}^2} \tag{25}$$

Which provides voltage at terminal

$$\begin{aligned} V &= -IZ = (-BX_c + jBR_L)\left(I_d + jI_q\right) \\ &= -B(X_c I_d + R_L I_q) + jB(R_L I_d \text{-} X_c I_q) \end{aligned} \tag{26}$$

The voltage equations can be reduced to:

$$\begin{aligned} V_d &= R_s I_d - X_q I_q = -B(X_c I_d + R_L I_q) \\ V_q &= R_s I_q + X_d I_d = B(R_L I_d - X_c I_q) \end{aligned} \tag{27}$$

The values of minimum capacitance required can be determined for self-excited synchronous reluctance generator under resistive condition is by rearranging Eq. (27) and eliminate I_d and I_q (**Figure 9**).

3.1.3 Finite element analysis of SRG

The present section includes the finite element validation of the proposed design of the SRG, as shown in **Figure 1(a)**. The finite element analysis (FEA) performance is evaluated as per the proposed design specifications. The maximum induced electromotive forces (emfs) in the stator winding of the 1 kW SRG with symmetric design are shown in **Figure 10**. The emf in the synchronous reluctance machine is given by Eqs. (16), which clearly indicates that if the effective q-axis flux is reduced, it leads to a decrease in emf. The results obtained through finite element analysis/ simulation with the excitation current of the machine is shown in **Figure 10**.

The peak values of the back emf produced at 1500 *rpm* corresponding to different magnetizing currents for the machine is shown in **Figure 11**. It is observed that, emf starts with zero voltage.

Figure 6 shows that the performance of the machine in motor and generator modes. From **Figure 6**, it is clear the average torque is as a function of square of stator current till the current of the machine is 9A. But, after 9 A the different between (L_d-L_q) is approximately constant, hence, the variation of average torque is observed to be linear.

Figure 8 represents the electromagnetic torque of the machine with variations of γ. Moreover, as the magnitude of average torque increases, the torque ripple also increases and viva-versa.

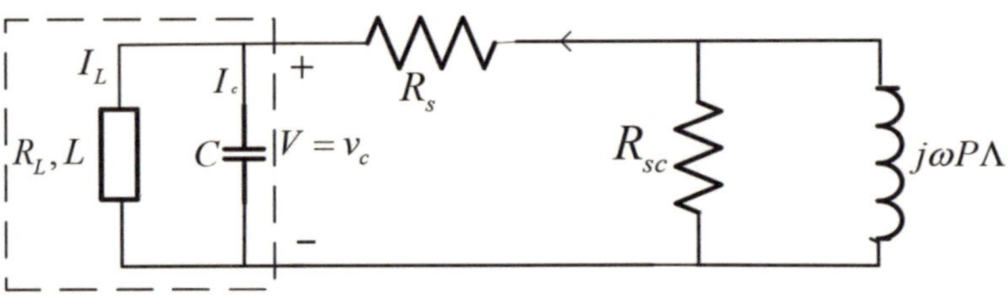

Figure 9.
Synchronous reluctance generator equivalent circuit with capacitor C, resistive R_L and inductive X_L loads ($V = V_d + jV_q$, $I = I_d + jI_q$, $\Lambda = \lambda_d + j\lambda_q$).

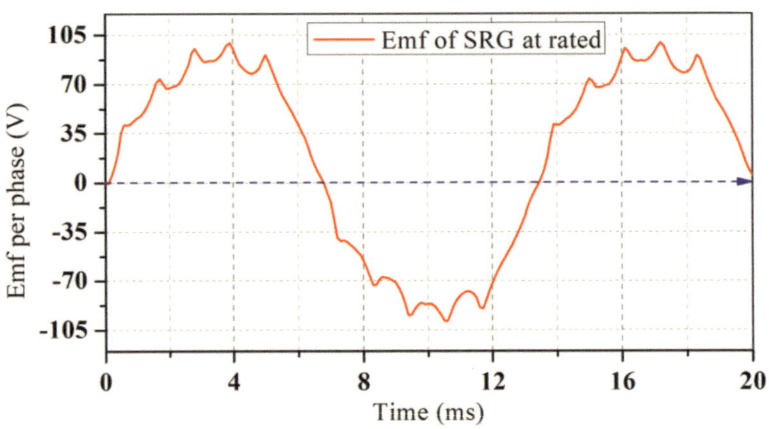

Figure 10.
Machine emf at given frequency.

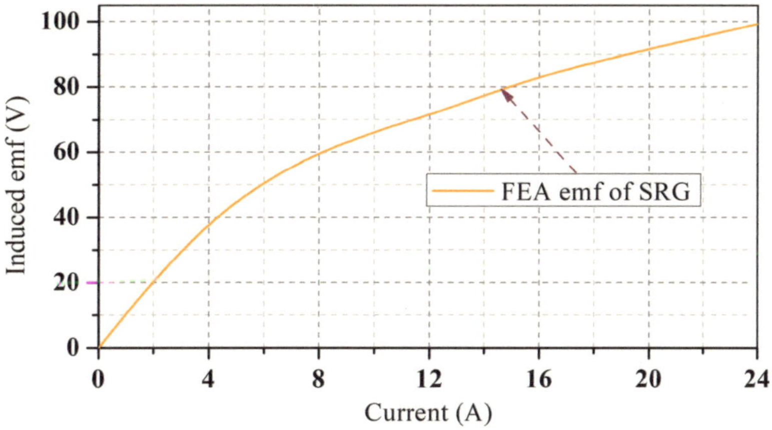

Figure 11.
Peak emf curves as a function of current.

Parameters	Analytical	FEA
Te	6.4 Nm	6.493 Nm
L_{ls}	0.80 mH	0.78 mH
L_m	17.42 mH	18 m H
L_d	16.9 mH	17.4 mH
L_q	1.73 mH	1.65 mH
T_r	—	9%

Table 5.
Analytical, and FEA results of the machines.

Table 5 provides the performance of self-excited SRG generator. From **Table 5** and **Figure 7**, the performance, flux line, and field density are observed. Overall, it can be said that the design of SRG is more robust and less costly. Since, cost effectiveness and robustness are major criteria for the suitability of generator for rural electrification application, SRG is more suitable for rural electrification. **Figure 7** provides magnetic field density distribution and flux lines of SRG generator.

Table 5 provided that the performance comparison of torque, linkage inductance, magnetizing inductance, d and q-axes, and ripple torque percentage. It can be observed that the analytical and FEA similar with small deviation.

4. Conclusion

The design and the modeling of synchronous reluctance generator for 6 poles, 1500 rpm, 1 kW, from wind turbine modeling specifications, are presented. The performance verses tip speed, mechanical power and torque verses turbine speed have been evaluated. The rotor design reducing q-axis inductance of this generator have been analyzed. Therefore, the torque ripple has been reduced. The relationship between generated emf voltage, and torque with the change of time are evaluated.

The effects of stator resistance on electromagnetic torque with variation of power angle have been considered. The design algorithm of reluctance generator are analyzed. Using finite element, the performances of the machine for field density, and flux line are also determined. With increase in current the performance of developed torque and generated voltage have been presented. The analytical and finite element results are evaluated and compared.

Appendices and nomenclature

ω	Generator rated speed (mechanical)
ω_e	Generator rated speed (electrical)
k_u	Fill factor
B_m	Air gap flux density
k_d	Ratio of L_d/L_m
k_q	Ratio of L_q/L_m
L_s/τ	Stack aspect ratio
V_w	Mean wind speed
J	Current density
ρ_a	Air density
P	Number of pole pairs
Ω	Turbine tip speed
E_{ph}	Emf induced per phase
G_r	Gear ratio
λ	Tip speed ratio
D_o	Stator air gap diameter
D_{in}	Stator yoke diameter
D_r	Rotor outer diameter
R_s	Phase resistance
R_t	Wind turbine radius
L_{ls}	Stator leakage inductance
I_d	d-axis current
I_q	q-axis current
X_d	d-axis reactance
X_q	q-axis reactance
n	Number of conductor turns per phase
l_g	Air gap length
λ_q	q-axis flux
λ_d	d-axis flux
L_q	q-axis inductance
L_d	d-axis inductance
L	Active stack length

Author details

Tefera Kitaba
Department of Electrical Power Engineering, Ethiopian Defense University College of Engineering, Bishoftu, Ethiopia

*Address all correspondence to: nasify2016@gmail.com

References

[1] Rahim Y, Fletcher J, Hassanain N. Performance analysis of salient-pole self-excited reluctance generators using a simplified model. IET Renewable Power Generation. 2010;**4**(3):253-260

[2] Nagrial M, Rahman M. Operation and characteristics of self-excited reluctance generator. In: Industry Applications Society Annual Meeting, 1988, Conference Record of the 1988 IEEE. USA: IEEE; 1988. pp. 55-58

[3] Elder J, Boys J, Woodward J. The process of self excitation in induction generators. In: IEE Proceedings B-Electric Power Applications. Vol. 130, no. 2. IET; 1983. pp. 103-108

[4] Boldea I. Reluctance Synchronous Machines and Drives (Monographs in Electrical and Electronic Engineering). USA: Oxford. 2002

[5] Wang Y, Bianchi N. Investigation of self-excited synchronous reluctance generators. IEEE Transactions on Industry Applications. 2018;**54**(2):1360-1369

[6] Maroufian SS, Pillay P. Self-excitation criteria of the synchronous reluctance generator in stand-alone mode of operation. IEEE Transactions on Industry Applications. 2018;**54**(2):1245-1253

[7] Murthy S, Malik O, Tandon A. Analysis of self-excited induction generators. In: IEE Proceedings C (Generation, Transmission and Distribution). Vol. 129, no. 6. IET; 1982. pp. 260-265

[8] Bianchi N, Fornasiero E, Ferrari M, Castiello M. Experimental comparison of pm-assisted synchronous reluctance motors. IEEE Transactions on Industry Applications. 2016;**52**(1):163-171

[9] Stipetic S, Zarko D, Kovacic M. Optimised design of permanent magnet assisted synchronous reluctance motor series using combined analytical–finite element analysis based approach. IET Electric Power Applications. 2016;**10**(5): 330-338

[10] Abdel-Kader F. The reluctance machine as a self-excited reluctance generator. Electric Machines and Power Systems. 1985;**10**(2–3):141-148

[11] Liu C-T, Luo T-Y, Hwang C-C, Chang B-Y. Field path design assessments of a highperformance small-power synchronous-reluctance motor. IEEE Transactions on Magnetics. 2015;**51**(11):1-4

[12] Chan T. Steady-state analysis of a three-phase self-excited reluctance generator. IEEE Transactions on Energy Conversion. 1992;**7**(1):223-230

[13] Bianchi N, Bolognani S, Bon D, Dai Pre M. Rotor flux-barrier design for torque ripple reduction in synchronous reluctance and pm-assisted synchronous reluctance motors. IEEE Transactions on Industry Applications. 2009;**45**(3):921-928

[14] Mohamadein A, Rahim Y, Al-Khalaf A. Steady-state performance of self-excited reluctance generators. In: IEE Proceedings B-Electric Power Applications. Vol. 137, no. 5. IET; 1990. pp. 293-298

[15] Holttinen H, Peltola E, Tuhkanen S. Renewable Energy Sources and Distributed Energy Production. Technology and Climate Change CLIMTECH 1999; 2002. p. 5

[16] Tefera K, Tripathy P, Adda R. Design and Modelling of Self-Excited SRG and FM-SRG for Wind Energy Generation. IET Renewable Power Generation Published by John Wiley & Sons Ltd; 2021. pp. 1-17

Chapter 15

Methods and Devices for Wind Energy Conversion

Janis Viba, Vitaly Beresnevich and Martins Irbe

Abstract

The chapter deals with the analysis and optimization of the operational safety and efficiency of wind energy conversion equipment. The newly proposed method of wind energy conversion involves flat blades or space prisms that perform translation motion due to the interaction with air flow. Air flow interactions with 2D moving prisms (convex, concave) are studied by computer simulation. Optimization of prism shape is made using as criteria maximum of generating force and power. Theoretical results obtained are used in the designing of new devices for energy extraction from airflow. Models of wind energy conversion devices equipped with one vibrating blade are developed (quasi translatory blade's motion model; model with vibrating blade equipped with crank mechanism). The operation of the system due to the action of air flow is simulated with computer programs. Possibilities to obtain energy with generators of different characteristics, using mechatronic control, have been studied. The effect of wind flow with a constant speed and also with a harmonic or polyharmonic component is considered. Partial parametric optimization of the electromechanical system has been performed. The serviceability and main advantages of the proposed methods and devices are confirmed by experiments with physical models in a wind tunnel.

Keywords: air flow, vibrating blade, energy conversion, computer simulation, experiments

1. Introduction

Various types of methods and devices are used for energy extraction from airflow. The operation principle of existing wind energy conversion systems is mainly based on air flow action on blades mounted on a special wheel and further transformation of air flow kinetic energy into the mechanical energy of wheel rotation [1, 2]. But as it turned out in practice, such a design does not provide the desired position of the blade against the air flow in relation to the rotating wheel. Its position is only optimal at certain wheel angles.

The interaction between the rotating plate and the air flow under different aerodynamic conditions is analyzed in [3]. The special orientation of a plana-shaped object during movement is realized in [4, 5]. The main disadvantage of these devices lies in a large number of blades. For example, it has been argued in [6] that turbine operating power efficiency decreases with an increasing number of active blades. This means that the front (airflow side) blades prevent airflow from accessing the rear blades.

Consequently, the interaction of the rear blades with the air flow differs significantly in the direction and size from the interaction of the front blades. This is because air vortices are generated in the system. Besides, it is known that increased efficiency of existing equipment can be achieved by increasing the radial grooves of the blades. However, such a solution has a negative effect on the use of wind turbines, as the final speed of the blade rotation increases and the generated noise becomes higher. In addition, it becomes possible to kill birds and other living things with a rotating flat blade.

This chapter discusses some new methods for describing wind interactions with rigid bodies and provides recommendations for wind energy conversion based on the use of flat blade translational motion excited by air flow.

2. New approach to the air flow interaction with a moving rigid body

The main focus of the present work is to investigate the stationary air flow interaction with rigid body and extend the interaction concept for non-stationary body-air flow interaction without requiring "space–time" programming techniques [5]. In accordance with the concept proposed, a space around rigid body interacting with the air medium is split into several zones (see **Figure 1**).

It has been found in theory and practice that the non-stationary interaction of air flow can be divided into two parts using the principle of superposition. For this purpose, the interaction can be considered within two zones: the frontal pressure zone and the rear intake zone. In addition, it is possible to separate slow movements from fast air particle movements (i.e., from Brownian chaotic particle movements).

Hereinafter, this approach is used to study air flow interaction with flat blades and space prisms that perform translation motion.

2.1 Object interaction with a windless air medium

The model of moving rigid body interaction with air medium is shown in **Figure 2**.

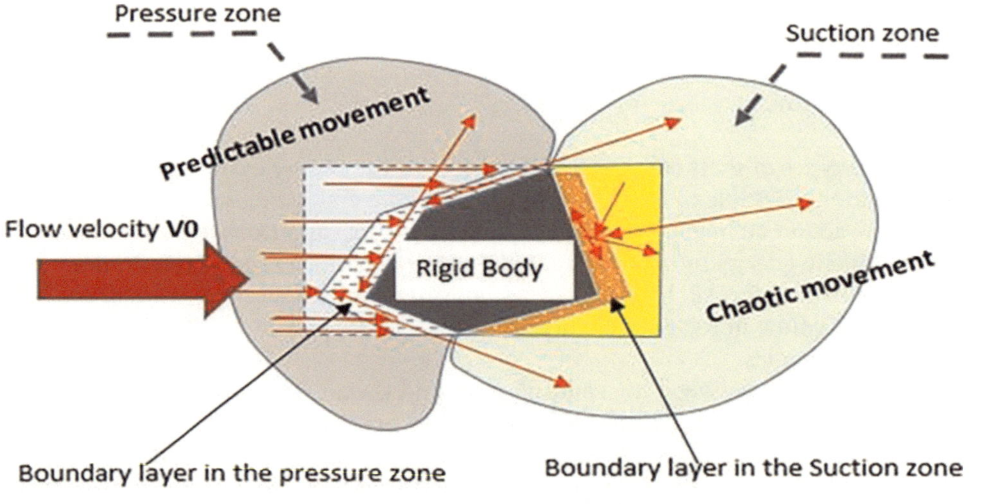

Figure 1.
Concept of zones (pressure and suction zones) for a rigid body immersed in an air flow.

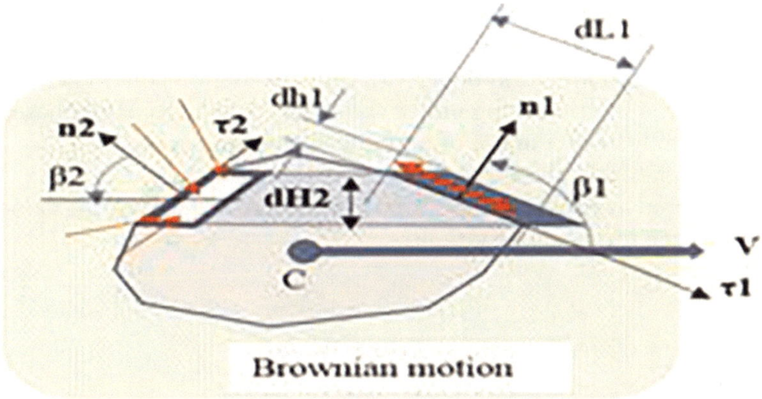

Figure 2.
Interaction of the rectilinearly translationally moving body (prism) with a windless air medium.

By applying the theorem of momentum change in the differential form [7] to a very small air element in the pressure zone and accordance with the superposition principle, the following system of equations can be received in the projection on the area normal n_1 before and after collisions (air–body interaction), taking account of Brownian motion:

$$\mathrm{m_{10} \cdot VB_1 - (-m_{10} \cdot VB_1) = -N_1 \cdot dt, m_{10} = VB_1 \cdot dt \cdot dL_1 \cdot B \cdot \rho} \quad (1)$$
$$\mathrm{p_{10}} = \frac{|\mathrm{N_1}|}{\mathrm{dL_1 \cdot B}},$$

where m_{10} is Brownian interaction mass; VB_1 is an average value of air normal velocity within the pressure zone; N_1 is a force directed along a normal to a small element of air medium; dt is an infinitely small-time interval; dL_1 is a width of a small element; B is a prism height in the direction perpendicular to the plane of motion; ρ is air density; p_{10} is atmospheric pressure in the pressure zone.

Considering body and air interaction at the windward side (pressure side), the following system of equations can be formed:

$$m_1 V \cos(\beta_1) - 0 = -\Delta N_1 \cdot dt, m_1 = V \cos(\beta_1) \cdot dt \cdot dL_1 \cdot B \cdot \rho, \quad (2)$$
$$\Delta p_1 = \frac{|\Delta N_1|}{dL_1 \cdot B},$$

where m_1 is a mass due to prism interaction with air in boundary layer; V is a velocity of prism; β_1 is an angle between velocity V and normal n_1; ΔN_1 is an additional normal force acting on a prism; Δp_1 is an increment of pressure in the windward side.

By the solution of the system of Eqs. (1) and (2), six unknown parameters can be found. From the practical point of view, the most required are parameters p_{10} and Δp_1, which can be determined by the following calculations:

$$p_{10} = 2 \cdot {VB_1}^2 \cdot \rho \cdot dt, \quad (3)$$

$$\Delta p_1 = \rho \cdot dt \cdot V^2 [\cos(\beta_1)]^2. \quad (4)$$

Besides, it is possible to apply a mathematical model similar to Eq. (1)–(4) in the suction zone (leeward side). However, the task is complicated a little due to the increasing number of momentum differentials in the suction zone. Therefore, it is suggested to find the solution using one or the other hypothesis. Hypotheses should be tested experimentally or by the use of numerical computer programs.

The first hypothesis. In the suction zone, pressure reduction Δp_{21} over the entire surface is considered as constant and proportional to the square of the velocity V in accordance with the following equations:

$$\Delta p_{21} = -\rho \cdot C_1 \cdot V^2, \tag{5}$$

$$p_{20} = 2VB_2{}^2 \cdot \rho \cdot dt, \tag{6}$$

where C_1 is a constant found according to the experimental or numerical simulation; VB_2 is an average air normal velocity in the suction zone.

The second hypothesis. It is assumed that in the suction zone, pressure reduction Δp_{22} over the entire surface is not constant, but is proportional to the square of the velocity V and also depends on the normal n_2 to the surface area and position angle β_2. Thus, the following equations can be obtained:

$$\Delta p_{22} = -\rho \cdot C_2 \cdot V^2 \cos(\beta_2), \tag{7}$$

$$p_{20} = 2VB_2{}^2 \cdot \rho \cdot dt, \tag{8}$$

The obtained Eqs. (3)–(8) can be used in the engineering analysis and synthesis tasks in the low-velocity range and for bodies that undergo rectilinear translation motion. For practical engineering calculations, it is recommended to adopt $VB_1 = VB_2$ for low-velocity ranges $V << VB_1$ and $V << VB_2$. Then it is assumed $p_{01} = p_{02} = p_0$, where p_0 is the mean atmospheric pressure around the given prism.

2.2 Stationary rigid body (prism) interaction with air flow

The model of airflow interaction with a stationary prism is shown in **Figure 3**.

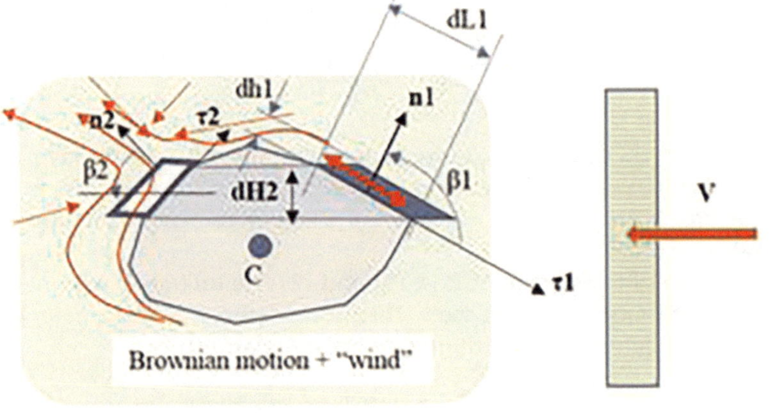

Figure 3.
Model of air flow interaction with stationary prism.

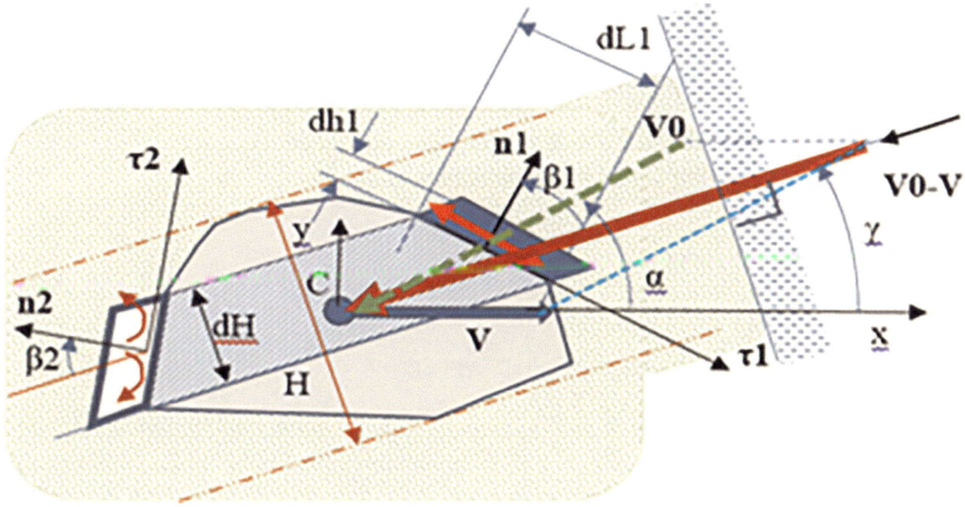

Figure 4.
Model of air flow interaction with moving rigid body.

Airflow interaction with a stationary prism (**Figure 3**) is dependent on the extra velocity and extra kinetic energy of air particles. However, by applying the interaction concept to relative motion, it is possible to use the Eqs. (3)–(8) in the engineering calculations of systems with air flow velocity.

2.3 Moving rigid body (prism) in an air flow

The model of air flow interaction with moving prism is shown in **Figure 4**.

In this case, the relative motion velocity Vr vector in the pressure zone must be recalculated by determining the angle γ from the elementary parallelograms with normal directions n_1 and n_2 (**Figure 4**). By projecting the vectors V and V_0 onto the x and y axes, the following formulas are obtained:

$$Vr = \sqrt{(-V_0 \cdot \cos\alpha - V)^2 + (-V_0 \cdot \sin\alpha)^2}, \tag{9}$$

$$\cos\gamma = \frac{-V_0 \cdot \cos\alpha - V}{\sqrt{(-V_0 \cdot \cos\alpha - V)^2 + (-V_0 \cdot \sin\alpha)^2}}, \tag{10}$$

where V_r is a relative velocity module; γ is an angle indicating the direction of the vector V_r of relative velocity; V_0 is a velocity of wind air flow; V is a velocity of prism in its rectilinear translation motion; α is an angle indicating the direction of the vector V_0 of air flow velocity (see **Figure 4**).

By the use of obtained Eqs. (3)–(10), it is possible to solve various technical problems of air flow and rigid body (prism) interaction. For example, it is possible to solve the problems of energy extraction from an air flow. Besides, body's shape optimization problem can be solved in order to obtain the desired effect along with motion control realization.

2.4 Model of air flow interaction with a perforated flat plate

Pressure distribution for a flat plate element with a rectangular cross-section is shown in **Figure 5**.

In accordance with the theorem of linear pulse change in the differential form [7], the following equations for the plate's pressure side can be written:

$$dm_1 \cdot V\cos\beta = dN_1 \cdot dt, \tag{11}$$

$$dm_2 \cdot V\sin\beta = dN_2 \cdot dt, \tag{12}$$

$$dm_1 = V\cos\beta \cdot dt \cdot dL_1 \cdot B \cdot \rho, \tag{13}$$

$$dm_2 = V\sin\beta \cdot dt \cdot dL_2 \cdot B \cdot \rho, \tag{14}$$

where dm_1, dm_2 are masses of elementary air flow particles with relative velocity V against inclined surfaces; dN_1, dN_2 are elementary impulse forces in the directions of normality toward the surfaces of the elemental area; β is an angle between elementary pulse dN_1 and air flow; dt is an elementary time moment; dL_1, dL_2 are elemental lengths of the surface; B is a width of the element, which is considered as constant in the case of a two-dimensional task; ρ is a density of air medium.

Using Eqs. (11)–(14), the change in pressure on the sides of the perforated plate can be expressed as follows:

$$\Delta p_1 = V^2\rho \cdot (\cos\beta)^2; \tag{15}$$

$$\Delta p_2 = V^2\rho \cdot (\sin\beta)^2. \tag{16}$$

The suction pressure in a small layer directly along the plate's lower edge is considered as constant and can be expressed with the following equation:

$$\Delta p_3 = V^2 \cdot \rho \cdot C, \tag{17}$$

where C is a constant determined experimentally or by computer modeling [5]. For subsonic velocity flow, the C value varies within interval $0 < C < 1$.

The model of air flow interaction with perforated plate is shown in **Figure 6**.

For the length L_3 of the perforated gap, the following condition is satisfied:

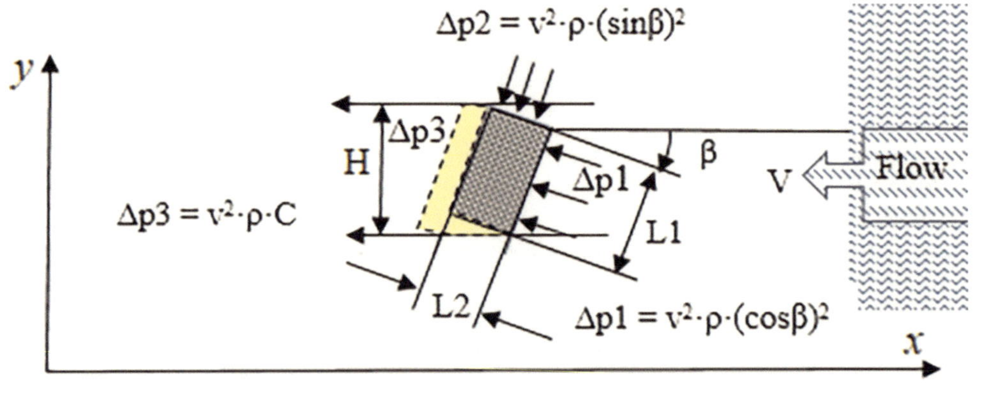

Figure 5.
Pressure distribution for a rectangular element of flat plate.

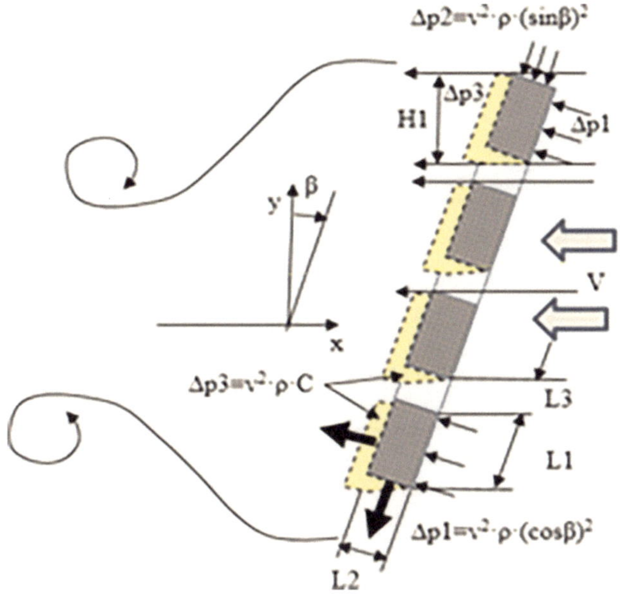

Figure 6.
Pressure distribution in a cross-section of a rectangular flat perforated plate, where L_1, L_2 are lengths of plate's edges, L_3 is a length of the perforated gap.

$$L_3 \geq L_2 \cdot tg(\beta). \tag{18}$$

Using the laws of classical mechanics for a two-dimensional flat plate [7], interaction force IF_x in the air flow direction (direction of the x-axis) can be determined by the formula:

$$IFx = -k \cdot B_1 \cdot H \cdot V^2 \rho \cdot \left[C + \frac{(\cos\beta)^3 + d \cdot (\sin\beta)^3}{\cos\beta + d \cdot \sin\beta} \right], \tag{19}$$

where k is a total number of elements between perforations; $d = L_2/L_1$ is a ratio of plate edges; $H = (L_1 \cos\beta + L_2 \sin\beta)$ is a dimension of the plate's element in the direction perpendicular to air flow. Another notation is the same as in Eqs. (11)–(17).

The mathematical model of perforated plate interaction with air flow is validated by computer simulation with the program Mathcad. Simulation is performed in application to translational motion of two-dimensional perforated plate in air flow with velocity V. Plate interacts with a linear spring with stiffness coefficient c and a linear damper with damping constant b (**Figure 7**).

Following the methods of classical mechanics [7], it is possible to determine relative interaction velocity V_r by the formula:

$$V_r = V + v, \tag{20}$$

where V is an air flow velocity; v is a velocity of a flat plate in the direction of *the* x-axis.

For the plate with the very small thickness ($\delta \approx 0$), the differential equation of its motion along the x-axis can be written in the following form:

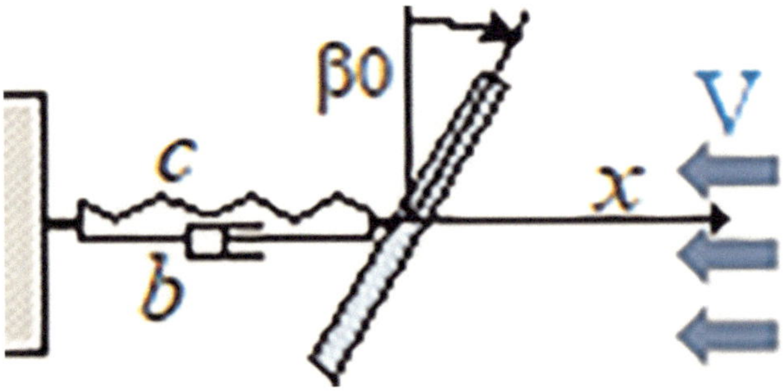

Figure 7.
Model of air flow interaction with perforated plate.

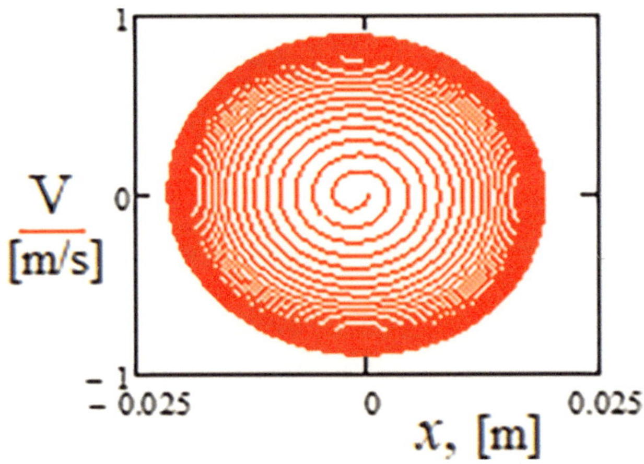

Figure 8.
Motion on phase plane "displacement x *– velocity* v*".*

$$m\ddot{x} = -cx - b\dot{x} - A_0(1 - a \cdot \text{sign}(\dot{x})) \cdot \rho \cdot \left\{C + (\cos\beta_0)^2\right\} \cdot (V + \dot{x})^2 \frac{V + \dot{x}}{|V + \dot{x}|}, \quad (21)$$

where A_0 is an average value of contact surface area of the plate; ρ is the air density; a is a constant of area variation; β_0 is plate angle against air flow; m is mass of the plate; C is an air flow and plate interaction constant.

Mathematical simulation of Eq. (21) was performed with program MathCad assuming the following values of main system's parameters: A_0 = 0.04 m^2; V = 10 m/s; ρ = 12,047 kg/m^3; m = 1.56 kg; c = 3061 kg·s^{-2}; b = 5 kg·s^{-1}; a = 0.5; C = 0.065; β_0 = π/6. Results of simulation for the perforated plate translation motion are presented in **Figures 8** and **9**.

From the graphs in **Figures 8** and **9**, it can be concluded, that stable oscillatory movement can be initiated in the aerodynamic system by the variation interaction area of the perforated plate. As it is seen from the analysis of the graph for generated power (**Figure 9**), the almost stationary oscillatory regime with maximal power P can be achieved after some cycles of a transient process.

2.5 Model of air flow interaction with a quadrangular convex prism

An analytical model of a quadrangular convex prism interacting with air flow is shown in **Figure 10**.

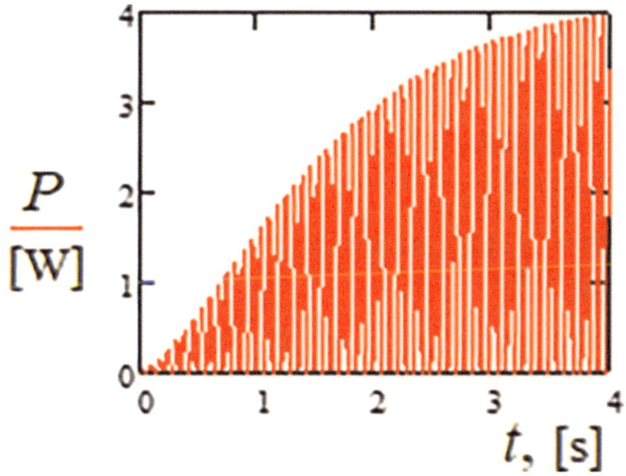

Figure 9.
Power P = b·v² *generated for the given time interval.*

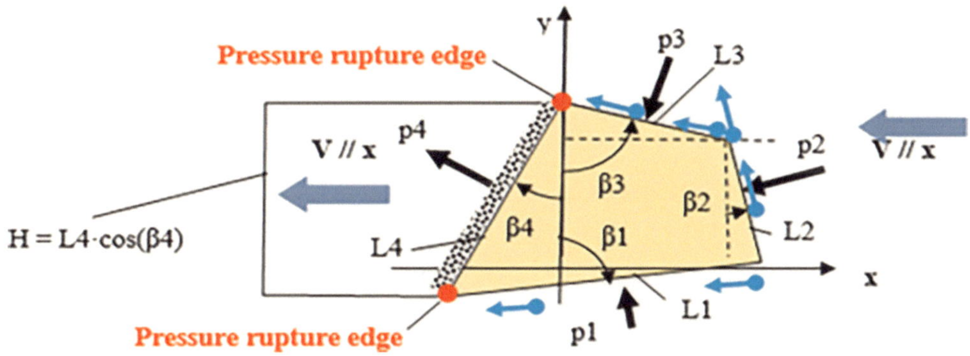

Figure 10.
Model of a quadrangular convex prism: L_1, L_2, L_3, L_4 *are lengths of edges;* β_1, β_2, β_3, β_4 *are prism's frontal angles;* H *is a height of prism; is a symbol of the direction of air flow on the surfaces of the prism (due to flow-prism interaction).*

By applying the theorem of air flow motion quantity change in the differential form [7], pressures p_1, p_2, p_3 on frontal planes of the prism (in pressure zone) can be expressed in the following form:

$$p_1 = (V\cos\beta_1)^2 \cdot \rho; \qquad p_2 = (V\cos\beta_2)^2 \cdot \rho; \tag{22}$$
$$p_3 = V^2\rho \cdot \left[(\cos\beta_3)^2 - C_{12}\cdot C_{23}\cdot \cos\beta_2 \cdot \sin(\beta_3 - \beta_2)\right].$$

where $\beta_1, \beta_2, \beta_3$ are angles of lateral orientation of prism sides relative to the air flow; C_{12}, C_{23} are constants for changing the flow rate along with the boundary layer at breaking points of the flow. For example, condition $C_{12} = C_{23} = 1$ means that the speed at the breaking points is not changed and is the same as at the beginning of the entire flow.

Accordingly, the pressure p_4 in the suction zone between the two broken edges can be determined by the formula

$$p_4 = C_4 \cdot V^2 \cdot \rho, \tag{23}$$

where C_4 is an air flow and prism interaction constant [5].

In calculations, it is necessary to take into account that Eqs. (22) and (23) are applicable to a prism that has curved surfaces in the pressure zone. For example, for the prism shown in **Figure 10**, the following angle relationships must be satisfied: $0 < \beta_4 < \pi/2$; $0 < (\beta_3 + \beta_4) < \pi$; $(\beta_3 - \beta_2) > 0$.

Using the Eqs. (22) and (23), the following projections of the interaction forces on the x and y axes can be obtained:

$$-F_x = V^2 B\rho \cdot \{L_1 \cdot (\cos\beta_1)^3 + L_2 \cdot (\cos\beta_2)^3 + L3 \cdot \left[(\cos\beta_3)^3 - C_{12} \cdot C_{23} \cdot \cdot \cos\beta_2 \cdot \cos\beta_3 \cdot \sin(\beta_3 - \beta_2)\right] + L_4 \cdot C_4 \cdot \cos\beta_4\}; \tag{24}$$

$$-F_y = V^2 B\rho \cdot \left\{L_1 \cdot (\cos\beta_1)^2 \cdot \sin\beta_1 + L_2 \cdot (\cos\beta_2)^2 \cdot \sin\beta_2 + L_3 \cdot \left[(\cos\beta_3)^2 \cdot \sin\beta_3 - C_{12} \cdot C_{23} \cdot \cos\beta_2 \cdot \sin\beta_3 \cdot \sin(\beta_3 - \beta_2)\right] + L_4 \bullet C_4 \sin\beta_4\right\}, \tag{25}$$

where F_x is a resistance force (along the direction of air flow); F_y is a lifting force (perpendicular to the direction of air flow).

When analyzing or optimizing forces expressed by Eqs. (24) and (25), the following geometric relationships should additionally be observed:

$$L_4 \cdot \sin\beta_4 + L_3 \cdot \sin\beta_3 + L_2 \cdot \sin\beta_2 - L_1 \cdot \sin\beta_1 = 0; \tag{26}$$

$$L_4 \cdot \cos\beta_4 - L_3 \cdot \cos\beta_3 - L_2 \cdot \cos\beta_2 - L_1 \cdot \cos\beta_1 = 0. \tag{27}$$

2.6 Model of air flow interaction with a quadrangular concave prism

An analytical model of a quadrangular concave prism interacting with air flow is shown in **Figure 11**.

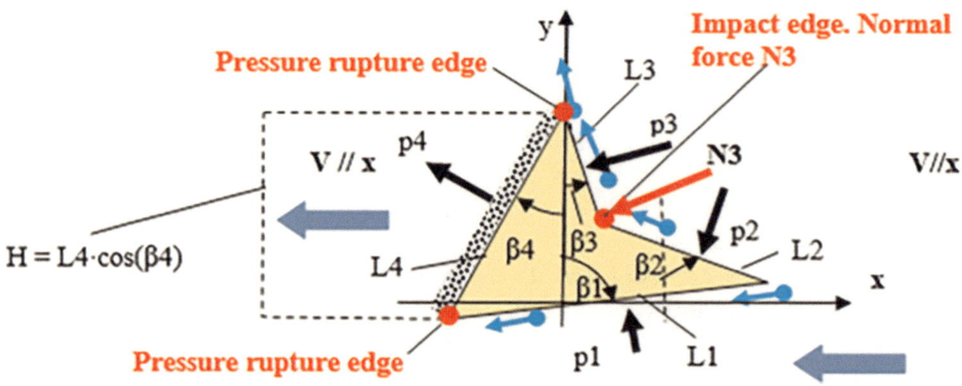

Figure 11.
Model of a quadrangular concave prism: L_1, L_2, L_3, L_4 *are lengths of edges;* β_1, β_2, β_3, β_4 *are prism's frontal angles;* H *is a height of the prism.*

In this case, the air flow impact force N_3 acts on the concave edge of the prism. The force N_3 is perpendicular to the edge with the length L_3, as shown in **Figure 11**. According to the boundary air flow motion change, when the direction of flow is varied from edge L_2 to edge L_3, the impact force N_3 is as follows:

$$N_3 = L_2 B \rho V^2 \cdot \cos\beta_2 \cdot \sin(\beta_2 - \beta_3) \cdot [0.5 + 0.5 \cdot \text{sign}(\beta_2 - \beta_3)]. \tag{28}$$

In the case of the concave prism, the following criterion additionally must be satisfied: $\sin(\beta_2 - \beta_3) \geq 0$. Eqs. (24) and (25) remain valid, only negative members should be excluded, since this part of the interaction is equivalent to N_3.

The resulting relationships (24), (25), (28) make it possible to analyze interactions of air flow with various prismatic forms, solving the tasks of optimization and synthesis.

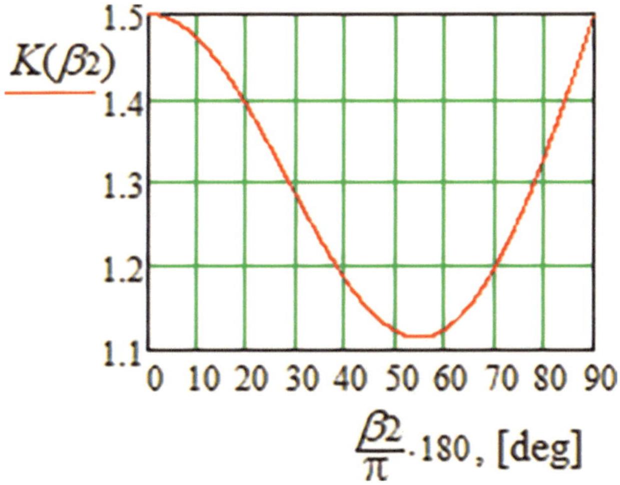

Figure 12.
Results of optimization for $C_4 = 0.5$.

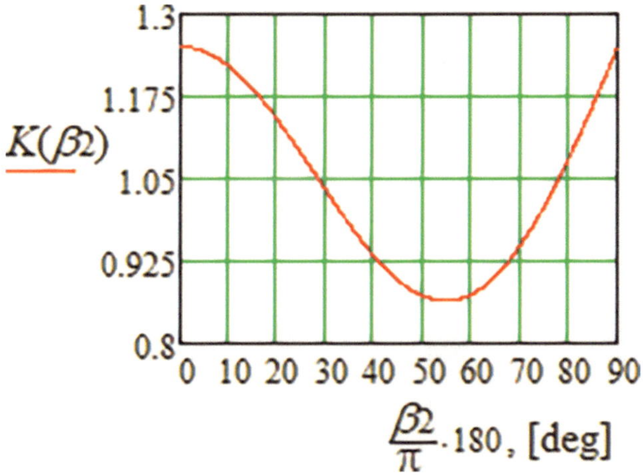

Figure 13.
Results of optimization for $C_4 = 0.25$.

2.7 Example of shape optimization for a quadrangular prism

Problem of prism shape (**Figure 10**) optimization is solved by computer simulation. The optimization criterion is resistance force F_x in accordance with Eq. (24) and taking account of limitations given by Eqs. (26) and (27). It was assumed that sides L_2 and L_3 are equal to the constant height H but for simplifications $\beta_1 = 0$ and $\beta_4 = 0$. Parameters V, ρ, B remained constant (were not varied).

Results of optimization for the criterion $K(\beta_2) = F_x/(V^2 B\rho H)$, which is a resistance coefficient in the direction of air flow, are presented in **Figures 12** and **13**.

As it is seen from the diagrams presented (**Figures 12** and **13**), it is possible to maximize or minimize a resistance force F_x by the variation of angle β_2. Qualitatively the same results are obtained for two different values of flow rate constant C_4.

3. Computer simulation of air flow interaction with simple form prisms

The interaction theory discussed above has been tested in computer modeling. Two-dimensional and three-dimensional problems are considered [8].

3.1 Air flow interaction with two-dimensional objects

Air flow interaction with rhombic and triangular prisms of various shapes was studied by numerical modeling. The aim of the study was to find out the reliability of the formulas obtained in the previous section in the description of air flow interaction with objects. The studied two-dimensional objects are shown in **Figures 14** and **15**.

The multiplication numbers under the prism drawings (**Figures 14** and **15**) indicate the position of the prism side (in angular degrees) relative to normal against the flow in both the pressure and suction zones. Software ANSYS Fluent was used to perform the numerical simulations. All the simulations were made assuming a constant air speed of 10 m/s.

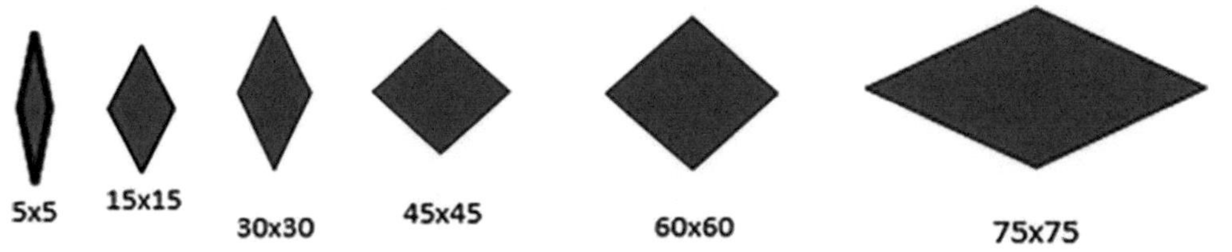

Figure 14.
Parameters of computer-studied rhombic section prisms.

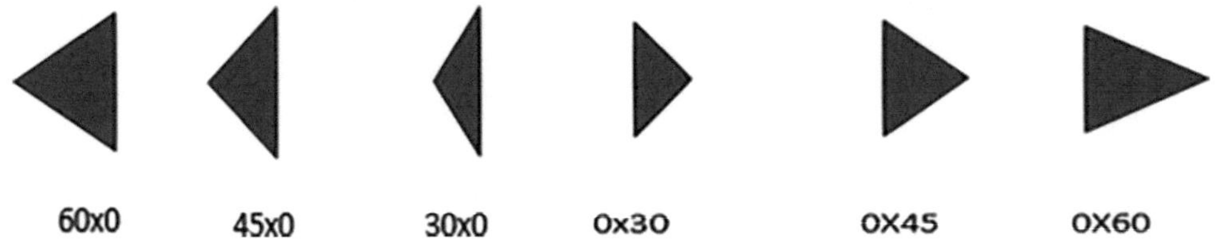

Figure 15.
Parameters of computer-studied triangle section prisms.

Results of numerical simulation are presented in **Figures 16** and **17** in the form of diagrams for pressure distribution around rhombic and triangles prisms.

Pressure and suction zones around prisms are shown in diagrams (**Figures 16** and **17**). Pressure distribution around the prisms is presented using different colors. As it is seen, the color of the suction zone is almost invariable. Therefore, it can be concluded that pressure in the boundary layer practically is almost constant.

3.2 Air flow interaction with three-dimensional objects

A four-ray star prism's interaction with air flow was simulated. A diagram for pressure distribution around this prism at a supersonic velocity of 1.8 Mach (equivalent to 612.5 m/s) is presented in **Figure 18**.

As it is seen from the diagram presented (**Figure 18**), even at high supersonic speeds, the pressure in the boundary layer of the suction zone is visually constant. This confirms the opportunity to apply the above considered analytical formulas for the analysis of flow-prism interaction at supersonic velocities.

Air flow interaction with full and perforated flat plates was studied in order to find out the physical nature of air medium in the suction zone. The distributions of streamlines in the suction zone for full and perforated plates are shown in **Figures 19** and **20**.

As it is seen from the diagram for the full flat plate (**Figure 19**), the vortices and bubbles are formed behind the plate in the suction zone. But for the perforated plate (**Figure 20**), the nature of the air flow in the suction zone is changed fundamentally. There is no vortices and bubbles downstream of the plate. This property should be

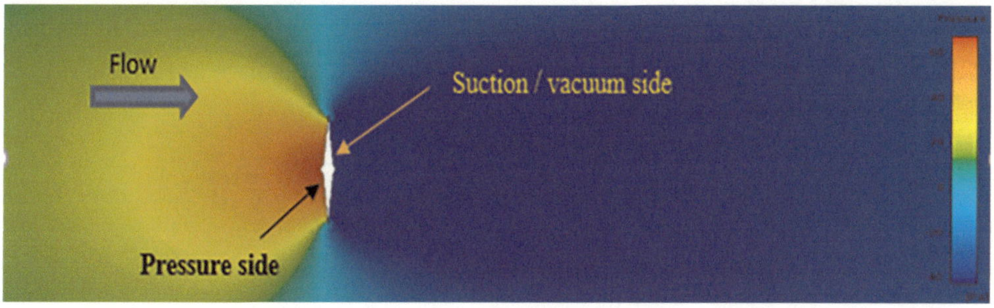

Figure 16.
Pressure distribution around the rhombic prism 5x5.

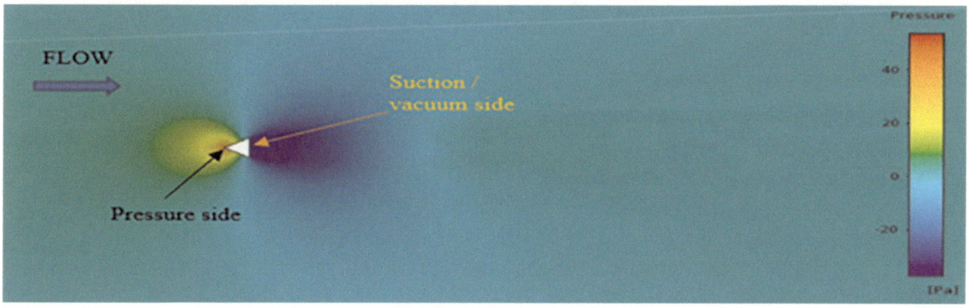

Figure 17.
Pressure distribution around the triangle prism 60x0.

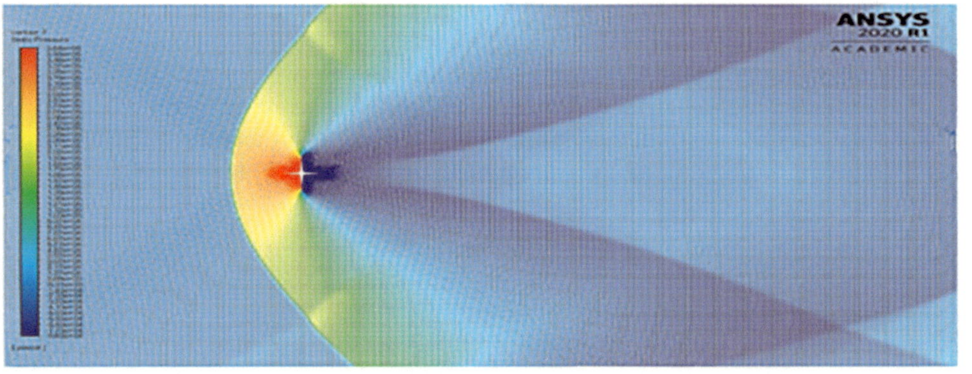

Figure 18.
Pressure distribution around the star prism at supersonic velocity.

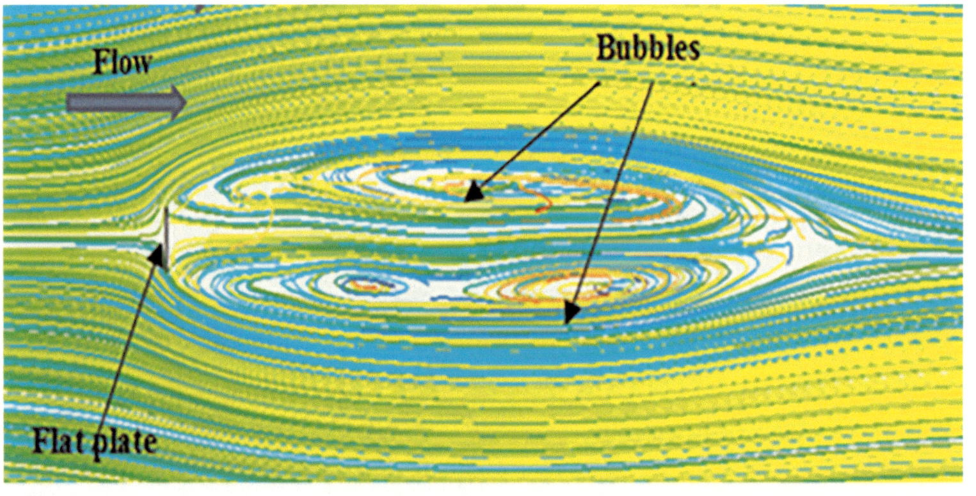

Figure 19.
Streamlines distribution around the full flat plate.

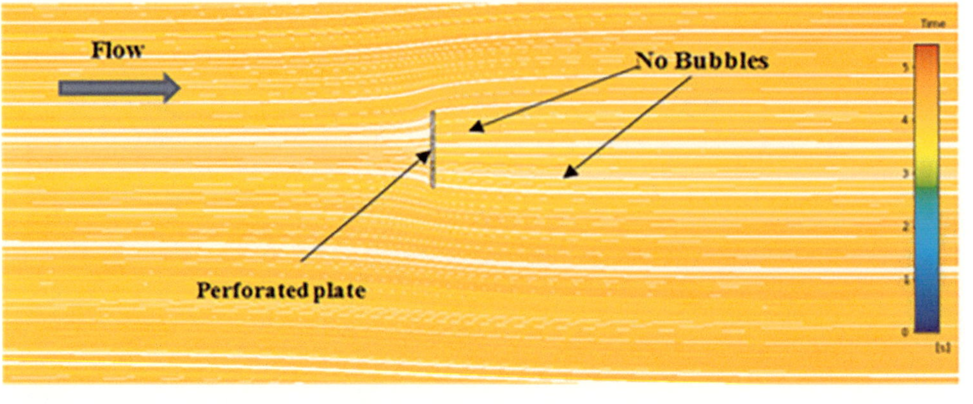

Figure 20.
Streamlines distribution around the perforated flat plate.

taken into account in analytical calculations by the reducing air flow interaction constant C in the suction zone.

The results of numerical modeling confirm that air flow and rigid body interaction phenomena can be analyzed within two completely different zones: the pressure zone and the suction zone. It was shown that pressure in the suction zone along the entire boundary layer is constant. In the pressure zone, the interaction has an analytical relationship, but in the suction zone, it is possible to supplement the formula with a constant parameter C. It is found that for the velocity of 10 m/s, the constant parameter for two-dimensional modeling is $C = 0.5$, but for three-dimensional modeling, it is reduced to about $C = 0.25$.

4. Experimental investigations in wind tunnel

Experiments were carried out in the Armstrong Subsonic wind tunnel, available at Riga Technical University. The main specifications of the wind tunnel can be found in [9].

4.1 Experiments with full flat plate

The object of study is a square flat plate with dimensions 0.159 x 0.159 m, which is about two times less than the dimension of the tunnel working section (0.304 m). The drag force is measured using the concept of balanced weights. The schematic diagram of experimental installation and process parameters is shown in **Figure 21**.

V_0 is air flow velocity; V_N is a normal component of air flow velocity; β is an angle of plate's normal position against the flow direction; F_x and F_y are horizontal and vertical components of the air interaction force; L_1 is a length of the square plate's edge; $L_2 = 0.005$ m is a thickness of the plate.

The main purpose of the experiment was to test the applicability of analytical formulas for calculation of drag force F_x (horizontal component of air interaction force). Experimental interconnection between drag force F_x and angle of attack (90° + β) is graphically presented in **Figure 22** (results are obtained for the constant air flow velocity $V_0 = 10$ m/s).

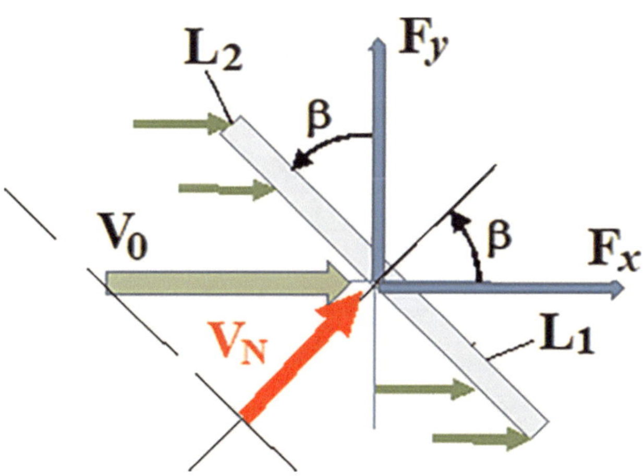

Figure 21.
Schematic diagram of experiment and process parameters.

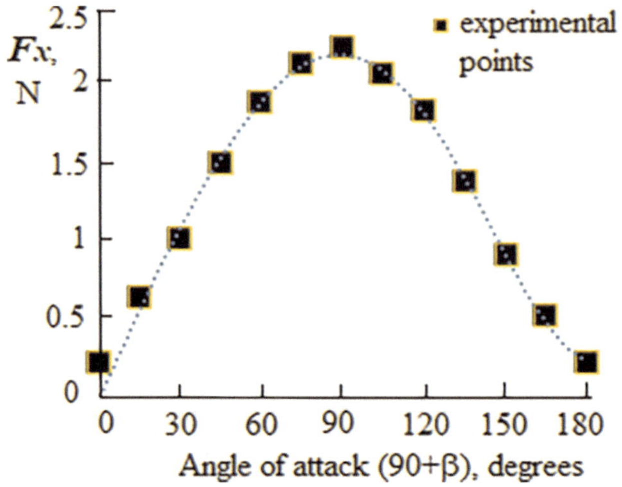

Figure 22.
Experimental results for drag force F_x *for the full flat plate.*

Analytically drag force F_x for a flat plate interacting with air flow can be determined by the formula [8]

$$F_x = \frac{H(\beta)\bullet B\rho}{2}\bullet\left[C + \frac{(\cos\beta)^3 + (\sin\beta)^3}{\cos\beta + \sin\beta}\right], \tag{29}$$

where $H(\beta) = L_1 \cos\beta + L_2 \sin\beta$.

Analytical curves $F_x = f(\beta)$, constructed by Eq. (29) at the three different values of constant C (0.125; 0.25; 0.50), are presented in **Figure 23**. For comparison, experimentally measured values of forces F_x are shown on this diagram, too.

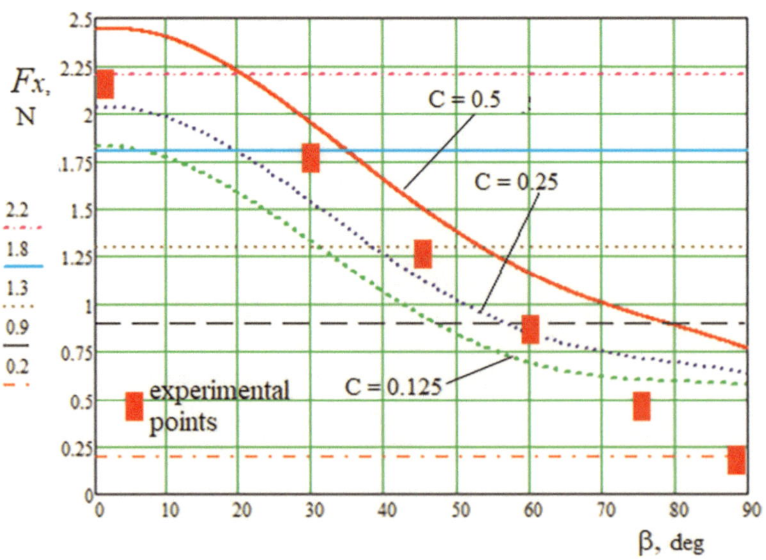

Figure 23.
Comparison of analytical and experimental results for drag force F_x.

Theoretical curves $F_x = f(\beta)$ agree qualitatively well with the experimental data (**Figure 23**), i.e., the curves have the same shape. But the quantitative difference is satisfactory and lies within the range from 12–25%. Such difference could be explained by the limited cross-sectional dimensions of the wind tunnel (0.304 x 0.304 m) in comparison with plate dimensions (0.159 x 0.159 m), as well by the operation principle of the tunnel (not pressing, but suction principle). Therefore, it has been experimentally proved that obtained analytical formulas can be used in air flow interaction calculations (tasks of analysis, optimization, and synthesis).

4.2 Experiments with perforated flat plate

Experiments were held with a perforated flat plate shown in **Figure 24**. During experiments, different orientations of perforated grooves were used: horizontal (as in **Figure 24**) and vertical. The velocity of air flow was constant and equal to 10 m/s.

Experimental interconnection between drag force F_x and angle of attack (90° + β) is graphically presented in **Figure 25**. Curves $F_x = f(90° + \beta)$ are constructed for the plates with horizontal (H) and vertical (V) orientations of perforated grooves. Additionally, results of analytical calculations of drag force F_x by formula (19) are shown (for the perforated plate with vertical grooves, assuming C = 0.5).

On the analysis of experimental results (**Figure 25**), it can be concluded that drag force F_x is always higher if grooves are oriented horizontally. This could be explained by the fact that there is an additional air flow interaction with the edges of perforated horizontal grooves. But in the horizontal position of plates (under the β = 90°), drag forces are the same both in the vertical and horizontal grooves orientation (and equal to the drag force for the full plate F_x = 0.2 N, see **Figure 22**). This is well understood because the perforation in both plates is covered, if β = 90°.

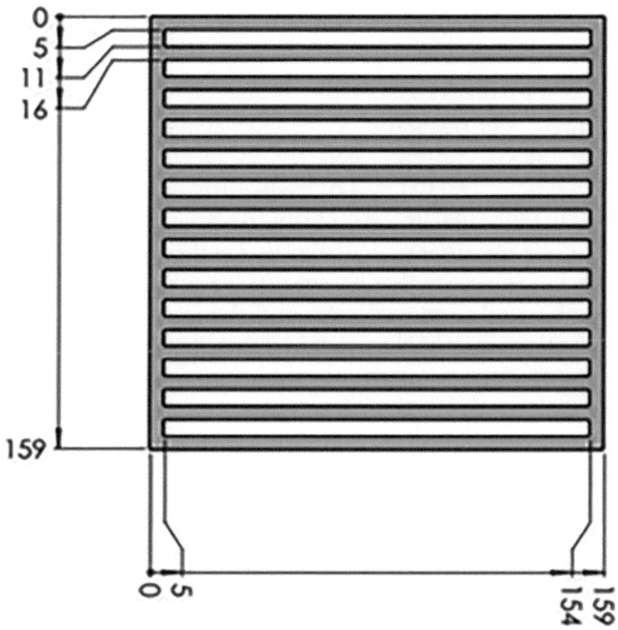

Figure 24.
The geometry of the perforated flat plate (all dimensions are in mm).

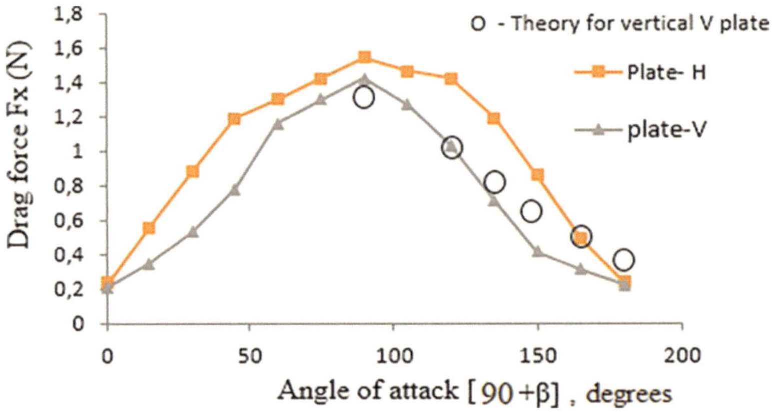

Figure 25.
Experimental values of drag force F_x *for the perforated flat plate.*

Results of analytical calculations of drag force F_x by formula (19) agree well with experimental data (see **Figure 25**). Therefore, the mathematical model obtained for perforated plate can be successfully used in air flow interaction calculations.

5. Models of wind energy conversion devices

The above results of the theoretical and experimental analysis are used in the designing of new devices for energy extraction from air flow. Models of wind energy conversion devices equipped with vibrating plates (disks) are developed.

5.1 Wind energy conversion device equipped with rotating perforated disk

A new model of wind energy conversion device equipped with working head made from two concentric circular flat plates (disks) with alternate flow sectors is synthesized (**Figure 26**). Disks are connected to each other at the center. Besides, the disk whose front area is subjected to the action of air flow has an ability to rotate freely over the other circular non-rotating disk. Both disks have the same surface area and identical sector perforations (holes). During the rotation of one disk, perforations are cyclically opened and closed, and due to this equivalent surface area of the working head is periodically changed in accordance with the given control action [10].

V is air flow velocity; x is a displacement of the disk in its translation motion; $-bV_x$ is a force of linear generator; $-cx$ is the elastic force of a spring; ω_0 is an angular velocity of rotating disk.

Control action for the variation of perforated disk's surface area A can be given in the following form:

$$A = A_0\left\{1 + \frac{\cos^{-1}[\cos(\omega_0 t)]}{\pi}\right\}, \tag{30}$$

where A_0 is a medium surface area of the disk per its one cycle; ω_0 is an angular frequency of harmonic control action. Area A variation function (30) graphically is shown in **Figure 27**.

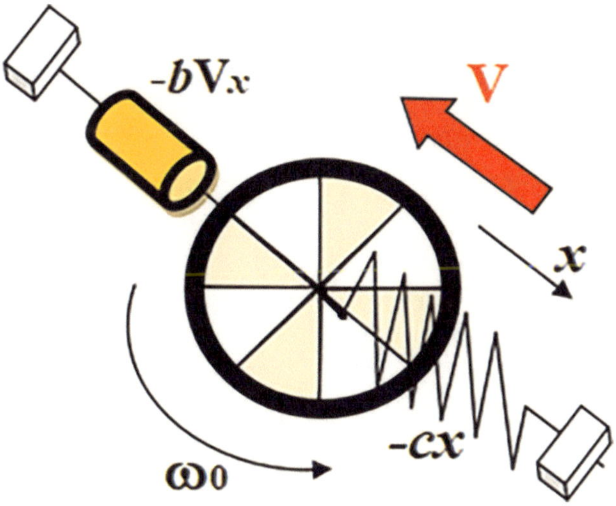

Figure 26.
Model of wind energy conversion device with rotating perforated disk.

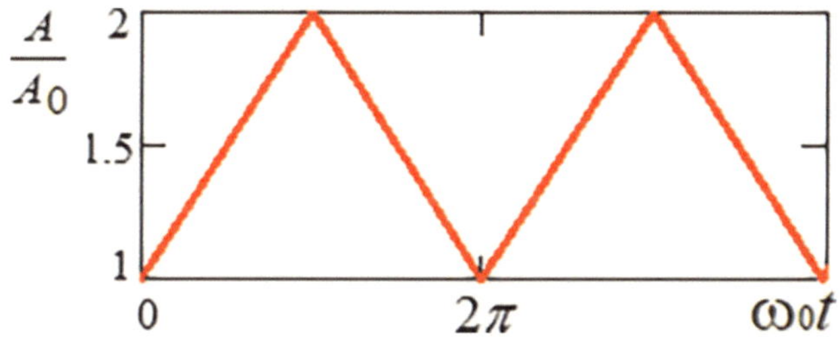

Figure 27.
Control action by the variation of area A of perforated plate.

Translation motion of the perforated plate in the direction of x-axis (**Figure 26**) under the control action (30) is described by the following differential equation:

$$m\ddot{x} = -cx - [F_0\text{sign}(\dot{x}) - b\dot{x}] + (1 + C)\frac{A}{\pi} \cdot \left[\cos^{-1}(\cos\omega_0 t) + \pi\right] \\ \cdot \rho(-V_0 - \dot{x})^2 \cdot \text{sign}(-V_0 - \dot{x}), \tag{31}$$

where m is a mass of perforated plate; c is stiffness coefficient of spring; F_0 and b are constants of linear damping generator; C is an interaction coefficient between air flow and plate; V_0 is an air flow velocity; ω_0 is an angular frequency of harmonic control action; A is a constant surface area of the plate; ρ is air density.

By the simulation with program Mathcad of disk motion under the Eq. (31), the optimization task was solved. It is shown that maximal power P through disk interaction with air flow is generated under the resonant condition $\omega_0 = \sqrt{c/m}$. The graph of generated power P versus time t for the V_0 = 10 m/s is shown in **Figure 28**.

As it is seen from the analysis of the graph presented (**Figure 28**), a stationary oscillatory regime with maximal generated power P can be achieved after some cycles of a transient process.

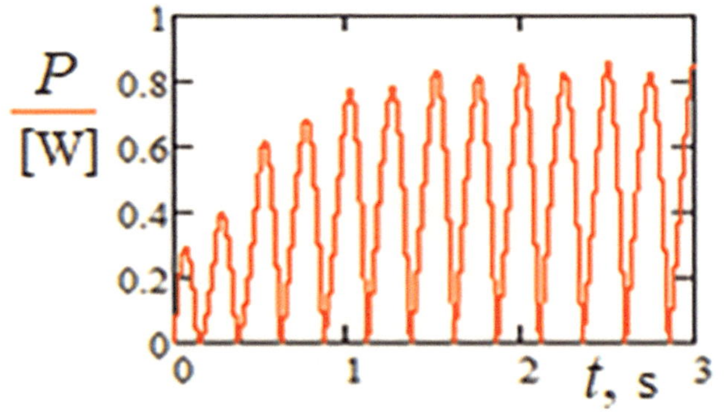

Figure 28.
Power generated due to air flow interaction with a perforated disk.

5.2 Air flow generator on the base of a closed track conveyor

The principal model of the wind energy conversion generator synthesized on the base of a closed track conveyor is shown in **Figure 29**. Closed track conveyor 1 forms a Central part of the generator, besides the track has an ability to move parallel to coordinate plane $x0y$. The conveyor is driven by an air flow with velocity V_0, acting on blades 2 in parallel to the $0z$ axis.

Power is obtained from a generator connected with rotor 3 at the one end (left or right) of conveyor 1 (**Figure 29**). Flat blades 2 is attached tightly to conveyor 1 with a rigid fastening element 4 (welded hinge). Besides, blades 2 are fixed at the angle α toward the x-axis. The model of generator has several flat blades 2. Due to the action of air flow V_0, the translation motion of blades 2 along conveyor's straight and circular sections (in final turns) is excited.

The three-dimensional design of air flow generator made with the program Solid Works is shown in **Figure 30**.

The generation of useful power in the proposed device (**Figures 29** and **30**) is due to the translation movement of the flat blades. Therefore, the wind flow load is uniformly distributed over the lateral surface of the flat blades. This provides a simple way to increase the operational efficiency of the device, which can be achieved by increasing the area A of the blade's lateral surface.

5.3 Air flow generator on the base of vibrating flat blade and crank mechanism

The model of the developed wind energy conversion device is shown in **Figure 31**. Flat blade 1 is a main element of the device, and it is attached to the rotating axle 2 by a cylindrical axial hinge. And symmetry axis z_1 of blade 1 simultaneously is a longitudinal axis of axle 2. Besides, rotating axle 2 is rigidly attached to slider 3, which has the ability of translation motion along the x-axis. Additionally, the translation motion of slider 3 is limited by elastic springs 4 and shock absorbers 5, but turning of the blade 1 around axis z_1 is restricted by a torsional spring 6 and a rotary shock absorber 7. The crank 8 is rigidly attached to the flat blade 1 perpendicular to its side surface. Additionally, there is a connecting rod 9, which opposite ends are hinged to the crank 8 and slider 10 of an electric generator. And slider 10 has the ability to move inside the electric coil 11 along the x_1 axis.

DOI: http://dx.doi.org/10.5772/intechopen.103120

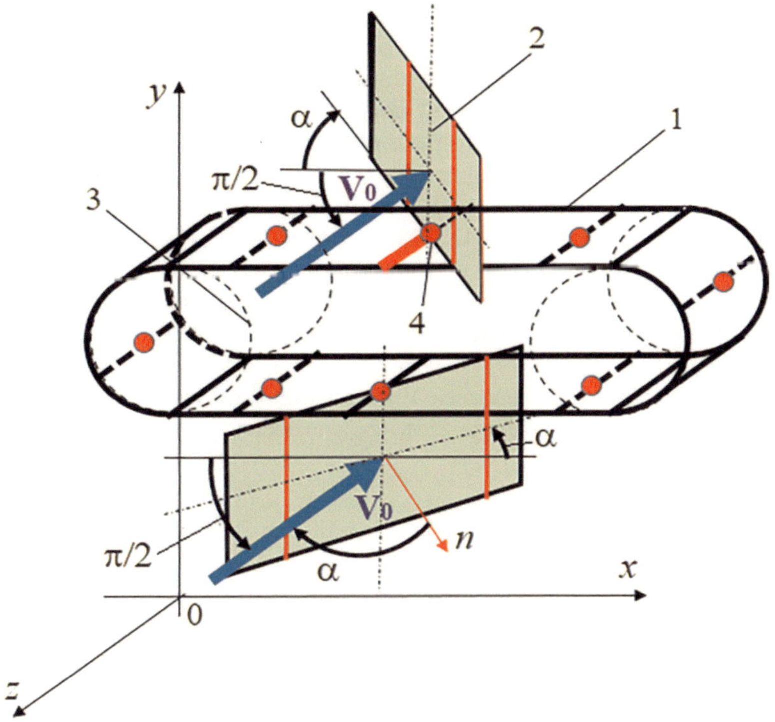

Figure 29.
Principle model of air flow generator on the base of track conveyor: 1 – closed track conveyor; 2 – flat blade; 3 – rotor; 4 – rigid fastening element.

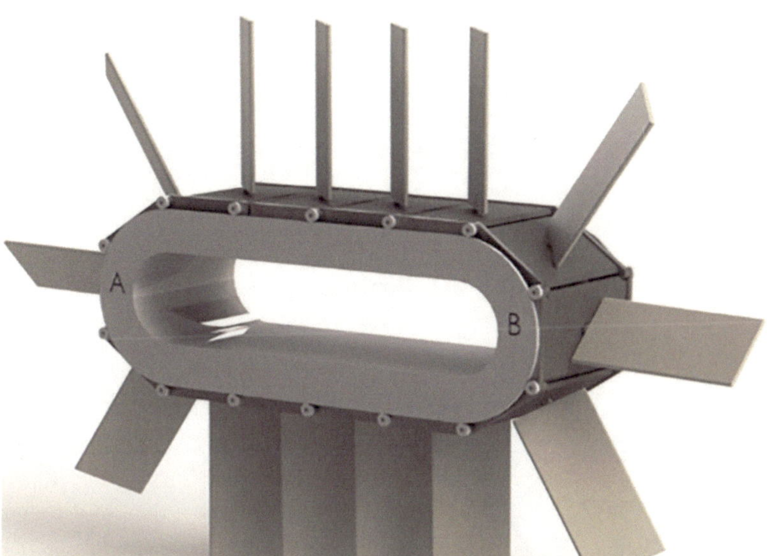

Figure 30.
Three-dimensional design of the air flow generator.

The operation of the wind energy conversion device starts from the position shown in **Figure 31**. It is assumed that wind flow has a speed of V_0 and is directed perpendicular to the x-axis. Due to the effect of wind flow on the side surface of the flat blade 1, a force N is formed in the direction of normal n (**Figures 31** and **32**). The action of the force N causes slider 3 to move to the right along x-axis. As a result, compressive force F_k is formed in the connecting rod 9.

The force F_k of the connecting rod 9 acts on the linear generator, consisting of a slider 10 and a built-in electric coil 11. This force holds the rotating flat blade 1 at the left rotary shock absorber 7 (**Figures 31** and **32**). At this device position, the turning angle α reaches the maximum value (α_0 clockwise). Then spring 4 is stretched, and the right shock absorber 5 is deformed until slider 3 stops in the right extreme position.

Then, under the action of the elastic forces, slider 3 moves back to the left. At the beginning of this translational movement, the connecting rod 9 is tensioned, and,

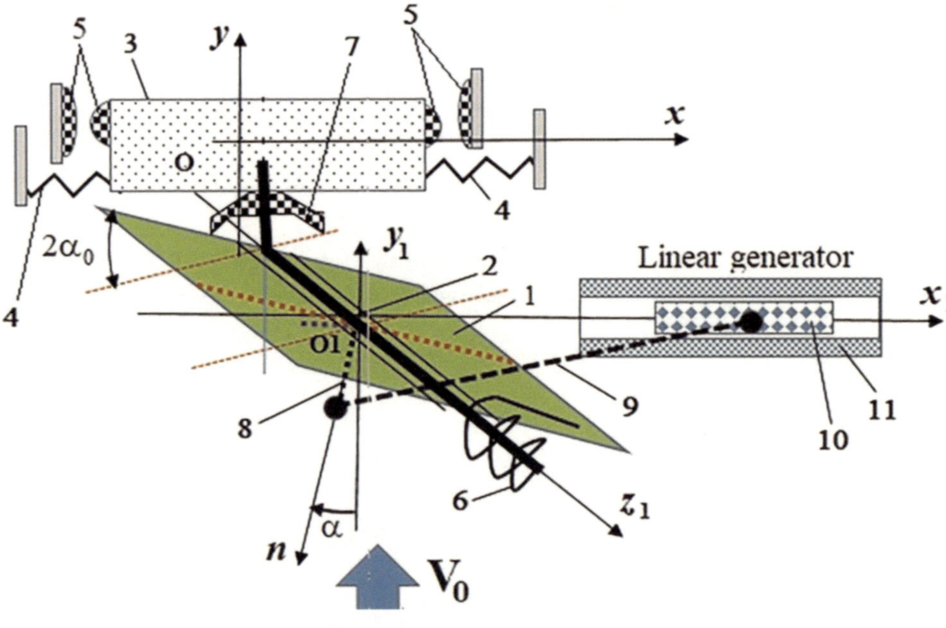

Figure 31.
Principle model of the wind energy conversion device: 1 – flat blade; 2 – rotating axle; 3 – slider; 4 – spring; 5 – shock absorber; 6 – torsional spring; 7 – rotary shock absorber; 8 – crank; 9 – connecting rod; 10 – slider of the linear generator; 11 – electric coil.

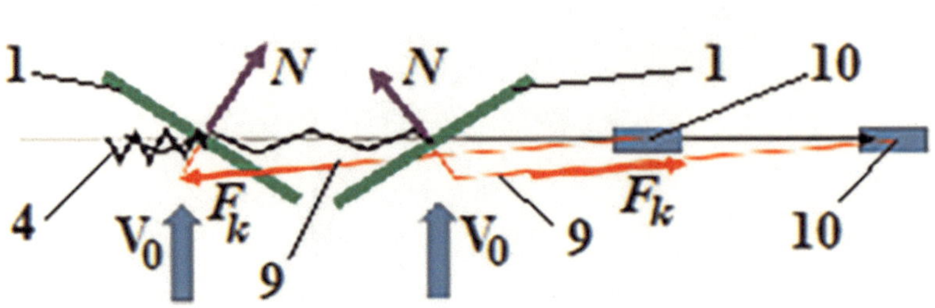

Figure 32.
Two extreme stopping positions of the blade during operation of the device: 1 – flat blade; 4 – spring; 9 – connecting rod; 10 – slider of linear generator.

consequently, the force F_k acts in the opposite direction. The flat plate 1 rotates counterclockwise about the symmetry axis z_1 and reaches the right rotary shock absorber 7. At this position, the angle of rotation α reaches a maximum value (α_0 counterclockwise). As a result, the normal n to the flat blade changes its position against the wind flow V_0. Therefore, a new force F_k pushes the flat blade 1 in the direction opposite to the x-axis. As a result, slider 3 moves to the left, deforms spring 4 and also the left shock absorber 5 until slider 3 stops in the left extreme position. The cycle then repeats as the compressive force F_k again begins to interact with the connecting rod 9.

During the generated cyclic movement, the generator's slider 10 moves backward inside the electric coil 11 along the x_1 axis. As a result, electrical energy is produced in the generator (alternating current is generated in the electric coil 11). This dynamic operational principle of a linear generator has been described in the literature [11].

6. Conclusions

Air flow interaction with flat blades and space prisms that perform translation motion was studied using the concept of zones (pressure and suction zones) for a rigid body immersed in an air flow. This method allows to solve problems of non-stationary body-air flow interaction without requiring intensive and laborious "space–time" programming techniques.

The air flow–rigid body interaction theory has been tested in computer modeling of two-dimensional and three-dimensional problems. Applicability of the proposed formulas for engineering calculations of interaction forces (drag forces) are confirmed by experiments with physical models of air flow devices in the wind tunnel.

New wind energy conversion devices are designed. The operation principle of these devices is based on the utilization of flat blades translation motion due to the interaction with air flow.

Author details

Janis Viba, Vitaly Beresnevich* and Martins Irbe
Riga Technical University, Riga, Latvia

*Address all correspondence to: vitalijs.beresnevics@rtu.lv

References

[1] de Renzo DJ, editor. Wind Power: Recent Developments. Park Ridge, New Jersey: Noyes Data Corporation; 1979. p. 347

[2] Dirba J, Levins N, Pugachevs V. Electromechanical Converters of Wind Energy (in Latvian). Riga: RTU Publishing House; 2006. p. 300

[3] Viba J, Beresnevich V, Noskovs S, Irbe M. Investigation of rotating blade for energy extraction from fluid flow. In: Vibroengineering Procedia. Vol. 8. Kaunas: JVE International Ltd; 2016. pp. 312-315

[4] Viba J, Beresnevich V, Irbe M, Dobelis J. The control of blades orientation to air flow in wind energetic device. Energy Procedia. 2017;**128**: 302-308. DOI: 10.1016/j.egypro.2017.08.317

[5] Viba J, Beresnevich V, Irbe M. Synthesis and Optimization of Wind Energy Conversion Devices. In: Maalawi KY, editor. Design Optimization of Wind Energy Conversion Systems with Applications. London: IntechOpen; 2020. pp. 125-141. DOI: 10.5772/intechopen.90819

[6] Eltayesh A, Castellani F, Burlando M, Hanna MB, Huzayyin AS, El-Batsh HM, et al. Experimental and numerical investigation of the effect of blade number on the aerodynamic performance of a small-scale horizontal axis wind turbine. Alexandria Engineering Journal. 2021;**60**:3931-3944. DOI: 10.1016/j.aej.2021.02.048

[7] Meriam JL, Kraige LG, Bolton JN. Engineering Mechanics: Dynamics. 8th ed. New York: John Wiley & Sons; 2016. p. 736

[8] Tipans I, Viba J, Irbe M, Vutukuru SK. Analysis of non-stationary flow interaction with simple form objects. Agronomy Research Journal. 2019;**17**(S1):1227-1234. DOI: 10.15159/AR.19.083

[9] Armfield. Engineering teaching and research catalogue – ISSUE 2 [Internet]. Ringwood, Hampshire, England: Armfield Limited; 2019. Available from: https://worlddidac.org/wp-content/uploads/2019/07/CAT5L-Jan2019-V6-live.pdf [Accessed: 23 December, 2021]

[10] Vutukuru SK, Viba J, Tipans I, Viksne I, Irbe M. Analysis of flat plate vibrations by varying frontal area to the flow. Jelgava: Proceedings of 18th International Scientific Conference "Engineering for Rural Development"; 22–24 May 2019, Latvia University of Life sciences and Technologies; 2019. pp. 1408-1414. DOI: 10.22616/ERDev2019.18.N147

[11] Bal VB, Gecha VJ, Goncharov VI, Ezhov EV. Linear electric reciprocating machines – application (in Russian). Problems of Electromechanics. 2015;**149**: 3-17 eLIBRARY ID: 34421985. Available from: http://jurnal.vniiem.ru/text/149/3-17.pdf